MINING Irish-American LIVES

Mining the American West

SERIES EDITORS: DUANE A. SMITH | ROBERT A. TRENNERT | LIPING ZHU

Boomtown Blues: Colorado Oil Shale | ANDREW GULLIFORD

Eben Smith: The Dean of Western Mining | DAVID FORSYTH

From Redstone to Ludlow: John Cleveland Osgood's Struggle against the United Mine Workers of America | F. DARRELL MUNSELL

Gambling on Ore: The Nature of Metal Mining in the United States, 1860–1910 | KENT A. CURTIS

Hard as the Rock Itself: Place and Identity in the American Mining Town | DAVID ROBERTSON

High Altitude Energy: A History of Fossil Fuels in Colorado | LEE SCAMEHORN

A History of Gold Dredging in Idaho | CLARK C. SPENCE

Industrializing the Rockies: Growth, Competition, and Turmoil in the Coalfields of Colorado and Wyoming | DAVID A. WOLFF

The Mechanics of Optimism: Mining Companies, Technology, and the Hot Spring Gold Rush, Montana Territory, 1864–1868 | JEFFREY J. SAFFORD

Mercury and the Making of California: Mining, Landscape, and Race, 1840–1890 | ANDREW SCOTT JOHNSTON

Mining Irish-American Lives: Western Communities from 1849 to 1920 | ALAN J. M. NOONAN

The Once and Future Silver Queen of the Rockies: Georgetown, Colorado, and the Fight for Survival into the Twentieth Century | CHRISTINE A. BRADLEY AND DUANE A. SMITH

The Rise of the Silver Queen: Georgetown, Colorado, 1859–1896 | LISTON E. LEYENDECKER, DUANE A. SMITH, AND CHRISTINE A. BRADLEY

Santa Rita del Cobre: A Copper Mining Community in New Mexico | CHRISTOPHER J. HUGGARD AND TERRENCE M. HUMBLE

Silver Saga: The Story of Caribou, Colorado, Revised Edition | DUANE A. SMITH

Thomas F. Walsh: Progressive Businessman and Colorado Mining Tycoon | JOHN STEWART

Yellowcake Towns: Uranium Mining Communities in the American West | MICHAEL A. AMUNDSON

MINING Irish-American LIVES

Western Communities from 1849 to 1920

Alan J. M. Noonan

UNIVERSITY PRESS OF COLORADO
Louisville

Published by University Press of Colorado
245 Century Circle, Suite 202
Louisville, Colorado 80027

Printed in the United States of America

The University Press of Colorado is a proud member of the Association of University Presses.

The University Press of Colorado is a cooperative publishing enterprise supported, in part, by Adams State University, Colorado State University, Fort Lewis College, Metropolitan State University of Denver, University of Alaska Fairbanks, University of Colorado, University of Denver, University of Northern Colorado, University of Wyoming, Utah State University, and Western Colorado University.

∞ This paper meets the requirements of the ANSI/NISO Z39.48-1992 (Permanence of Paper).

ISBN: 978-1-64642-250-0 (hardcover)
ISBN: 978-1-64642-251-7 (ebook)
https://doi.org/10.5876/9781646422517

Library of Congress Cataloging-in-Publication Data

Names: Noonan, Alan J. M., author.
Title: Mining Irish American lives : western communities from 1849 to 1920 / Alan J. M. Noonan.
Other titles: Mining the American West.
Description: Louisville : University Press of Colorado, [2022] | Series: Mining the American West | Includes bibliographical references and index. | Text in English and Gaelic.
Identifiers: LCCN 2022017913 (print) | LCCN 2022017914 (ebook) | ISBN 9781646422500 (hardcover) | ISBN 9781646422517 (ebook)
Subjects: LCSH: Irish Americans—West (U.S.)—History—Sources. | Irish American copper miners—West (U.S.)—Sources. | Hard rock mines and mining—West (U.S.) | Gold mines and mining—West (U.S.) | Frontier and pioneer life—West (U.S.)
Classification: LCC E184.I6 N66 2022 (print) | LCC E184.I6 (ebook) | DDC 979/.0049162073—dc23/eng/20220606
LC record available at https://lccn.loc.gov/2022017913
LC ebook record available at https://lccn.loc.gov/2022017914

Cover photograph: "Going to Work, Wallace, Idaho," ca. 1909. Public domain image. Library of Congress Reproduction # LC-USZ62-107002.

To my parents,

Bernadette and Seán Noonan,

ever a light in the dark.

Contents

Acknowledgments

This book represents an effort to recover some lost or forgotten voices from the past and to develop a clearer vision of the story of Irish mobility and communities in the US. Just as the people in these stories traveled far and wide, I followed in their footsteps visiting archives, libraries, and people across America looking for traces of their existence. You hold in your hands a book that is the product not just of my research and writing but also of many helping hands of family and friends too numerous to thank in full, though any errors in this work are mine alone.

As a Glucksman Government of Ireland Fellow at New York University's Glucksman Ireland House, I received much help from Joe Lee, Marion Casey, Hilary Mhic Suibhne, and Anne Solari. As a Kluge Fellow at the Library of Congress, both Mary Lou Reker and Travis Hensley and my colleagues at the Kluge Centre were welcoming and generous with their time, including Kristen Shedd, Arun Sood, Kaveh Abbasian, Amy Bride, Marenka Odlum-Thompson, Samira Mehta, and Shaden Tageldin. My thanks to my colleagues at University College Cork, including Donal O'Driscoil and William Mulligan Jr. My journey to visit various archives across the American West would not have been possible without the help of Michael Larkin, Karl Duesterberg, Matt Frey, Johanna Höög, Duke Day, James Walsh, Joshua Pollarine, Beki Parham, Dan Church, Curtis the truck driver, and Bud and Cindy Carter. Thanks also to Jack and Zara Noonan, Timothy Guilfoyle, David Bass, and April Braden. Patrick J. Barrett provided quality Irish translations, and his linguistic expertise and support of the project were much appreciated. Dozens of archivists and librarians helped me during my research, including those at the National Archives, Washington, DC; Library of Congress, Washington, DC; Bancroft Library, University of California, Berkeley; California State Library, Sacramento; California Historical Society, San Francisco; Butte-Silver Bow Archives, Montana; Kathleen Durfee at Coeur d'Alene's Old Mission State Park, Idaho; Cork City Archives, Cork, Ireland; Boole Library, University College Cork, Ireland; Alice Phelan Sullivan Library, San Francisco; Downey Historical Society, Downey; Douglas County Historical Society, Gardnerville, Nevada; Cudahy Library at Loyola University, Chicago; Wisconsin Historical Society, Madison; University of Colorado Library, Boulder; Nevada Historical Society, Reno; National Library, Dublin; Public Record Office of Northern Ireland, Belfast; Montana Historical Society Archives, Helena; Historical Society of Pennsylvania, Philadelphia; Idaho Historical Society, Boise; Mansfield Library at the University of Montana, Missoula; and Bobst Library, New York University. Special thanks to Julie Monroe at the University of Idaho Special Collections, Paddy Fitzpatrick at the Irish Migration Centre in Omagh, Jacquie Oboes at the Museum of St. Mary's in the Mountains in Virginia City, and William Breault at the Diocese of Sacramento. I also want to thank Rita O'Neill and Damien Keane for their wonderful mentoring during my early education. Theresa and Curtis Noonan's kindness and generosity was a pillar in the foundation of this work, and I cannot thank them enough for everything.

David Emmons bears much responsibility for inspiring me to follow the path of historical inquiry, and I offer him my deep thanks for his years of kind friendship, a kindness that he did not extend to the space within the four walls of the handball court. Kerby Miller was extremely generous in sharing his time and remarkable collection of Irish migrant letters with me when I visited him at the University of Missouri. Andy Bielenberg expertly supervised my successful doctoral dissertation that formed the research basis for this book and has supported my work since then. David Brundage has kindly encouraged my work through the years. Sara Goek generously offered her help on the early structure and orientation of this work. Michael O'Connell guided me in the digital use of census records. Sincere thanks to Rachel Walker for her friendship and support over the years. I am grateful to Emily Robideau who helped clarify my vision in the text as I prepared the final manuscript; her encouragement of the project helped push it over the line.

Thanks to Charlotte Steinhardt, Rachael Levay, Cheryl Carnahan, Dan Pratt, Laura Furney, the reviewers, and the staff at the University Press of Colorado for their help in preparing this manuscript for publication.

Many, many deep thanks to Colin, Aoife, Emma, and Declan and especially to my nan, Mary. My final deepest thanks go to my parents, Bernadette and Seán, to whom this work is dedicated. Their boundless love and support over the years helped me to continue and complete this project.

MINING
Irish-American
LIVES

Introduction

Fuaireas-sa litir ó bhrathair gaoil,
Dul go tapaidh anon thar toinn,
Go raibh ór go flúirseach le fáil anso,
Is ná feicfinnse choíche lá cruaigh ná bocht.
I got a letter from a relation
Telling me to hasten across the sea,
That gold was to be found in plenty there
And that I'd never have a hard day or a poor one again.[1]

"The only place in Ireland where a man can make a fortune is in America."[2]

In the spring of 1883, a troupe of twenty-one made a prospecting expedition to Baja California from the mining town of Tombstone, Arizona. This was perhaps the largest effort of its kind and included local experts such as Bill Hogan. The expedition members were blessed with a rumor that a dying Mexican's last breath told them the specific location of a great fortune or gifted them a map, depending on the version of the story one read. The fact

https://doi.org/10.5876/9781646422517.c000

that the expedition failed is unsurprising; many did, and for many different reasons. What is surprising is that the leader of the group, described by an Indian agent and newspaper editor as someone whose "energetic, courageous, and self-sacrificing life was an inspiration on a wide frontier during half a century" and who was of "frank manner . . . self-reliant spirit, [and who had an] emphatic and fascinating Celtic brogue," was a woman—Nellie Cashman.[3]

The image of a foreign woman leading a mining expedition challenges some preconceived notions of a stereotypical miner. If asked to imagine the stereotype, people would likely envision a man panning gold from a river. The overwhelming majority of miners in nineteenth-century America were men, were foreign-born or second-generation Americans, lived in towns rather than in isolated rural areas, and were likely to be Irish. These generalizations do not tell the whole story in the knotted and complex history of how the many different people and groups interacted with each other.

Mining was foundational for the industrialization of the United States and the emerging Irish-American identity. The opportunity to acquire wealth prompted one of the largest migrations in American history, the California Gold Rush, in turn spurring the settlement of the West with the possibility of employment and thereby solidifying US territoriality. It continued to change vast regions through settlement and development, not least by its transformation of the landscape that only vaguely hints at the innumerable miles of tunnels dug underground. Many in the emerging workforce were Irish who marched each year in St. Patrick's Day parades as a demonstration of their identity and as a message to wider society of their communal strength. They built and supported Catholic schools, hospitals, and churches staffed with compatriot clergy and other religious workers; sponsored nationalist causes; repatriated vast sums of money to relatives in Ireland; and carved out a place for themselves in the emerging urban West. Their communities were oft-maligned with nicknames such as Poisonville or were referred to as a "black beating heart" by those who viewed the ideal of American western society as pastoral, Protestant, and native-born.[4] These towns represented a vital local market for goods and services, especially agriculture; as Charles Dickens wrote, "It would be hard to keep your model republics without them, for who else would dig, and delve, and druge."[5] His observation held true across the American West where the greatest engine of change—mining—established the urban frontier, prompting the rapid expansion of rail networks across the vast landscape and through the formidable terrain of the Rockies, further propelling demographic changes.

To fully comprehend the ways the American West changed, it is necessary to look at the varied ethnic makeup of this landscape during the nineteenth and early twentieth centuries. The Irish constituted a unique and vital element of that workforce, and their experiences are littered across a diverse range of surviving sources. We should look beyond the cities of New York, Boston, and indeed Butte in our efforts to craft a fuller appreciation of the varied experiences of Irish-Americans. An occupational survey enables us to closely investigate these mining towns across a broad area where we can compare and contrast the events and show how they intersect. Historians have long acknowledged a distinctive Irish culture in America. Using hundreds of letters, Kerby Miller's pioneering *Emigrants and Exiles* details the distinctive experience of the Irish in the United States. More recently, David Emmons's *Beyond the American Pale* goes a step further than Miller's argument that this separation from the dominant Protestant culture put the Irish at a disadvantage in the real and imagined American West.[6] His work builds on and challenges both of these views, arguing that Irish Catholicism could be both a blessing and a curse, defined both inside and outside the group. It was a blessing in that it could allow immigrants to retain their cherished beliefs notwithstanding their mobility, empower them to face down adversaries, and enable the repeated creation of community in spite of the boom and bust nature of mining towns. It was a curse in that it could isolate them from others, including possible allies; render them susceptible to targeted attacks from opponents; and expose them to claims of disloyalty to the United States and to Anglo-American society. The chapters in this book build on those earlier explorations and find that the stability and conflict that defined the mining regions of the nineteenth and early twentieth centuries were most often shaped by the Irish presence and the varied groups that supported or opposed them. Beyond letters, newspapers, and records, the testimony of their lives shows how they elected to retain their Irishness and moreover how they determined to reshape their world and influence it, crafting it into Irish-America. By extension, it was encounters with others that formed the lens through which we can see and understand the ethnic dimensions of workplace relations and business frameworks expressed in the invisible social contract.

To scratch the surface of an Irish miner's life is to reveal the multilayered complexity of their identity. Nellie Cashman's family fled her home in Midleton, Cork, at the outset of the Great Famine. She grew up in Boston and went west over the Panama crossing to San Francisco before leaving the city with her mother to work in Virginia City and Pioche, Nevada; Tucson, Arizona; and a number of other infamous boomtowns. Together, they worked in boardinghouses before establishing their own in Pioche. In 1872 the state mineralogist reported to the state senate that Pioche "scarcely ever had a parallel [for] lawlessness and horrid murders, which have scarely ever had a parallel in the history of this coast."[7] American observers relished lurid accounts as the Cashman family lived happily in the excited bustle of the town. It was not fear that caused them to move to the next opportunity; rather, it was the declining economic fortunes of the mines and thus of Pioche itself. Cashman left boardinghouse work and decided to join a group prospecting in the frozen north. She led a rescue expedition and earned the sobriquet "the Angel of Cassiar." Afterward, she opened a restaurant at Tucson in 1879 but left to join the rush to Tombstone where she partnered with other women to open two businesses. One reporter commented, "She is as adventurous in pushing forward to a new region as any nomadic miner," but in many ways, that was what she was.[8] She aggressively collected donations for the construction of Catholic hospitals, schools, and churches; but she was equally dismissive of women's Progressive reform groups as Protestant snobbery, "clubs for catty women and false standards of living."[9] She saw America through Irish Catholic eyes, stating that "all enjoy the liberty of pursuing the road of wealth and happiness according to the dictates of his own conscience" by flying the American flag on British holidays, collecting for the Land League, and calling her hotel "the American."[10] She cared for her sister's five orphaned children for three years and then, as Tombstone began to fade economically, placed them in Catholic institutions before continuing her wandering with the next mining rush, again defying social and occupational expectations.[11]

Cashman demonstrated the importance of her heritage through her unremitting support of and loyalty to Catholicism, but this was just one of many ways of expressing her multifaceted Irishness. Ethnic identity could subtly or overtly influence people's lives and choices. It could affect where they lived, where they prayed, where they socialized, who they befriended, the way they interacted with others, and the way people from other ethnic backgrounds—including Anglo-Americans—interacted with them, positively or negatively. Unsurprisingly, they, like the larger body of Irish-America of which they were part, were not a uniform mass; their Irishness reveals itself as measureless and heterogeneous, best understood on a sort of spectrum where the varied hues of green and orange of the Irish-born and their descendants in America show themselves to be as diverse as the lives they lived. Some of them fell away from their Irishness and wholly adopted an American identity, though the numbers who did this and the degree to which this was possible for migrants and their descendants will remain a source of fruitful historical discussion for some time.[12]

Whatever its rather fuzzy boundaries, there is no dispute that Irish Catholicism became the de facto identity of the group regardless of where on the spectrum of Irishness or religiosity individuals placed themselves, and both the Roman Catholic Church and the Nationalist label became central in forging the physical structures and imagery that bound Irish-American communities together. Irish Protestants, in contrast, mingled and merged seamlessly into the Anglo-American population, practically disappearing within a generation. Unlike Irish Catholics, there is no evidence that Irish Protestants suffered discrimination from Americans; nor is there evidence that they formed a separate culture or community. Irish Protestants used the term *Scotch Irish* or *Scots Irish* in mug-shot histories such as the *Progressive Men of Montana*, not as an effort to form their own identity but rather to distance themselves from the prejudice against Catholicism and the ways it was tied to popular perceptions and stereotypes of geographic Ireland, in particular the association with poverty. Thus, when individuals used the term *Scotch Irish*, they did so to vouch for the religious and class pedigree of their ancestors. Irish-born Catholics and their children—those who would form the identity known as Irish-American—remained a distinct group from white Anglo-Saxon Protestants and continued to visibly demonstrate their ethnicity even if Irish Catholics had ample opportunity to leave it all behind and fully "assimilate."

Other subtle differences existed within this group, the most important of which is the class distinction between the lace-curtain / middle-class Irish and working-class Irish. There was also a depth of local association. On a broader scale, there was the Fardowner and Corkonian rivalry, which stretched back to the United Irishmen rebellion and beyond.[13] More narrowly, allegiance to parish was intense and commonly appears in any description of an Irish person by themselves in their recollections or by others in obituaries or on headstones (see chapter 3, this volume). In mining, other divisions within the group were related to occupational skill levels corresponding to three primary types of Irish miner. The first was the skilled miner. Often, they entered the mines at a young age, grew up to be as skilled as their famed Cornish counterparts, and had similar pride in their occupational skill. The aristocrats of the mining occupation, they were found in significant numbers in Virginia City, Leadville, and Butte. They were generally lifelong miners and rarely left the occupation unless compelled by death or infirmity. The second type was the temporary miner, sometimes called the "ten-day miner." They mined for a while but alternated between mining and other jobs, usually depending on economic circumstances.[14] Often, these were experienced miners, but they had no deep attachment to the occupation, viewing it as a job rather than a lifelong career. The third category was the placer miner. Unlike hard-rock miners, they rarely delved deep into the bowels of the earth and had limited mining skills. They were searching for the rich diggings and usually sold off their promising claims for someone else to work and began their hunt anew somewhere else. They panned for gold in the rivers and are remembered fondly in the popular imagination as the classic '49er Argonauts. The categories could be porous; some skilled miners became placer miners and some, such as Cashman, developed mining skills after a lifetime of prospecting. Some hard-rock miners joined the gold rushes and mingled briefly with these placer miners, but the surface deposits in the American West were rapidly combed in a few frantic years; as hydraulic mining became widespread and deeper mine shafts were needed to reach gold and silver deposits deeper underground, the demand for the skills of the hard-rock miners increased.[15] The prospect of longer-term employment made the decision to continue gambling on the next big strike less likely and attractive, although miners sometimes kept spare mining equipment in case of a particularly promising rush nearby.

As the diversity of ethnic attachments and occupational divisions suggests, there is no single story of Irish miners. A wide spectrum traversed the American West in the nineteenth and early twentieth centuries to dig, haul, drill, and blast an income from the earth. They worked in every manner of mining, as famed gold-rush prospectors, as hydraulic miners washing hills away, and as copper miners in the honeycombed depths under the city of Butte. A small few struck the mother lode and became immensely wealthy. Irish miners migrated not only from Ireland but often several times across the US. They traveled seeking jobs, comrades, and wives, remaining in one location from a few weeks to a few decades before moving on. Perhaps the most famous example of a *spailpín fánach*, a wandering laborer, was Donegal-born Michael MacGowan. He tried his hand at copper and gold mining, wandering in both an occupational and a geographic sense, and recorded his movement across America and eventually back to Donegal in the memoir *Rotha mór an tSaoil* (The Great Wheel of Life).[16] The title tied the Irish proverb on the turning fortunes of life to his mobility.

The late 1840s witnessed two formative periods in Irish and American migration history: the Great Famine and the Gold Rush. Omitting either of these watersheds would leave a huge gap in the story of the Irish, the Irish in America, and the development of the American West. This book closely examines the emerging mining industry and shows micro-frontiers of opportunity opening, evolving, and—sometimes as quickly—closing. These frontiers formed parts of a system whereby Irish miners and laborers traveled looking for jobs, establishing communities, often raising families, and sometimes finding a manner of stability in the most uncertain industry in the nineteenth century. The mid-nineteenth century marked the advent of large-scale industrial mining, while the arrival of mechanization and strip mining in the late nineteenth century was akin to a long twilight for the skilled miner.

By the early twentieth century, mechanized mining was a very different occupation from its antecedent. Large-scale mining swept aside smaller, less profitable mines—at first locally, then nationally. This final act in the present story marks the waning point of Irish influence in the mining towns. Irish-American mining communities faded as the company-worker relationship that sustained them broke down and was replaced by corporations whose loyalty lay solely with their shareholders rather than the owner's ethnic group. This move, in turn, encouraged a largely Irish-American trade union leadership to build a more inclusive tent and move away from the skill- and ethnic-based distinctions that had often fractured and weakened earlier mining unions.

Upward mobility from blue-collar mining to white-collar labor was most often a multi-generational journey for the Irish. Whereas statistics in literacy or fluency in English or Irish remain difficult to establish, every Irish migrant would have encountered the English language by the time they arrived on America's shores. Most Irish were fluent or bilingual, and this gave them significant advantages in labor relations.[17] Later, as opportunity declined, their children's rigorous, largely Irish-Catholic education represented the flip side of the same coin of American mobility.

The most significant decline in opportunity was represented by the concurrent creation of a professional managerial and engineer class, which, in turn, solidified the stratification of employment in the mining industry, closing possible avenues of career and class advancement for miners once dependent on circumstances as tentative as ethnic favoritism by mine owners (see chapters 5 and 6, this volume). Earlier mining histories lumped miners of all skill levels into one occupational category or suggested skill distinctions based entirely on ethnic identity, muting their heterogeneity.[18] The Irish as a group show how mistaken this historical approach is, since although many were placer miners and "ten-day" miners, there were also many skilled miners. A reason for this error has been the ease with which historians classed the Cornish as the premier miners in the US. Historians have tended to ignore their privilege (i.e., preferential promotion of more acceptable Anglo-Protestants by mine management and discrimination against other ethnic groups) and confuse their long-standing mining heritage with the fallacy that the Cornish were the only skilled miners. They interpret long-held historical prejudices as historical realities instead of contextualizing each place with its unique amalgamation of people. For example, a wheelbarrow was often referred to as an Irish buggy by miners, as an insult to the Irish and to associate them with the less skillful "mucker" position in the mines. Conversely, narrow tunnels were nicknamed "'Cousin Jacks,' because only Cornishmen were supposed to be able to work in them."[19] As the American miner who wrote those details noted, however, his Irish comrades did work there "and they taught me how to do it too."[20]

The diversity of the Irish mining experience coupled with their enduring adherence to their religious and cultural identity meant that those historians who had previously lumped them together did so without an adequate understanding of the paths the Irish trod on their way to the American West.

A mining history that avoids this trap is David M. Emmons's monograph on the Butte Irish wherein he details the unique position of Irish miners in a single American mining town. However, treating Butte as a node rather than the sole focus illuminates the twisted paths of fortune trod by the migrant community, even in towns near that green beating heart of Montana, such as Marysville.[21]

Historian Frederick Jackson Turner defined the early historiography of the American West by postulating that the frontier experience reinvented the immigrant as American; thus, in this framework, the Irish are mashed into the broader body of white immigrants and effectively erased from the history of the American West.[22] This muting of ethnic distinctions was integral to the early mythology of the American West, with linear or overly reified narratives including cowboys versus Indians and the progress of civilization through Manifest Destiny, acting as a veneer to justify the territorial expansion west and simultaneously portraying its population as exclusively American—by definition, white and Protestant. This Turnerian school has been thoroughly eclipsed by New Western history, an approach that made important strides in correcting the historiographical lens. This effort focused on the underwritten history of the exploited, beginning with American Indians and later expanded to include women, Latinos, and Asians. Yet this approach has sometimes failed to explore the very real fissures and fusions between and within cultural groups and subsumed the Irish under the broad terms Anglo, British, or white.

The ways the Irish saw their experiences in the American West were different from the way others viewed them and were informed by their own history. Exemplifying the distinctive Irish perspective were Fr. Eugene O'Connell and Michael MacGowan. For example, O'Connell served the California missions in the 1850s, where he blamed the decline in the Indian population on relentless exploitation at the hands of "rapacious agents." He wrote mournfully, "What a people that race of the 'red man' might have become . . . he has no country anymore," and in a resigned Christian hope of a hereafter he added, " 'tis well he will have a grave and a Father beyond it."[23] Decades later, in the far-off state of Montana, MacGowan was working in a silver mine at Granite Mountain when he noticed the strong tensions between Indians and encroaching Americans. He contextualized the origins of the conflict as a result of the actions of a "greedy white man . . . with friends at court or a *planter* without conscience."[24] The use of the word *plandóir* (planter) rather than *ionnaitheoiror* (settler) deliberately linked the plantations in Ireland with the contemporary struggles of the Indians in the West. In the next few lines he reinforces this parallel:

> The Indians that were left here and there were in a bad way and we had a great deal of pity for them—the same thing had happened to ourselves home in Ireland. We knew their plight well. We understood their attachment to the land of their ancestors and their desire to cultivate it as well as their wish to keep their own customs and habits without interference from the white man. We were interfering with them I suppose, as well as everybody else but at least some of us sensed that if they were wild itself, it was not without cause.[25]

MacGowan simultaneously paralleled the Irish history of dispossession with the experiences of Indians and distinguished the Irish from "the white people" (*ón mhuintir bhána*), a cultural and ideological divorce from the racial ideas of whiteness. He further admitted their partial culpability, tempering it with a resigned sympathetic note contextualizing the Indians' present difficulty as imposed privation and their hostility as justified.[26] He was not looking at the situation as the average Anglo-American would, with the Indians as irredeemable savages or a ground-up by-product within the wheels of progress.

The Irish brought their unique worldview with them on their long journey through the territorial and industrial expansion of the US. The chapters in this book trace the veins of these communities that spread across the region, revealing how the people organized their lives, their relationships, and their ties beyond the places they lived. The transnational, sometimes multi-generational migrants traveled from Ireland, often through Britain, to the eastern United States and then to the American West. Direct lines of migration perpetuated by social networks, for example, from the Beara Peninsula to Butte, played an important role in the emergence of tight-knit neighborhoods such as Corktown in Butte. Britain is also an oft-ignored and important part of the staggered and indirect migration story of Irish miners to the US. They often traveled through British ports, frequently spending time working in mines in Scotland, England, and Wales to earn money for the onward journey.

Irish migration was a web stretching across the world and was not limited to a single town or city. With notable exceptions such as David Thomas Brundage's *The Making of Western Labor Radicalism: Denver's Organized*

Workers, 1878–1905 and Gunther Peck's *Reinventing Free Labor: Padrones and Immigrant Workers in the North American West, 1880–1930*, the limited city or town or state view has been the standard. Even across the broad historiography of the American West, a cursory search for the term *Irish* in the indices of the hundreds of history books on the American West or the dozens of books in the subcategory of mining leaves the distinct impression that the Irish melted into the background of American history rapidly and effortlessly.[27] One edited collection opens with the clarion call "European immigrants are the forgotten people of the West," and certainly the complexities and diversity of the European immigrant population of the West have yet to be fully recovered from the mythologizing period that forged the first histories of these places.[28] The research in the following pages arose out of these explorations and, in turn, reveals in a direct way that the Irish cannot be divorced from any part of the American West. Correspondingly, mining cannot be extracted from the history of these places—its presence lingers on through the imprint of a thousand abandoned mine shafts and ghost towns on the landscape to the story of the communities now living in these regions.

The reasons the influence and breadth of the Irish were forgotten in historiography varied based on the period and the field of study. The subcategory of mining preoccupied itself with the technological and operational aspects of the industry. For example, explorations of the subtleties of ethnic identity usually extend only so far as to mention Cornish miners, largely because of their notoriety as the premier skilled hard-rock miners in the nineteenth century and the famed mining traditions of Cornwall. Greater focus on the mixed composition of the workforce and a wider awareness of the diversity of mining experience distracted from the economic narrative, one that seemed to operate on the premise that the quantity of material mined, rather than the workers who mined it, defined the story of the extractive industry. This led to an overemphasis on technological changes that sidelined the transformations in the mining workforce, the most important of which occurred at the end of the nineteenth century when eastern and southern European peoples overtook British, Irish, Chinese, and German residents as the largest immigrant groups—a change that further subsumed the Irish into the Anglo or British category for the sake of a simpler historical narrative.

Yet a simple narrative does not easily accommodate cultural distinctions or the story of Irish miners in the American West, either in the first part detailing the often thin occupational line between miner and mucker or laborer or, later, in telling the story of how the Irish interacted and worked with other ethnic groups, in particular the complicated relationships between them and the Chinese, Finns, Italians, and others.[29] These distinctions are prerequisites for any attempt to understand the development of trade unionism in the American West. Like others, miners' day-to-day lives were determined more by their social circles and their interactions with others than by the technology of their occupation, and it is impossible to understand why some would be in favor of trade unionism or opposed to workers' organization without the context of personal identity and experiences of circumstance. Stability, better wages, and safety were defining motivators, broadly encapsulated as opportunity; but in many cases this was a secondary consideration compared to the close bonds of family, friends, and faith that tied them to each other and gave their lives meaning.

Industrial history requires a more inclusive approach and nuanced framework that builds on the work of the New Western school in which we can understand the unique place of the Irish in the American West. Cultural allegiances, seen most obviously through Irish immigrants' continued allegiance to Catholicism and the persistence of ethnic divisions, contradict the very core of the frontier hypothesis—the reinvention of immigrants as Americans. The Irish fashioned their communities into ones that fostered their Irishness even in remote mining towns, often by choosing Irish wives, Irish friends, and Irish associations. The temporary nature of mining and the corresponding shifting ground of opportunity for employment and steady wages regularly meant that these community bonds had to be forged and re-forged many times over the course of multiple migrations.

Their persistence in reestablishing an organized Irish presence is remarkable. Emmons notes the dozens of mining towns in the American West that are included in newspaper subscription lists as having collected funds to support Irish nationalist causes.[30] Some are explored in the following chapters, while others have yet to have their full histories written; still others have left so meager a historical record that an odd reference in some surviving newspaper remains the only trace of what was a thriving Irish congregation. The diversity of experiences helps us see similarities in the struggle for identity and place, the dynamic relationships with allies, and the reasons the Irish faced the adversaries they did.

Despite repeated dislocation, community anchored many to their sense of self and society and played a vital role in miners' lives. The men and their income provided the foundation for what emerged, but the structure was built by women, priests, nuns, and children without whom Irish communities were evanescent. Although hundreds of smaller outposts in the form of mining camps and prospectors dotted the landscape, most mining during this period was an urban occupation because the workforces required to work the larger, deeper mines necessarily gave rise to sizable towns. The population density proved alluring to Irishmen, who disliked the isolation of American agriculture and remembered the fresh trauma of the Great Famine, *an Gorta Mór*. The examples of Virginia City and Butte (chapters 3 and 6, this volume) suggest that Irish women shared the same preference for urban frontiers rather than rural ones. If women appear as secondary actors in portraying the life of these mining towns and communities, it is not intentional but instead represents the scarcity of firsthand accounts from the figures coupled with the occupational dominance of males in mining. The limited surviving miners' letters, fraternal records, and company records are almost silent when it comes to the role of women in nineteenth-century mining towns. Parish records such as the internment book of Smartsville, the patient logbook of Virginia City, and census records (chapters 2 and 3, this volume) reveal some aspects of their lives but represent the barest of starts in an effort to comprehensively detail women's and families' roles in ethnic communities throughout this mining diaspora.

This project began with the discovery of a series of Irish emigrant letters in an early version of the Irish Emigration Database, which I found on a research visit to the Irish Migration Centre in Omagh. Further research led me to Professor Kerby Miller who generously granted me access to his vast collection, which included many letters that contribute greatly to our understanding of the life of Irish migrants in the American West. In the context of the period, the surviving letters are a fragment of what was once a global communication network. They offer a fascinating window into the thoughts and feelings of the Irish scattered across the world and the types of information they thought it was important to share.[31] Newspapers provided another major source of information, and both tools relied on migrants' ability to write and read or at least to know someone who did. As literacy increased, so too did the migrants' reliance on them for information; the letters often

Figure 0.1. The arrival of an emigrant letter was an exciting occasion for family and friends in Ireland. The letter was a source of consolation, advice, and warning often treasured by relatives. In the painting a young girl reads a letter from America aloud to her family. James Brenan, *News from America* (1875). Crawford Art Gallery, Cork, Ireland.

asked relatives in Ireland to send them newspapers so they could read the local and national news. This shrinkage of the world through regular news correspondence and faster communication also encouraged the formation of Irish societies and fraternities dedicated to causes dear to the migrants' hearts, such as the Ancient Order of Hibernians, whose limited surviving records are utilized in chapters 3 and 6.

A consistent theme across this broad chronological and geographical span is the importance of ethnic organizations and forms of protest in the struggle for a fair living. The Irish in these mining towns had the same exceptional organizational and political savvy that made them so powerful in the major cities of New York, Boston, Chicago, and San Francisco. Sometimes these organizations were either ethnically or religiously exclusive, but a draft note from Marysville, California, reveals that there were some efforts to reach out to the wider community and explains the causes of hostility, if it existed. The note sought the organization of all Irishmen of Marysville, regardless of religious denomination, into an association for the promotion of Irish causes, and in this case ethnic and national identity trumped religious loyalty. The note also contained a warning against "wolves in sheeps [*sic*] clothing"; the writers of this document had experience dealing with agitators, spies, and agent provocateurs.[32] A further consistent theme reinforcing Irish suspicions was the predatory behavior of vigilantes and businesses. The most personal spaces were vulnerable by definition and became targets for infiltrators. The widespread evidence from Pinkerton and Thiel detective reports provides proof that such fears were fully justified. These sources also provide a wealth of intimate information about miners, albeit filtered through prejudiced eyes and ears. The detectives frequently report dialogue overheard between workers and hint at how the Irish perceived their fellow workers, themselves, and their position in society. When used carefully, they offer a remarkable view into life and labor for these groups during this period.

Pre-migration experiences also form a crucial part of these people's story. The Irish brought with them an intimate familiarity with opposition to powerful systems of authority through various means, political and otherwise, drawing on strong traditions of agrarian agitation—a set of grievances established by the trauma of recent Irish history and reinforced by their own emigration. They recognized themselves as a single drop in the vast swell of economic emigrants, who were also making sense of their own dislocation and identity. In the US they became foremost agitators for workers' rights and consistently resisted company control of miners' pay, communities, and lives. When mine companies targeted the Irish or unions (the terms were often interchangeable), agitation increased as a response to their actions.

Evidence of anti-Catholicism directed at Irish-run Catholic institutions and fraternities is presented in chapter 3, while local anti-Irish discrimination, most conspicuous in chapters 4 and 5, represented a coordinated mobilization of business and government forces against an ethnic group. Mine companies intentionally engaged in exploitative practices and encouraged ethnic friction and violence to strengthen their position and dominate communities, both politically and socially. Management's short-term goals were only partially focused on lowering wages and maximizing profits; as newly discovered documents reveal, their worldview focused on consolidating power, the mechanisms of which were infused with specific ethnic prejudices. In other words, company documents reveal how these officials perceived the Irish and prove that many labor disputes were based on bigotry rather than any natural friction between workers and owners.

The importance of ethnic identity and the uniqueness of Irish miners in nineteenth- and early twentieth-century America is a constant refrain in the primary sources of the period. Their lives expose a distinctive web, with hubs concentrated in mining towns and reaching across America and worldwide back to Ireland. The history of Irish involvement in the US mining industry details a relentless effort to earn a fair income and form a community despite repeated setbacks, often created by Anglo-Americans in positions of power in mine companies. Management stoked up and armed nativist groups in an attempt to divide towns and pit communities against one another.[33] The following chapters set out to explore the degree to which Irish-American experiences differed from those of other ethnic groups and demonstrate how Irish-Americans interacted with those groups within the backdrop of each location. Comparisons between these local case studies highlight the complexity of the story of Irish miners in the American West and the unique space they carved for themselves in many varying contexts. Irish men, women, and children shared broad interactions with other peoples, including Asian immigrants, European immigrants, Indians, and native-born Americans. The rich tapestry of cultural experiences expressed in these Irish identities formed the backdrop for these complex encounters as the Irish sought to survive and thrive in the spaces they built for themselves in mining areas.

To be an Irish miner in the American West was to be both uprooted and transplanted many times and a wandering laborer whose home and community would always be transient—ephemeral in a way. And yet the time and effort spent building and rebuilding those homes and communities again and again was not futile—it sustained them, most obviously in their identity, while also strengthening them. The links forged demanded respect from others, and this local power was linked to other Irish networks as far distant as the mythologized homeland, inviting them to follow and deepen the legacy. This is hinted at on the scattered headstones describing the parish of origin, in the Irish counties listed in hospital books next to Irish patients, in the malevolent letters written among American mine managers about their Irish workers, and on statues of St. Patrick in the many Catholic churches scattered throughout those vast expanses. Being Irish could ensure that one would find employment in Butte but not 60 miles away in Marysville. It could offer companionship within unions or fraternities but court hatred from vigilantes and masons. It created a sense of community but placed the Irish and their communities beyond the American pale.[34] Inevitably, a degree of accommodation and an element of friction were the consequences. The trails and trials of the Irish in the mining frontiers of the American West illuminate some forgotten historical stories and place them as an important piece of the wider puzzle: that of the Irish diaspora.

1

Varied Hues of Green

Wherever a hole, a pit, or a shaft was dug in America during the nineteenth and early twentieth centuries, chances were good that the person hewing the rock was Irish. At a glance, mining might seem to have provided a stable, settled life for workers near the mine entrance, but mining was often a sporadic occupation that was dependent on the quantity and quality of resources in the earth, the fickle demands of markets, and the whims of management. Mines closed when a seam pinched out and could open as quickly when new ones were discovered. According to John Brophy, "Miners of those days had to lead a gypsy life."[1] Mine managers could fire their entire workforce or import immigrants in an effort to drive down wages and hobble labor unions. Strikes for better conditions and wages resulted in some of the bloodiest labor conflicts in American history and remind us that underground presented just one of the many dangers miners faced in their day-to-day lives. In the case of Irish-born miners, movement was integral to their life and safety was a calculated risk beginning with the first step out the door of their homes and their journey from Ireland. The road to the American West often

https://doi.org/10.5876/9781646422517.c001

led through Britain, Australia, and the Atlantic seaboard; mobility meant many journeys within a vast region. Their life and work encouraged movement from one place to the next, chasing one opportunity after another in an effort to improve income and status.

It is difficult to trace the exact geographic movements of Irish miners and their families due to the diffuse and multifaceted nature of their migration. Economics constituted one major determinant of their movement, but the slow passage of information and the time spent traveling meant the reality they faced upon arrival did not always align with the descriptions of friends and relatives. They sometimes stayed in one location for weeks, years, or even a generation before moving on to what they hoped or had been informed might be greener pastures. As individual migrants or when traveling with relatives and friends, the Irish relied on information from other migrants through letters, newspapers, and word of mouth. The growth and disappearance of earlier Irish mining communities in Ireland, Britain, and the United States—for example, in Pennsylvania and the Upper Peninsula of Michigan—were linked to the later emergence of Irish communities in Virginia City (Nevada), Leadville (Colorado), and Butte (Montana). These towns, home to the largest Irish mining communities in the American West, offer a variety of signposts as to the different paths migrants took, but before looking at these locations individually, this chapter will assess overall numbers and broad paths of migration.

Our best source of information on where the Irish settled is the US Census records. There are three important figures to consider when looking at Irish mining in a particular town or region. The first is the total number of Irish-born residents in the state or territory. This gives a general idea of whether the region is popular with Irish migrants. These figures can be obtained from the US decennial census reports, which, apart from the notoriously unreliable 1850 California census, are a rich source of information. The second figure is the number of Irish involved in mining. This figure is not simply those listed as "miners"; many more workers would have been involved in support roles for the mines, such as carpenters, blacksmiths, and machinists.[2] As such, they are part of the narrative but tend to be set aside when it comes to calculating the core miner workforce.[3] When speaking of Irish miners, it is important to consider the parallel number of less skilled or unskilled Irish laborers. Muckers, who cleared the blasted rock and put it into the rail trams, and trammers, who oversaw the trams and movement of rock and ore out of the mine, were two such types of workers. In many cases they worked under Irish miners who vouched for and supervised them, but these men, working underground, were not classed as miners in the statistics.

To more accurately reflect those missed in the occupational focus, the final figure is the most important: the size of the Irish community itself, including men, women, and children. This figure encompasses both Irish-born residents and those of the second generation who together form the core of Irish-America in this period. Comparing the Irish-born to the larger Irish-American category gives an idea of the stability of the community and helps contextualize the influence and interaction of the Irish with people of other ethnicities in these towns. Most censuses offer figures for occupation and birthplace, but in the US in the nineteenth century, only the 1880 census records include the nativity of a person's parents; this information allows us to establish the size of the national communities or ethnic groups. Due to the inordinate amount of work required to manually compile the individual returns on a computer and crunch the figures, historians have only been able to do this previously on a very small scale. The recent digitization of the 1880 US Census by the Integrated Public Use Microdata Series at the University of Minnesota, Minneapolis, means we can compile every individual response to parents' birthplace to create a far more accurate estimate of the size and stability of ethnic groups in the America West in the late nineteenth century.

Although it is only possible to retrieve accurate estimates for the size of entire Irish communities in the 1880 census, we can still establish a rough sense of the increase in and decline of Irish presence in the several important mining centers detailed throughout the chapters of this book by counting only the Irish who were born between 1850 and 1920. The greatest concentration of Irish miners was found in Storey County, Nevada, home to the silver mines of the Comstock Lode and the town of Virginia City; in Lake County, Colorado, where Leadville developed into another large silver mining center; and in Silver Bow County, Montana, with its Irish-dominated copper mine city of Butte. These regions were home to thousands of Irish miners and, in turn, supported families, other Irish workers, and Irish businesses through the wages the miners earned. Altogether, the Irish communities in these mining towns represented a major portion of the total population in mining towns in the American West.

Irish-America

While born-in and nativity figures are useful, they are not the most accurate reflection of the size of the Irish communities in the American West. For this we need to include those whose parents were also Irish. Even the use of this more accurate measurement fails to capture the totality of these communities, however. Evidence of this is more clearly seen in the traces they left throughout the region in the form of churches, schools, hospitals, the names of streets, and overgrown graveyards. The establishment and growth of Catholic congregations in mining camps and towns and the accompanying infrastructure that sprung up wherever the Irish settled provide further evidence of a parallel culture distinct from other ethnic groups, including Americans. One highly visible strand of this identity is the Irish adherence to Catholicism. The other strand was cultural, most visible through the formation of Irish organizations that furthered their ability to maintain links within their immediate community, across America, and in Ireland. Both strands were intertwined and spread across the globe in the form of transnational links that proved important conduits for Irish mobility and sources of comfort and familiarity during periods of transition. The braided strands of identity, religious and cultural, helped these migrants create what became known as Irish-America.

Other Irish-born residents did not subscribe to this identity. Most Irish Protestants in mining towns allied with, identified with, and acted so closely akin to native-born Americans that they can be considered a substrata of that group. Only occasionally did they choose to separate themselves with the label *Scots Irish*, and when they did so it was mostly to clarify that the land of their birth or ancestry could also be home to Protestants. The overall number of Irish Protestant miners in the US was small, probably a low single-digit figure.[4] This estimate is based on the small numbers of Protestant miners in the north of Ireland at places such as Coal Island and at Castlecomer in Kilkenny.[5]

There was also a limited degree of apostasy for the Irish who chose to assimilate entirely into American society, their surname being the sole indicator that they had any links to Ireland. Certainly, there is a degree of difficulty in assessing identity over time. The cross-generational cultural transfer within families whose parents were of different ethnicities remains largely unexplored in the nineteenth century, so figures for second-generation "Irish-Americans" in this book, in spite of heavy revision upward for the size of the Irish-American community in the West, actually tend toward conservative by tallying only those for whom both parents were Irish-born. For the vast majority of the second-generation Irish in 1880, this was the case and they were ethnically Irish or, as they called themselves, Irish-American. Despite the use of far more accurate figures than ever used previously, this effort to prevent overestimation means that the figures may represent a slightly conservative estimate of the size of the entire Irish-American community.

For the vast majority, the creation of a hyphenated identity was a very important development for the Irish-born who sought to accommodate loyalty to their Irishness with loyalty to their new homeland, and it can be seen in the growth of both Catholicism and Irish fraternities in the American West. Furthermore, this development enabled the children of Irish immigrants to proudly identify with their distinctiveness, even if their families had left Ireland a generation or more earlier. This new identity often manifested itself through efforts to help those in Ireland or a longing to return to the motherland. Other miners noticed the Irish tendency to reminisce and romanticize about home. The Scottish mining bard Davie Robb wrote "Reflections of an Irish Emigrant" in 1908:

Away o'er the sea is the land of my fathers,
Far over the ocean my heart turns to thee;
But fond recollections of childhood ne'er wither,
For green in my mem'ry is Ireland to me.[6]

Robb was born to a Scottish mining family in Staffordshire, England, and migrated to Ohio in 1902.[7] Robb wrote these lines either because of a personal familiarity with Irish music and culture or in an effort to evoke the popular theme of Irish longing for Ireland. This well-established motif can be seen in contemporary Irish sheet music titles, including "Erin Is My Home," "Come Back to Erin," "Exile of Erin," and "Almost Home."[8] The popularity of such songs contributed to the popularity of this Irish stereotype.

Irish miners used many labels to define themselves—Irish, Catholic, worker, miner, unionist, Irish-American, Democrat—whereas Irish Protestants adopted the term *Scots Irish*. The failure to appreciate the differences between "ethnic group" and "race" leads historians into difficult territory, as happened in Philip J. Mellinger's otherwise excellent book *Race and Labor in Western Copper*, where he terms a simple ethnic alliance between Irish and Cornish miners in Bisbee, Arizona, a "nativist" alliance. The following sentence illustrates this apparent confusion with ethnic terms: "the Protestant Anglo-Irish commingled, and to a limited extent, they were socially connected to the Catholic Anglo-Irish, while the non-Anglo-Irish were both physically and socially separated from both of the other groups."[9] Mellinger's use of the hyphenated term *Anglo-Irish* is problematic because it subsumes the diversity of Irish immigrant culture and language under the phrase Anglo-Irish, one that has particular context and meaning for Irish studies.[10] Perhaps most important, it was not a term Irish miners used in reference to themselves. The author's main point, that it would have been impossible to ally if the Irish and Cornish had no intercommunication, is valid. His case study reflects similar events repeatedly shown in the present work: that the shifting systems of ethnic alliances and social exclusion were powerful tools of labor and communities in the nineteenth and early twentieth centuries.

Those in the second generation understood their partially muddled position in society, as can be seen in the poem titled "A Scoto-Irish American's Protest" by John F. Kearney, published in *Miners' Magazine* in 1903.[11] This use of multiple identities doubled as a pun on identity and an attempt to encourage worker solidarity by overcoming ethnic divisions.[12] Kearney's humorous use of three identities to describe himself should underline his keen awareness of its importance for both himself and the workforce. The importance of identity and culture was intimately understood by the working class, as further exemplified by Mother Jones—who was actually the Irish-born Mary Harris, who adopted a Welsh pseudonym in a sophisticated mimesis to bridge ethnic divisions by cloaking herself in the artificial identity of another. Kearney and Mother Jones were just two among many who attempted to unite disparate cultural groups under a class umbrella. These distinctions between the various ethnic groups would have had important consequences for where the Irish lived, who they socialized and prayed with, and how their children would see themselves and negotiate their place in the United States.

The question of identity was a mediation that began back in Ireland but came sharply into focus when a person moved from their home. An example of this complex process and an inherent statistical weakness of historians is the traditional over-reliance on "born-in" figures in official records as a basis

Table 1.1. Irish-born and Irish parentage miners in select US states, 1880[a]

State	*Irish-Born*	*Irish Parentage*
California	2,689	3,365
Colorado	2,252	3,872
Nevada	1,375	1,692
Montana	702	923
Idaho	433	550
Arizona	412	598
Utah	385	554
South Dakota	317	543
Oregon	135	189
New Mexico	105	157
Washington	85	115
American West	8,890	12,558
Total US	28,065	45,990

[a] Only two states in the American West are not included in the table because of small figures: Texas and North Dakota. Texas had four Irish-born miners and six others with Irish parentage. North Dakota had two miners who were Irish-born and one other with Irish parentage.

for determining the ethnic composition of mining towns. Furthermore, despite extensive writing about and substantial research into the existence of a distinctive Irish community's presence in British mining towns, historians generally display a total lack of awareness of the impact these Irish people might have had on figures for all migratory miners from Britain and Ireland.[13] This blind spot in the history of miners, particularly of Irish miners, derives from a historiographic tradition that fails to look at Irish migration in a transatlantic or transnational context. Anecdotes such as those given above point to the problematic nature of "born-in" statistics. For children born to Irish parents and raised in Irish communities in British mining towns, we can solve this problem by reassessing their inclusion as English, Welsh, or Scottish and thus bring greater clarity to the British category.[14] As mentioned, there were legitimate technical reasons for historians' inability to discover the numbers of British-born "Irish" miners, but thanks to the recent digitization of the 1880 US Census, we can now establish accurate parentage figures.

The number of Irish-born miners enumerated in US Census data (table 1.1) represents the Irish in one obvious sense but provides no evidence of the size of the multi-generational ethnic group or the entire Irish community. Worse still, such figures lead to a vast overestimation of the Anglo-American population in the mining towns, as the Irish who were generational migrants moving through Britain or across the US would subsequently be counted as English (or Scottish or Welsh) or American in the census. To eliminate this error, it is useful to rely on the individual's parentage. This enables us to establish ethnically Irish individuals who were born in, for example, Scotland,

England, Pennsylvania, or Michigan under the column "Irish Parentage." A comparison of these two metrics shows us that Irish-born miners were more likely to travel to the American West than were those of Irish parentage (31.7 percent versus 27.3 percent). The explanations for this vary. It is possible that the Irish mining communities in the East and Midwest were more established or perhaps too poor to move (we know from the census that most miners there were coal miners), which, in turn, limited their movement to the American West. It is also possible that they were mobile but chose to migrate within their home state to closer opportunities rather than completely uproot their families again.

The information presented in table 1.1 includes only the 1880 census and therefore presents a snapshot of a rapidly changing environment. Nevada's Comstock Lode was fading in importance, whereas Montana's copper mines were in their infancy. Likewise, Idaho was experiencing a lull between the end of placer mining and the development of industrial mining. Meanwhile, Colorado's mining boom had just begun, prompting the growth of Leadville. A quarter of all Irish-born miners in the US in 1880 lived in four states—California, Colorado, Nevada, and Montana. Pennsylvania remained home to the largest number of Irish miners, with over 40 percent of all Irish-born miners (11,430) in the US—most of whom worked in the anthracite fields in the eastern part of the state in Luzerne, Schuylkill, Lackawanna, and Carbon counties. Many Irish were moving from Pennsylvania to the mines in the West (see chapter 5, this volume), blurring the distinction historians traditionally made between coal miners and hard-rock miners. In particular, we can see evidence of this in the migration path from Pennsylvania to Colorado, where 927 miners had been born in Pennsylvania to Irish parents, proving that the difference between the categories of miners was overstated. Skilled miners could get work regardless of their mineral sub-specialization, and ethnic links—Irish following Irish and Irish hiring Irish—were important factors in this movement.

The census also allows insight into where in the US these miners were coming from by looking at the state of birth for the Irish miners, as seen in the Pennsylvania figures. Most of those of the second generation whose occupation was "miner" came to the American West from states that had significant coal mines, such as Illinois (595), Missouri (392), Ohio (372), and Iowa (328).[15] Some of those from Illinois probably came from Chicago—where families may have paused to work for a time before pressing on—rather than the coal mines. An illustration of such movement is the fact that the largest recorded birthplace of Irish-American miners living in Colorado was New York (1,451), suggesting the importance of New York City, not necessarily as a destination for miners but as a stopover point where the Irish could earn enough to continue their multi-generational journey onward.[16] Somewhat surprisingly, fewer miners came from the copper mining state of Michigan (314), where there was a significant Irish population but apparently less onward mobility than was the case in any of the aforementioned coal states.

In cities such as Chicago and New York, Irish miners relied on their ethnic networks to sustain them. Mostly, these were family networks, as in the case of Séamus Feiritéar, who stayed with his uncle Pádraig Feiritéar in New York before moving on to Butte, and Nellie Cashman, the "Angel of Tombstone," who grew up in Boston after her family fled the effects of the famine in Cork.[17] These nodes had their own Irish communities and institutions that helped emigrants. For two Irish gold miners, Thomas Conway and James Egan, the Emigrant Savings Bank in New York provided a means to safeguard their money on their way to and from the goldfields.[18] The bank's Irish leadership gave it an implicit mark of trustworthiness in the eyes of other Irish.[19] The establishment of the Hibernia Savings and Loan Society by John Sullivan in San Francisco in 1859 gave the Irish on the West Coast a nearby option for protecting their money beyond sewing their gold into their clothes or hiding it in their boots.

Again, the "Irish-American" figure here only includes those individuals whose parents shared the same birthplace. This standard, applied to the statistics from the 1880 census alone, avoids a double counting of individuals of mixed ethnicity and also mitigates the natural fall-off that would happen with any ethnic group over time (i.e., those who no longer identify with their Irishness to any notable degree). There would be a degree of overestimation in labeling everyone who had one parent born in Ireland as Irish-American and assuming that a child would actively identify with their Irishness.[20] There is much less risk in doing so when both parents are Irish. If we use the census "born-in" figures (table 1.2), we overestimate the numbers of ethnically American and Canadian miners by 30 percent and 56 percent, respectively, compared to when we include the "parentage" figure (table 1.3).[21]

An important reason for this jump in the total number of Irish miners is that the figures include the second generation born in Scotland, England,

Table 1.2. Miners' birthplace figures, American West, 1880[a]

Place of Birth	*No. of Miners*	*Percent American West*
Ireland	8,892	8.9
US	41,450	41.4
China	21,763	21.7
England	9,713	9.7
Canada	3,794	3.8
Germany	3,331	3.3
Mexico	1,572	1.6
Scotland	1,553	1.5
Wales	1,412	1.4
Sweden	1,192	1.2
France	1,128	1.1
Italy	1,029	1.0
Other	3,406	3.4
Total	100,235	100.0

[a] The nationalities listed account for 97.5 percent of all miners in the US in 1880. Steven Ruggles et al., Integrated Public Use Microdata Series: Version 5.0 [dataset], University of Minnesota, Minneapolis, IPUMS, 2010 [hereafter IPUMS].

Table 1.3. Miners born in the United States and parentage figures, American West, 1880[a]

Parentage	*No. of Miners*	*Percent*
Irish	12,559	12.5
American	31,187	31.1
Chinese	21,748	21.7
English	10,877	10.9
German	4,261	4.3
Canadian	2,436	2.4
Scottish	2,214	2.2
Mexican	1,687	1.7
Welsh	1,596	1.6
French	1,290	1.3
Swedish	1,225	1.2
Italian	1,051	1.0
Other	8,104	8.1
Total	100,235	100.0

[a] The ethnic groups listed in this table account for 92.0 percent of all miners in the US in 1880. IPUMS.

Wales, and Canada to Irish parents before coming to the US. This figure is established by looking at the nationality of parents for people of British birth in the US. A mere 1 percent of Welsh-born miners in the US have dual Irish parentage (versus 5 percent non-Welsh parentage overall), whereas 11 percent of Scottish-born miners have Irish parentage on both sides (making up most of the 13 percent non-Scottish parentage overall). England sits between both provinces, with 4.2 percent of English-born miners in the US having Irish parentage on both sides (5.5 percent overall).[22] This is consistent with the assessment that Irish miners disliked or were dissuaded from living in Wales, despite its proximity to Ireland, because they came to believe that Scottish and English mining towns were more hospitable places to live and work.[23]

The Mining Frontier

Mining was a crucial occupation for immigrants in the nineteenth and early twentieth centuries and was vital to the continued industrialization of the United States. In 1870, almost a third of all workers in the American West were employed in the mining industry; while this number declined in subsequent decades, it was the industry most responsible for creating the main urban centers and populating the region's vast expanses. Frederick Jackson Turner wrote, "The unequal rate of advance compels us to distinguish the frontier into the trader's frontier, the rancher's frontier, or the miner's frontier, and the farmer's frontier."[24] While he meant this in terms of civilization's staggered progress, it remains useful to assess separately how the mining frontier developed, due to its unique history and its consequential impact on the surrounding regions. Historians have long grappled with the difficulties presented by fragmented frontiers and migration patterns and of disparate social and geographic mobility. Migration was generally focused on departure or arrival, treating emigration and immigration almost as two distinct phases rather than as part of the same process of broad mobility.

A different but related problem emerges in the historiography of British and American mining, with the dominance of single town or regional case studies, which obscures the movement of the Irish miner and his family by focusing on a permanence that was atypical for the majority of the Irish. As historian David Fitzpatrick noted about the Irish in Britain, "By the crude indicators of socio-geography, they were settlers; by their own testimony, they were frustrated transients; by the omissions from their testimony, they were displaced persons."[25] Many other letters reflect this sentiment. Irish letters, such as the Mullany letters (chapter 5, this volume) that traveled back and forth within the US or those that traveled around the globe such as the Orr family letters (chapter 2, this volume), demonstrate the interconnected nature of the disparate frontiers that stretched across continents. Accounts, exemplified by John Brophy's interview in the 1950s (see below), further emphasize the impact these links had on encouraging the Irish to migrate onward, in particular from Britain to the US, in the hopes of improving their economic and occupational position.

A limitation in mining historiography has been an effort to maintain certain occupational orthodoxies. For example, in Graham Davis and Matthew

Goulding's recent work, the authors criticize the conventional geographic limitations used in miners' history—a valid point—yet they continue to adhere to specific occupational definitions by using the term *hard-rock miners*.[26] Such an approach downplays evidence that many of the presumed skill barriers between hard-rock and coal miners (particularly those who mined anthracite) were muted as ethnic identity influenced employment patterns in mining towns throughout the American West, in particular Virginia City, Nevada, and Leadville, Colorado (chapters 3 and 4, this volume). In terms of migration, the vaunted hard-rock mining skills supposedly required proved less important than being Cornish or Irish or Irish-American. Simultaneously, the possibility of employment, improved conditions, and the security of an established Irish community encouraged Irish chain migration, such as between Cork and Virginia City and between Avoca, Pennsylvania, and Butte, Montana (chapters 3 and 6, this volume).

Nellie Cashman was certainly singular, but she was not the only Irish woman in mining. The census shows other scattered examples: an unmarried woman, age 25, in Washington County, Utah; two unmarried women in Eureka County, Nevada (ages 17 and 40); one unmarried woman in Deer Lodge County, Montana (age 27); two women in Lake County, Colorado, one single (age 19) and the other with her husband (age 39); and two single women in California, one in Calaveras County (age 34) and the other in Placer County (age 30). Census data show that none lived with relatives except the woman who was married, leading to the supposition that these women decided to strike out on their own and try their luck in the male-dominated occupation. Were these women skilled miners and, if so, how did they learn, perhaps from a male relative or friend? Did they enter mining out of desperation? In the other states the Irish immigrated to, this appears to have been the case because the few women in mining were either very young or very old, as, for example, in Pennsylvania and Michigan. Rather than acting out of desperation, the women mining in the West, as was the case with Cashman, were part of the gambling, prospecting class of miners, trying to strike a claim that might make them economically independent and wealthy—an attractive opportunity in a society with few such avenues for them. This case is strengthened by the fact that none of these women miners were in the developed hard-rock mining towns of the American West (the census classed them as living in rural conditions) but were instead at the periphery where they could grubstake and try to strike the mother lode.

Another factor playing into this isolation may have been that hard-rock miners refused to allow women underground due to superstitious beliefs, fearing that women brought bad luck to a mine.[27] While there were indeed a small number of Irish-born women in mining, a number are likely hidden behind the narrow occupational definition "miner" and other Irish women may have worked in other capacities in the mining industry, perhaps as muckers, and as such were classed as laborers. Still, in the case of the American West, only California has double-digit numbers of Irish women listed as laborers, with San Francisco accounting for more than half of all those numbers.[28] Women's opportunities for work in the economy were most readily available either as housekeepers or at boardinghouses, hotels, or laundries—which, coupled with the support structures of relatives, fraternities, unions, and religious charities and the aforementioned superstitious intolerance for women working underground—meant that the choice of working in or around a mine was available only for a few exceptional individuals.

An article published in *The Era* magazine in 1902 attempted to challenge the prevailing prejudice against women in mining, stating that "a few clever women have discovered that this knowledge is by no means beyond the grasp of the feminine brain." Two examples were Irish women Miss Delia A. McCarthy and Mrs. Mary Kent.[29] Mary Kent owned a mine near Western Pass and several claims across Lake County. McCarthy had more extensive investments, and her comments are some of the few recorded by an Irish businesswoman in the mining industry. McCarthy was listed as president, treasurer, and general manager of the Cooperative Mining and Milling Company of Cripple Creek, as having hired a foreman to manage her mines at Cripple Creek and Idaho Springs, and as the secretary of the Bonacord Company at Empire. Apart from her impressive array of titles, she actively monitored her properties, traveling from Denver through rugged, mountainous terrain to inspect their progress. Yet she described her "hardest undertaking . . . [as] to persuade these gentlemen that a woman can know anything about mining."[30] In spite of the prejudice she faced, she was convinced that mining was "a most promising field for women, if they will only go into it with clear-eyed investigation, educating themselves for it as they would for any other pursuit."[31] She utilized sexist attitudes to her advantage, loftily claiming that she never lost a dollar in mining stocks by asking "common miners" questions about investments, locations, and prospects. Since they saw little threat from her, they were "always ready to answer any question"; thanks to their "unconscious assistance," she claimed she knew more about the property than "its own board of directors."[32] The article described her as having an abundance of "faculty," likely intended as an understatement of her determined attitude.[33] Both Delia McCarthy and Mary Kent were outliers in a few ways; they were involved in mining, were business owners, and directly confronted contemporary misogynism in their public lives.[34]

Yet Irish women did not need to enter the mines to have a vital role in sustaining the regions' Irish-American mining communities. Wives, members of religious orders, nurses, boardinghouse owners, domestic workers, teachers, and sex workers were all integral to the existence of larger urban spaces. The ratio of Irish sex workers to their overall ethnic group was lower vis-à-vis that of other ethnic groups, but some rose to prominence as brothel keepers, such as Mary Welch and Molly Burdan.[35] Whether their identity helped or hindered them in that role is difficult to discern, since few left personal accounts and their experiences are most comprehensively recorded in titillating newspaper articles that play up the tropes of the soiled doves for their readership instead of presenting a full and honest account from the women's perspectives. Some Irish women became renowned labor organizers; the most famous woman labor agitator of the period was Mary Harris, aka Mother Jones.[36] Armed with firebrand speeches and cloaked in the guise of an innocent grandmother, she used societal gender roles to her advantage and earned adulation from "her boys," the miners, who created the famous song "She'll Be Comin' Round the Mountain" to commemorate her relentless efforts to support their strike in the face of threats and repeated expulsions from the state during the Labor War in Colorado.

Generational Changes

The 1911 *Report of the Immigration Commission* (commonly referred to as the Dillingham Commission) details generational changes in occupations for various ethnic groups in the US.[37] Table 1.4, drawn from the Dillingham Commission, gives a general idea of the numbers of men mining in the US and their nationalities. The more exciting aspect of these statistics is that we

Table 1.4. Generational occupation differences among miners, 1900[a]

	First Generation (born abroad)		*Second Generation (born in US)*	
Group	*Number*	*% of Group*	*Number*	*% of Group*
Irish	22,892	3.2	28,421	2.6
English and Welsh	44,918	10.2	25,099	5.7
Scottish	9,740	7.5	6,198	5.6
Germans	19,038	1.5	16,887	1.1
French	2,945	5.7	1,013	1.8
Italian	25,465	9.2	534	3.1
Hungarians	26,550	30.0	394	10.2
Austrians	28,854	18.9	709	4.9
Poles	14,024	7.7	1,292	5.0
Russian and Finnish	7,585	4.0	196	1.3

[a] United States Joint Immigration Commission, *Report of the Immigration Commission*. An important addendum to the Dillingham Commission's E&W and Scottish figures is that they fail to account for the ethnically Irish portion of these groups—that is, the English-, Scottish-, or Welsh-born with Irish parentage. These figures also include quarrymen.

can use them to broadly track generational shifts in occupation. As with all statistics, it is important to keep their context in mind.

For example, if we consider the Italian figures, they might at first appear to show a spectacular abandonment of mining between first- (25,465) and second-generation Italians (534), but the figures are only telling us that large-scale Italian migration to the US was a recent phenomenon. This is reinforced by the 276,438 first-generation Italians who worked in all occupations in 1900, while there were only 16,986 second-generation Italians.[38] Newer immigrant populations such as the Hungarians, Poles, Austrians, and Russians (the latter figure includes Finnish people, as Finland was still a province in the Russian Empire at this time) also did not have a large second-generation group at that point. Mining companies sometimes advertised in these countries and imported these nationalities as a new source of cheap labor to undermine unions or to serve as strikebreakers.

Comparing the different groups is not an easy task. Only three groups can be accurately compared with the Irish: the English and Welsh (E&W), the Scottish, and the Germans. Each of these categories had a similarly long history of mining migration and arrived in the US in significant numbers in the mid-nineteenth century. The E&W and Scottish show the greatest decline in numbers of those mining between the first and second generations. British Protestants were the most readily accepted immigrant group entering the United States. More specifically, Welsh and Scottish immigrants avoided manufacturing and mechanical labor, with their lower pay and large immigrant workforce, and instead gravitated toward services and agriculture.[39]

The Irish, like the Germans, show a generational percentage decrease in mining, representing a gradual shift away from manufacturing and mechanical pursuits toward employment in trade and transportation. Irish-America, somewhat contrary to popular historiography, also had a significant percentage employed in agriculture; a figure from another part of the Dillingham Report shows that this figure actually increased, from 13.6 percent for the first generation to 16.5 percent for the second generation.[40] The huge decrease in Irish laborers from the first generation (22.3 percent) to the second (10.2 percent) marks an unparalleled generational leap in social standing, reflecting Irish communities' emphasis on education. The Scottish figure, by contrast, went the opposite direction between the two generations, from 5.7 percent to 6.2 percent, with no discernible reason as to the cause.[41]

Somewhat at variance with their generational mobility, the Irish were also the only ethnic group with a higher absolute number of miners in the second generation than in the first. This might represent a solidification of an Irish mining class that would have been less interested in upward mobility—further evidence of a divide between the lace-curtain Irish and other Irish—or an indication of a greater desire to remain in mining. Butte was probably the locus of many in this cohort. The city grew in response to the booming demand for copper, which coincided with the apex of Irish control over employment in the city in 1900. In the 1890s and 1900s, the Irish community was able to offer a degree of stability and a high wage, which certainly appears to have made mining a serious consideration for the children of the Irish-born.

The Twilight of Ethnic Mining Communities

By the turn of the twentieth century, increasing mechanization, the development of a college-educated engineer and manager class, and fading employment prospects in the mines cast a long shadow over the remaining Irish mining communities in the American West. Simultaneously, education enabled the upward and outward movement of the children of Irish miners. The Irish tended to see mining more as an opportunity to earn a high wage than as a cultural vocation. It was a difficult and dangerous one, to be sure, but that was a risk a miner took over the years of toil to secure a decent income for himself and his family and a better future for his children. Embedded in this movement to other occupations appeared to lie a seed of awareness over the high price wrought by this work. There was an awareness that it was an exchange; as they took something from the mines, the mines took something from them. Parents therefore generally discouraged their children from mining. The parents used education and mobility to present other opportunities for their children, and they fanned out to nearby towns or cities, some with Irish enclaves. Little evidence suggests much sentimentality regarding the work of mining itself.

Reflecting on the movement of the Irish from mining to different occupations in the twentieth century, John Brophy commented:

> There has been the dispersion and the attraction of new industries that have drawn them and tens of thousands of other miners who migrated to this country, so that the second and third generation[s] left mines and went into other fields . . . I think that diffusion in this case has been very, very good. It's been good for the people themselves and it's been good for the spread of trade union organizations . . . In fact, [miners] were part of the hard core of experience around which present day union organization [was] centered and achieved its great success.[42]

Certainly, historians of mining regions have generally perceived the word *decline* in a negative light, but Brophy interpreted this change as a movement—not a "decline" but rather a "diffusion," in which the old ethnic communities melted away and reformed in other occupations in the form of more general organized labor. In typical Irish fashion, Brophy did not sentimentalize the change in occupation. Mining might have been an important job for the Irish, a dangerous, good-paying job, to be sure, but it was just a job. For the Cornish, it was much more intimately tied to their identity. Superstitions such as the mischievous Knocker, who warned miners just before a cave-in, can be compared to the Leprechaun, but the former's existence was entirely underground whereas the latter's was not. The Irish imagination was bent entirely against the claustrophobic conditions of the mines.[43]

The poem *"Amhran na Mianach"* "The Miner's Song" could never have been written by a proud Cornish miner and stands in stark contrast even to Brophy's distant view of progress:

Gurb aindeis an tslí é chun beathadh
D'aon scaraire rábach tréan;
Mar beidh tú chómh doimhim sin fé thalamh,
Is ná feicfir an ghrian ná an rae,
Ach do choinneal bheag chaoch ar do hata
Ar maidin fáinne 'n lae.

And if you ever come amongst them
You'll remember the point of my story;
It's a miserable way to earn a living
For any strong, vigorous fellow;
For you'll be so deep in the ground
That you'll never see the sun or the moon,
But only the light of the dim little candle in your hat
At the break of day.[44]

For the poet, Seán Ruiséal, his work in the mines of the American West was a distinct trauma that many Irish may have suffered. It makes it all the more surprising that the large numbers of Irish found in mining would be willing to endure such psychological trauma in pursuit of better pay.

He was not alone in his pessimism. Seamus Ó Muircheartaigh worked in many mining towns in the American West, including Butte, and he highlighted the loneliness of this lifestyle in his poem *"Mo chiach mar a thána"*:

Sin mar a chaitheas-sa tamall dem shaol,
Ó bhaile go baile gan toinnte ar mo thaobh.

That's how I spent part of my life,
Going from place to place, with no company at my side.[45]

The two poems portray the same loneliness and depression caused by the fact that the occupation failed to fulfill their pre-migration expectations of a good wage and decent working conditions: *"Bhíos chun capall a cheannach dom mhάthair is uan"* "I was going to buy my mother a horse, and a lamb as well."[46] All his good intentions, his hopes of helping his mother, were stolen by mining; he was cheated of his dignity by an exploitative economic system.

The inherent transience of mining communities that arose from working a limited resource does not explain why so many Irish chose to travel constantly from place to place to work as *spailpíní fánacha*—wandering laborers. While some moved between occupations, between mining and logging, for example, most miners remained working at what they knew best. Their paths of seasonal migration took them north to Butte and Coeur d'Alene during the summer and back to the warmer climes of Arizona and New Mexico during the winter, constantly migrating in a large shifting circle.[47] One local in Burke, Idaho, when asked about the numbers of "tramp miners," estimated that there were "maybe two or three hundred miners" in Burke alone who would stay at the two large hotels, work for two weeks, and then move on.[48] There was no mention of violence or other disruption associated with them or with this process.

These men were often referred to as "hobos," but the present meaning of the word differs from its nineteenth-century usage. Evelyne Stitt Pickett writes, "[Hobos] between 1870 and 1910 were seldom ignorant misfits or begging bums who stole for sustenance, but rather working men caught in an endless round of cyclical employment."[49] These were men who worked in a range of dangerous settings for the little they had in life, and they were proud of that fact: "There is a pride among them in their work, similar to that of every free born American citizen."[50] The rootless hobos were in some ways freer than those native-born Americans referenced, and many of these hobos were legal American citizens. Exact figures are impossible to ascertain but many were Irish-American, including the itinerant poets Ruiséal and Ó Muircheartaigh. Both directed their pessimism not just at mining or the transient lifestyle but also toward their employers' disrespect for them as working men: "No genuine hobo wants to be addressed 'Come, boys,' which is the general expression used when [referring to] working native or local labor. Experienced foremen call the hobos 'men.'"[51] Their sensitivity over the specific terms *boys* and *men* emphasized an expectation of respect from employers and a keen awareness of the ways masculine language and class shaped the occupational experience.

Partnered with Irish-American hobos were other hyphenated Americans with roots in western or northern Europe, and the term *hobo* became the foremost overarching identity for this diverse group, representing an emerging sense of brotherhood and helping them to differentiate themselves from more recent "second wave" immigrants from eastern and southern Europe.[52] Employers detailed the differences between hobos and the more recent "European laborers," rating hobos as more skilled and harder-working "jack of all trades" workers. Because of their mobility, they were considered less reliable (and more difficult for the company to control) except in the case of emergencies: "European laborers are usually 'quitters' in emergencies. Not so with the hobo. Emergencies are his delight."[53] To be a hobo or indeed a hyphenated American created a degree of suspicion among many Americans, derived from a false equivalence that by refusing to commit to a single place geographically, they were also refusing to commit their political loyalty to one nation: the United States.

Over the course of the early twentieth century, a growing social stigma became attached to transience, although it was not perceived as such by Irish-American hobos or their relations. Michael "Joe" McGuire, the eldest son of a Grass Valley gold miner (see chapter 2, this volume), worked in the silver mines of Virginia City, the copper and gold mines of Montana, and Idaho's lead mines before he joined the Alaska gold rush in 1889. He eventually worked in Tonopah, Nevada, where he earned the nickname "wild McGuire." Every time he sold a mine claim he "drank and whored until the money ran out."[54] He eventually died in a gunfight in 1918 at the Empire Mine, less than two miles from his family home.[55] The family remembered how the "firstborn had disgraced the family name," and they believed "Joe's wildness and his tragic death contributed to [his mother's] passing."[56] The McGuire family's shame regarding Joe did not stem from his footloose nature—others in the family were similarly mobile—but instead was centered on his sinful carousing and drunkenness. Similarly, Bill Keating, an Irish-American balladeer, referenced how others viewed his movements as commonly accepted: "Said Pete McAvoy, 'Here's Bill Keatin' the scamp.' / Just back, Pete supposed, from a million-mile tramp."[57]

Although mine companies frequently imported workers, they disliked worker transiency because it undermined company control of the workforce. A sedentary, established workforce gave the company greater leverage against workers if they wanted to lower wages or raise prices at the company store. A mobile workforce weakened the effectiveness of blacklists and could quickly sap the mines of their skilled workforce, particularly during a strike. There was also the possibility that traveling miners might have been union organizers or radicals. Furthermore, there was a cultural discrimination against miner mobility. As Roger Bruns points out, the distrust of hobos was reinforced by the belief that the hobo lifestyle opposed the natural American order because it "worship[ped] idleness," thereby turning the Protestant work ethic upside down.[58] This also helps explain how businesses' scare-mongering tactics against hobos proved so successful. When the journalist Frederick Wedge heard that the Industrial Workers of the World (IWW) was bent on overthrowing the government, he decided to infiltrate the radical labor organization. He secured a union card and masqueraded as an IWW member, living the hobo life and traveling the railroad. He was arrested for possession of an IWW card and the "Little Red Songbook."[59] After his experiences, he concluded that the hobos were honest and hardworking, further adding that he now believed the Wobblies cared more about improving workers conditions than about any sort of revolution.

Police and detective agencies nurtured a public fear of vagrants through a variety of means, including distributing wanted signs emblazoned with eye-catching headlines such as "Look out for Tramps," thereby connecting vagrancy with criminality and poverty in the popular consciousness.[60] These actions reinforced the emerging distinction between the nineteenth-century migration involved in the mythological winning of the American West and the twentieth-century "sedentary" closing of the frontier. In addition, it now marked movement as respectable and vagrancy as criminal, meaning that mobility—both geographic and occupational—was defined by class and ethnic identity. Surveillance, so heavily relied on by government and businesses against hobos, was likewise used by other, sometimes surprising, organizations. In the mining town of Randsburg in Southern California, a Thiel Detective Agency spy noted that unions also kept track of wanderers: "There were some strangers in town whom he [Thomas McCarthy, a union leader] kept track of, and who were of a class of 'hobos' that drift into mining camps around pay day."[61] The hobo could do little to challenge his increasing demonization by companies and unions. If work conditions became too harsh, a disgruntled hobo would simply up and leave rather than strike. If public fear and local authorities or organizations made life unbearable, he left, knowing he risked his life if he stayed.[62]

One of the final nails in the coffin of the wandering miner came about as an unexpected side effect of slowly improving industrial conditions due to the mine companies' general introduction of medical screenings for new workers: "Until they got these doctors, checkups and that eliminated half of them just real quick because they couldn't pass a physical. That broke the circle of the tramp miner."[63] With their lungs scarred from inhaling dust in the mines, these older men were more wrecks than relics of a bygone era. The few surviving poems and letters reveal how they internalized their economic insecurity and partitioning from mainstream society with a sense of failure and isolation. One historian, who conducted interviews of miners throughout Nevada, recounted:

> An eighty-one-year-old Irishman lived in a small yellow railroad shack . . . His pension check was signed over to the proprietor of a small store who supplied his basic grocery needs. During his almost sixty years in Nevada, the lonely son of Galway had worked in every county in the state. He had finally grown old, and with neither a wife, nor relatives, nor close friends, he was quietly awaiting death in the railroad cabin. After patiently answering the interviewer's questions he explained, "This is the first time in my sixty years in America that anyone has shown the slightest interest in who or what I am."[64]

The historian who wrote this did not offer the man the dignity of being named, but his name was Frank Casey, and he had worked in Butte and other Irish mining towns before finding himself at the end of his lonely path.[65] It is difficult to imagine the depth of this man's loneliness; despite the Irish webs of family and friends stretching across oceans and continents, the diaspora had many gaps wide enough for people to fall through and be forgotten.

A fresh examination of the statistics used to track mobility helps us refocus on the forgotten and rediscover overlooked primary source material. The ability to glimpse the past more accurately allows us to see more clearly the paths trod (such as that of Séamus Feiritéar from Kerry, through New York to Butte), and the connections forged (enabling Nellie Cashman to become a miner) or sundered (as they were for Frank Casey). This deeper context for understanding ethnic formation and the definitions of identity, in turn, helps contextualize the roots of societies, organizations, and communities, as well as their waning.

Although it is difficult to find a mine that did not have an Irishman working in it in the nineteenth and early twentieth centuries in the American West, the cities of Butte and Leadville most strongly retain their historical and cultural memories of once thriving Irish mining communities. They may be unusual in this regard, but many other Irish mining communities existed throughout the American West at this time. The history recounted in the following chapters is an attempt to illuminate some of the forgotten stories of those whose only historical record is often the weathered piece of granite detailing their name, date of birth, nationality, and—unlike the American headstones they stand beside—their home parish in Ireland. These relics are some of the remaining physical markers but are not the only traces that remain of the pervasive Irish element in mining towns of the American West; the most obvious are the many Irish living in the United States today who count their ancestors among those who lived and labored in those places, knowingly or not.

2

Digging Lumps of Gold

Tens of thousands of Irish joined the gold and silver rushes across the American West. Their paths of migration brought them by land and by sea, from the Atlantic and from the Pacific.[1] Irish migration to the shores of California at the time of the famed gold rush was contrasted both by the fear of leaving a famine-stricken homeland and the nervous hopes of arriving in a land that promised wealth and opportunity. The transition over the period of sixty years, from placer mining to established settlements to the advent of corporate mining, shifted the individual experience illustrated below by the accounts of Irish emigrants such as Michael McGuire and John Orr into a community-oriented one such as that described by Fr. Andrew Twomey. Groups, then camps, then towns emerged or disappeared, depending mostly on the longevity of the resource; this ephemeral aspect of mining continued to haunt the occupation from these early days. The ways Irish gold-rush miners positioned themselves within and distinct from the burgeoning California society formed the foundational context for later encounters throughout the American West. The final transformation in California was demonstrated by

https://doi.org/10.5876/9781646422517.c002

the emergence of mining towns such as Randsburg, whose social and political organization orbits the relationship among labor, the townspeople, and the corporation. In each of these situations, the Irish experienced a diversity of challenges witnessed through accounts in both contemporaneous sources and songs that detail how they were perceived by other groups.

While the largest gold rush occurred in California, it remains difficult to estimate the numbers of Irish who flocked there in the early years. One priest passing through the gateway to the goldfields estimated that the Irish made up one-third of San Francisco's total population in 1850.[2] The foreign-born population figures for California in 1850 indicate that the Irish were the fourth-largest foreign-born group, behind the Mexicans, English, and Germans (table 2.1), but they experienced the second-largest growth in exact numbers during the 1850s. By 1860, the Irish population lagged only slightly behind the Chinese population. In 1870, the Irish-born were easily the largest component of California's foreign-born population.

Despite this numerical advantage, there are few surviving Irish accounts of these early rushes. Michael McGuire was one of the few Irish Argonauts to leave a written account of his journey. Enticed by reports of vast wealth in this new El Dorado, he left behind his seven-months' pregnant wife, Mary, at their new farm outside Philadelphia to travel to the goldfields. She waited patiently for word of his journey west, knowing he was traveling through the Isthmus of Panama, and he paid for his passage by guarding a shipment of supplies from a merchant who was a friend of the family.[3] Newspapers reported the frightful diseases striking down many travelers on this route, and details of Indian and bandit attacks on miners in this new land doubtless fueled Mary's dread. She received one letter announcing his arrival in San Francisco, followed by a second in which he explained his plans to journey into the goldfields. After that, she heard nothing.[4]

Months passed and Mary feared the worst, but then, in 1850, Michael unexpectedly returned to Philadelphia. He had earned a small fortune in California and traveled back home aboard a mail ship. The gold he carried was sewn into his boots in case he was robbed. His successful expedition and the lure of continued prosperity prompted the family to uproot and settle in Marysville, California, the following year.[5] The money he earned granted him the freedom to choose where he lived, and the novelty of California proved enticing for him and his family.

Table 2.1. Foreign-born population of California 1850–1920, select nationalities[a]

Birthplace	*1850*	*1860*	*1870*	*1880*	*1890*	*1900*	*1910*	*1920*
Ireland	2,452	33,147	54,421	62,962	63,138	44,476	52,475	45,308
China	660	34,935	48,790	74,548	71,066	40,262	36,248	28,812
England	3,050	12,227	19,202	24,657	35,457	35,746	48,667	58,572
Germany	2,926	21,646	29,699	42,532	61,472	72,449	76,305	67,180
France	1,546	8,462	8,063	9,550	11,855	12,256	17,390	20,387
Italy	228	2,805	4,660	7,537	15,495	22,777	63,601	88,502
Japan	0	0	32	133	1,224	10,264	41,356	71,952
Mexico	6,454	9,150	8,978	8,648	7,164	8,086	33,444	86,610
Portugal	109	1,459	2,495	4,705	9,859	12,068	22,427	24,517

[a] Cambell Loosley Allyn, "Foreign Born Populations of California 1848–1920," master's thesis, University of California, Berkeley, 1927, 33. Data compiled from the published edition of US Census Office, *Census of Population* for 1850–1920. 1880 Irish parentage figure from IPUMS. The 1850 US Census for California was notoriously unreliable. See Ralph Mann, *After the Gold Rush: Society in Grass Valley and Nevada City, California, 1849–1870* (Stanford, CA: Stanford University Press, 1982), appendix 1.

A young Protestant man named John Orr had a very different experience in his travels and wrote to his parents in County Down after the arduous journey: "I would not go back the overland rout [*sic*] to the States for less than £100 a year for life, there has been a great deal of sickness on the way, the latter part of the emigration suffered severely from cholera on the Platte afterwards from fevers and dissentery [*sic*], and on the latter end of the rout [*sic*] hundreds died of scurvy."[6] Orr took up mining but wrote to his parents on June 14, 1850, that his income from the previous five months amounted to only 125 pounds. While this was done without any capital, it fell "far short of the expectations we entertained before coming here."[7] After this initial disappointment he decided to try cutting hay instead: "if we can get 50 Tons each we can make at least $150, per Ton clear profit and perhaps more, for feed is a scarce commodity 3 months from this time."[8] While they did turn a profit, again they failed to match their hugely optimistic estimates.[9] Orr noted young male miners' propensity to gamble their earnings at Monte tables and the explosion in Sacramento's population over the past year, from 500 to 10,000. He also vividly described the effect of severe winter flooding, causing thousands of oxen to drown while "several houses . . . floated off their foundations and turned sideways to the streets."[10] On August 17, 1850, he wrote to his sister encouraging her and other relatives to write and not hesitate over how much it cost him to respond, "for it is nothing in California."[11]

By then he seemed less pessimistic about his wages, writing that "£100 a year is thought first rate at home, I have made it in a month here" and assessing that "if a person could do that & stick to it at home he would soon get rich but expences [*sic*] are very high here but not in proportion to what they are at home."[12] While he longed for news from family and friends, he never stated any desire to return home.[13] Accommodations on the frontier came at a premium, and most people made do with primitive conditions. Orr wrote:

> Boarding in the city is from $15 to $25 per week. Lodging extra. I have seldom had anything softer than the soft side of a plank or the ground with a Buffalo skin on it for a bed since I left Chicago, I don't think I could sleep as well in a bed now as I could on the floor.[14]

He also wrote that he had never suffered a cold while he was there but mentioned in passing a bout of diarrhea that lowered his weight 21 pounds, to 10 stone (140 pounds). Orr was well aware that disease posed a danger in California: "Cholera has appeared here though not to a great extent yet, some 8 to 12 cases a day in a city of 10,000," but it did not greatly worry him. He happily wrote of the many Irishmen he noticed in California, and, more specifically, "Portaferry and the neighbourhood are well represented here," indicating possible chain migration—neighbor following neighbor.[15]

On Sunday morning, November 3, 1850, John Orr was bedridden with diarrhea and cramps. His friends tried to treat him with laudanum and camphorated spirits but he vomited them up, after which they realized the seriousness of his illness and fetched a doctor from Sacramento.[16] A few days later his business partner, E. E. Griggs, wrote to his parents, "It is in sadness of spirit that I communicate to you the melancholy intelligence of the death of your affectionate son, John M. Orr. He died of cholera, on Sunday last, after 18 hours [of] illness."[17] He spent the rest of the letter consoling them in Christian tones and in the final paragraph asked the family how they wanted to manage the $750 worth of property and goods that belonged to their dead son. In the interim, Griggs put a friend of the family who lived in California, Thomas Warnock, in charge of the possessions until they wrote back.[18] Warnock was placer mining with his brother on the Makellome River, and

Orr had written in previous letters about Warnock's bitter disappointment at his lack of letters from family and friends after his emigration.

When the news of Orr's death reached his elderly parents in Portaferry in January, it devastated them. "It came on us like the bursting of a thunder-storm," they wrote back to Griggs. They remembered that when their son told them of his plans to travel to California, their worst fear was not that he might die; it was imagining him alone and afraid: "We pictured to ourselves his hardships and privations, and especially if attacked with sickness there would be no kind hand to attend him, or no kind voice to sooth him."[19] They wrote of the comfort Griggs's letter gave them, adding that it was only from "your sincere, feeling, affectionate Christian letter, that aided by the Grace of God we were restored to some kind of Christian composure."[20] Meanwhile, Warnock's anger over the lack of communication from home appeared to have been misplaced, as Reverend Orr asked Griggs to pass on the message that back in Ireland there was a "great anxiety about him" since they had only received one letter from him.[21]

Beneath Warnock's anger over the lack of contact welled a longing to return home, and Griggs replied to Orr's parents on December 12, 1851, that "a strong invitation from his friends could induce him to return to Portaferry."[22] Griggs also wrote that Warnock wanted to return home but was haunted by guilt over the circumstances of his departure: "He feels reluctant to return in consequence of having left contrary to the wish of his father, and frequently reproaches himself for the act."[23] He ended the letter with local news that "miners have generally made handsome wages during [the] past season; some have done remarkably well; and some are discouraged and consequently homesick." He concluded: "The population of this country is rapidly changing; newcomers with their families are adding a pleasing and moralizing feature to society. The business of Quartz mining is engaging the attention of capitalists and many steam power machines are being put in operation; but I think this is a very uncertain business."[24]

The final letter from Griggs, dated May 15, 1852, contained the payment of £160.10.6 ($800.50) to Orr's parents and information that Warnock had traveled 200 miles north of Sacramento to Shasta, presumably following a gold rush to the area. Griggs mentioned the increasing movement across the Pacific, detailing the migration paths that brought Irish gold miners from California to Australia and back again, rush following rush. "It appears that her majesty's dominions are not destitute of gold; and if the reports from Australia are true, a great rush will be the result; in fact many have already left this, for that land of promise."[25] Comparing and contrasting economic fortunes was a central theme of these letters, and in his letter to the grieving parents Griggs could not help but slip back into that habit. He ended his letter to Reverend Orr with a note that they had put a headstone at the grave of John M. Orr as a "last sad tribute of respect to the lamented dead, our mutual friend and companion," a small token that surely brought further comfort to the grieving parents 5,000 miles away.[26]

The lurid accounts of easy wealth meant that most believed their sojourn would be brief, though other Irish such as Orr and McBride became permanent residents of California for one reason or another. Finding their riches and keeping them proved a more difficult challenge than expected for most, and letters home were often filled with excuses for delayed returns. William Hayes wrote that his investments and a lengthy mining expedition that lasted several months were the reasons he failed to send regular correspondence or return to his family. Scrawled on one of his letters home was perhaps the beginning of a letter to him, written by his angry wife or a relative, "You wrong your wife and child in this letter . . . at any time you could get a living here and I have told you so, I wrote to you when in New York to come back and you would not."[27] Newspapers of the time had many sections seeking information about people, usually husbands, from whom contact had ceased. Dangers existed out west, but many of the missing likely absconded and began anew there.

Relatives usually traced the cause of this resettling to the temptation of wealth—"I hope the digging of gold has not changed your heart as we must say it has changed others."[28] This was a snide reference by the writer, an aunt, to emigrant Margaret Mehen, who went to Australia and married an Irish gold miner, John Maher.[29] Her family's bitterness was rooted in Margaret's perceived "ungratitude to us and their unthinking feel to us," and they jealously mentioned a neighboring family whose relative went to the Australian goldfields and who, more important, sent his mother twenty British pounds.[30] Far away in Ireland, divorced from the fickle nature of fortune on the mining frontier, they based their impression on infrequent letters, the stories of neighboring families, and newspaper reports. Margaret and John left Australia to join the gold rush to California in 1849, but he died a mere nine months after their arrival. She recounted that the experience "left me quite alone in the Mountains where there was not one White human but myself."[31] Her use of the term *white* seems to highlight her isolation and not necessarily the disappearance of her cultural identity, as not long after this she married Peter Mehen, an Irish-born merchant, in Sonoma—again showing the intense draw of the Irish to other Irish. After the family in Ireland reestablished contact, they noted conspicuously that Mehen was "immensely rich."[32] Her uncle Patrick Phelan, who had stayed in Australia to mine gold, wrote to Margaret Mehen to tell her that her grandmother was crying over her every day, "and if you should write home send them home a few pounds, for your grandmother and aunts loved you dearly."[33] Relatives were not above deploying an occasional emotional blackmail if it suited their needs.

The number of Irish-born in California peaked in absolute numbers in 1890. However, as a percentage of the total population, they had peaked two decades previously, at 10 percent of the total population. Their early influence was strongest from the 1850s to the 1890s, with their proportion of the population dwindling quickly thereafter, to a mere 1.3 percent by 1920 (table 2.1). The gold rush initially drew large numbers of different nationalities, but many of these newcomers did not work in mining, choosing instead the booming trade towns that supplied the mining camps across California. The Irish and Chinese proved particularly susceptible to the lure of mining and made up a huge number of placer miners in the American West.

The Lure of Gold

Goodbye Muirsheen Durkin sure I'm sick and tired of workin'
No more I'll dig the praties, no longer I'll be fooled
For sure's me name is Carney I'll be off to California
Where instead of diggin' praties I'll be diggin' lumps of gold

Irish folk song "Muirsheen Durkin"

The popular perception of the gold rush, the miners, and their effect on society can be detected in a range of sources, songs, stories, and newspapers that all tapped into a widespread fascination with the mass migration and the exciting possibilities linked to the "new" land. Irish society perceived emigration in a negative light, as highlighted by the tradition of American wakes, but the gold rush injected a uniquely optimistic and excited mood into migration, as seen in songs such as "Muirsheen Durkin." The song bookends the hopeful start of a migrant's journey with another song, "Spancil Hill," that details the crushing disappointment of a life uprooted. Taken together, these two songs represent an Irish experience of the gold rush life that was far from uncommon.

"Muirsheen Durkin" portrays an excited Irishman and soon-to-be-emigrant for whom the grim life of laboring was offset only by courting throughout Cork and Kerry, a hint of the joyful mobility he was continuing through emigration. The upbeat tune matches the optimistic lyrics of the piece and provides a powerful alternative vision to the darkness of the American wake, in which a large farewell party, akin to a funeral wake, would be held for the person about to emigrate.[34] It ended with the tearful final embrace of the parents and their child, both of whom knew they would never meet again. Such Irish traditions and emigration ceremonies were sorrowful occasions, since the emigrants were leaving their family and friends and might never see them again, but accompanying this sadness was a conflicting sense of excitement as they set out on their journey. The song references the Irish attachment to their American destinations, and the importance of contact with home is shown in the line "the next time you will hear from me is a letter from New York." New York represented both a logical port of entry for those coming from Ireland and an important hub for those traveling. The song also describes the emigrant's determination to return home after his adventure: "And when I return again, I never more will stray." Here the promise to return and stay is revealed to be an integral performative element in leaving and represents an attempt at reassuring both speaker and listener of the person's measure of true success and faithfulness to their home and identity.

Conversely, the tone in "Spancil Hill," about a gold rush miner at the end of his life, expresses a deep loss of hope and time. Both songs reveal the hope to return to the Promised Land of home. Indeed, the dream in "Spancil Hill" wherein the emigrant returns home expresses a spiritual experience: "And when our duty did commence, we all knelt down in prayer / In hopes for to be ready, to climb the Golden Stair. / And when back home returning, we danced with right good will, / To Martin Moilens music, at the Cross of Spancil Hill." The revenant reminisces deeply about his friends, his family, and the girl he left behind as he mourns the passing of those long years away from home, and a powerful sense of dislocation from a familiar place follows him as he notices the changes in the people there: "I went into my old home, as every stone can tell, / The old boreen was just the same, / and the apple tree over the well, / I miss my sister Ellen, my brothers Pat and Bill, / Sure I only met strange faces at my home in Spancil Hill." A feeling of futile longing runs throughout the song, and it ends with the haunting image of the Clareman waking from his dream to find himself aged and still "in California, far, far from Spancil Hill." Only in his dreams can he return to this place; he is a man lost, out of place and out of time.

Whereas the two songs differ in tone, they reflect a common theme in placer mining letters—initial hope and excitement eventually gives way to disappointment and a sense of becoming trapped in America, unable to return or even unwilling to do so because of shame over failing to strike it rich. As James Gamble wrote in a letter to his brother back home, "You speak about coming home. I would if I had the means to come and live there without working."[35] It was not enough for an emigrant to return, if such a thing was possible after years or decades away. They were under tremendous pressure to fulfill expectations, both their own and society's. Returning required them to have something to show for their years spent abroad and their determination not to scrape out an existence back in Ireland. An earlier letter from Gamble while gold mining in Calaveras County, California, reinforces this point: "I would not like to come unless I had money enough to support me without hard labour."[36] "Muirsheen Durkin" and "Spancil Hill" deal with the two sides of the mining rush, before and after, the eager fortune hunter and the disillusioned exile. An example of this duality was Jack Powers, who, after amassing a small fortune of over $10,000 at the diggings, managed to increase his haul to a staggering $175,000 through gambling. Declaring that he had enough to live on and that "my poor old mother shall never want for anything," he took the ferry from Sacramento to San Francisco. During the short journey, Powers lost all but $15,000 and, perhaps more disastrous, his will to return to his mother in New York.[37]

Newspapers around the world advertised the sensational gold strike in California. In Ireland the *Belfast Newsletter* reported, "The astounding reports of the new found mineral wealth in California have already given birth to innumerable schemes for emigration to that golden region."[38] Fantastic details of possible wealth accompanied dire warnings in these accounts. A contemporary assessed the draw of the goldfields as follows:

> Considering the arduous toil and unceasing drudgery of a gold-miner's life, the unending scraping, delving, washing, cooking, etc., it is one of the most wretched of all occupations. Yet withal it is attended by circumstances affording a continual and intense excitement. The constant suspense—when the turn of a shovel may bring forth a fortune to the lucky individual who wields it; after even the longest and most unsatisfactory labor; the feverish search for rich locations; the never relinquished hope that a blow of the pick, a stroke of the spade, will expose to daylight the treasure that is to make him independent forever.[39]

John S. Hittell considered the gambling fervor the main cause of the various rushes through the 1850s: "It was a bet of $1,000 against $20,000. Who can say that the chances were more than twenty to one against the river? And the chance of living through such and other excitement as that of California in '49—the fun would be worth a fortune almost."[40] "Almost" was right. Some did look back on the experience as a grand adventure, but writers often forgot the serious psychological toll these wild fluctuations in fortunes took on those involved. Frequently, their hopes were dashed when the mother lode turned out to be practically worthless. For the Irish-born in California, strangers in a strange land, the pressure to succeed was especially intense and was tied to another temptation: to never return and instead to settle in this new land.

Those who repatriated to Ireland held on to their time as Argonauts with a degree of pride. The Irish censuses of 1901 and 1911 detail many "gold miners," and some specify further details, as in the case of Farrel Reilly whose occupation was listed as "Miner California, U.S.A. (Returned American)."[41] His pride in being a miner at that time and in that place is obvious, and the categorization of him as a "Returned American" implied a lot. He had been transformed by his experience and was no longer solely Irish. Similarly, Edward Sword's details cover both his and his wife's occupational entry in the census: "Retired Gold Miner and Living on Income."[42]

Social networks helped spread news, and many heard reports of gold finds from friends or comrades (figure 2.1). Patrick Catherwood from County Down would become one of the rare Irish Protestant miners. While still in Ireland, he wrote to a cousin at the goldfields asking for information. Patrick was less enthusiastic to travel at this point than another cousin, Thomas, also still in Ireland, who told the emigrant's mother he would go to be with the

A DIGGER'S HUT.

Figure 2.1. A digger's hut. Prospectors lived in ramshackle huts that were as easy to build as they were to abandon. Charles Ferguson, *The Experiences of a Forty-Niner during Thirty-Four Years' Residence in California and Australia*, 277.

cousin "if God would spare him yet."[43] In the same letter Patrick wrote of another local man, James Gibson, who had six sons at the mines in Placer County, California, a cluster of relatives supporting each other in a difficult environment. Patrick wrote that it was not just the temptation of high wages that drew his relative's attention, "The money does tempt him," but also the belief that you could earn such money quickly: "he thinks that he would have the full sack before that he would be long there with you."[44]

Pat Magill emigrated in 1866 at age twenty to join his brother who was working on the railroad. While working near Pueblo, Colorado, he met an elderly man named "Dad" Burns who learned of Pat's interest in mining and advised him on prospecting leads. His words were unlike the typical stories: "Differing from the 'usual lost mine' story, 'Dad' did not claim any fantastical discovery, he simply said there was 'plenty of gold in that country.'"[45] Although the site was being worked when he arrived, he successfully got a lease and later bought a farm in Routt County, Colorado, with his proceeds.

Befitting his Protestant background, he joined the Masonic fraternity and raised a family with Myrel Palmer, the daughter of another mining prospector.[46] These networks of letters between families and relatives stretched from the American West through cities on the East Coast and back to Ireland and Britain. Despite the vast distances, they served migrants well and pointed them toward promising opportunities.

Letters also offered the possibility of escape for wanderers who grew tired of the lifestyle. Irish Argonaut John Williamson wrote to his brother in England stating his desire to leave both mining and America behind. He inquired, "What wages do they pay in the shops in Manchester for a good workman at the anvil & what chance is there to get work?"[47] Others remained optimistic for much longer and communicated their sentiments to relatives or the media. One man wrote to newspapers in Ireland saying "I believe there is no other employment half so exciting and hopeful as gold digging. Although you meet thousands in the mines, who perhaps have been here two years, and yet have not cleared expenses since their arrival, yet they generally all feel sanguine and confident."[48] Whereas Williamson was not haunted equally by optimism and the sunken cost fallacy, it would prove a more difficult barrier for most to overcome, perhaps illustrating a degree of willful delusion among those who had sacrificed so much to get there.

Emigrant letters expressed despondence regarding the passing of time as well as economic worth. A late arrival to the gold region, John McTurk Gibson, traveled the overland trail to Placerville, California, and met several miners going the opposite direction who "fully confirm previous reports from the mines and all agree in saying it is the greatest humbug ever got up in any age or country."[49] Another report came from Sam Dillon's brother who apparently worked all winter on the richest claim, earning a meager $1.50 a day.[50] The song "California as It Is" highlighted that the real measure of failure experienced by many miner migrants was not the loss of money; rather, it was the sense that they had wasted their lives:

I've been to California and I haven't got a dime;
I've lost my health, my strength, my hope, and I have lost my time.
I've only got a spade and pick, and if I felt quite brave,
I'd use the two of them there to scoop me out a grave.[51]

News and Views

Letters remained the most popular way for families to keep in touch with each other during the nineteenth century. Emigrant letters were such a common feature of life that they were printed in newspapers as sources of information and as inspiration for comedic parodies. These parodies point to popular stereotypes in the UK and the US of the Irish as stupid and naive and characterize them as having a pathological desire for contact with the homeland. Their need for contact was real (as shown by the many letters that form the basis of this book) and indicative of a vital system of communication that provided the Irish with a direct and sustaining link to their families and their culture. The letters back and forth informed this transnational network of potential opportunities and dangers and reinforced their sense of identity in a strange land among strangers.

One parody that mocked this system of communication was printed in the *Belfast Commercial Chronicle* on April 15, 1850, a spoof titled "An Irishman's Letter from California," written by a "Terence Finnegan" to "Biddy" in Ireland. It read, "Accushla, sarching (searching) for goold (gold); but a body might as well look for new pitayies (potatoes) in Thriffalgar-square (Trafalgar Square)."[52] The piece mocks the Irish accent throughout, helpfully providing the intended word next to the phonetic transcription and simultaneously undermining the character's fluency and literacy. It goes on to detail a pun (a quart as the measurement of liquid) as a humorous misunderstanding: "You may have read in the papers that the diggers goold [gold] in quartz [quarts]; but don't believe it, Biddy. I'll be on my oath none of them ever found a pint of it."[53] Terence concludes the spoof pleading with Biddy to ask his friends to hold a raffle and send him the fare back to Ireland.[54] His request for money is meant to be a preposterous inversion of the typical flow of resources from America to Ireland in the form of remittances. Superficially, this letter offers a typical example of the insulting caricature of the Irish that was popular during the period. But the theme of homesickness still shines through, as it does frequently in genuine emigrant letters.

An example of a more serious emigrant letter was printed in the *Belfast Newsletter* on February 15, 1850, as a more intimate way of reporting news from distant locations by using a firsthand account. The person who sent it was the chief mate of a ship in California, and the account boasted of the vast wealth easily available in that state: "An old friend of the captain's came on board the other day, and stated that in four months he had realized 9,000 dollars at the 'diggings.'" The letter ends on a note of warning, however, that these earnings could be easily lost: "He dined on shore that evening, brought his gold with him, got into a gambling house, and was cleared of every ounce of it."[55] As family news, this was treated as a trustworthy first-hand account, although the question might be asked, why would a person want their letter to be printed in the newspaper? The local notoriety and status that came with having something printed seems to have been enough of an incentive for individuals.

The popular view of the Irish miner in American society can be further illustrated through sheet music songs from the period, which frequently utilized the stereotype of a naive and unsophisticated Irish person. One such song, titled "Lament of the Irish Gold Hunter," dwells on the rapid transformation of the American West and the promise of wealth that was drawing so many to the goldfields. One verse of the song describes the cost of travel: "There's lots of change up here, Mary, / Though you'll find none in me / For I spent the whole that I was worth / In coming over the Sea."[56] The innocence of the Irish gold hunter is juxtaposed with his religiosity and poverty, portrayed as two sides of the same coin, as he thanks Mary for her parting gifts of food and drink. After the phrase "I bless you for . . ." the song lists beef, cheese, sausages, and beer and reveals that they spoiled during the trip, making a mockery of his prayer and her wasted goodwill. The Irishman's figure of devotion is Mary, a subtle jab at the Irish Catholic devotion to the Virgin Mary, which is thereby linked to very literal corruption acting as a metaphor for that of the Catholic Church.

In another part of the song, the recurring theme of the Irish person yearning for home makes an appearance, though in a peculiarly specific reference: "For tho' there's bread and work for all / I would a good deal rather, / Die in old Ireland once a week, / Than live here altogether," echoing the Irish immigrant toast *bás in Éirinn* (to die in Ireland) and lending the song greater authenticity.[57] This reference furthers a sense of longing and highlights the reality that despite all the difficulties the Irishman in this song has overcome, he has not found happiness in employment and sustenance; there is a tacit acknowledgment that personal relationships are what make life worth living.

Neither of the characters in the "Irishman's Letter" and the "Irish Gold Hunter" experienced any serious danger; however, many contemporary observers issued direct warnings to Irish migrants about the risks of the goldfields, even before they left Ireland. As early as January 15, 1849, newspapers warned, "No doubt many fortunate individuals will suddenly acquire vast riches, but in the aggregate disappointment and, in all probability distress, will be suffered by the thousands which the accounts of these gold regions have within the past couple of months caused to emigrate thither."[58] Fr. John O'Hanlon followed suit in his guidebook for Irish immigrants to the United States, writing, "for the one person that succeeds, hardly ten will acquire what they would consider a competency, and many we are assured, hardly make their board and clear ordinary expenses."[59] These accounts did little to stem the tide of emigrants who wanted to try their luck; casting the contest as an exciting gamble versus a mundane, stable job was not as powerful a disincentive as these authors may have believed.

Concerned groups also attempted to warn the public through vividly illustrated drawings, such as *The Slave of Gold* (figure 2.2). This illustration was published in the New York–based *National Police Gazette*, a titillating tabloid aimed at men.[60] In the first panel, a man pans for gold and has "A Find," giving credence to the stories of wealth in this new land. The next panel shows the same man drinking, with two curvaceous women sitting on each of his knees and another behind him during "A Spree in 'Frisco." His money now spent on debauchery, he returns a broke prospector and is "Waiting for a Grub Stake," that is, a promising lead whereupon the local merchant would extend him needed supplies on credit. His luck runs out, and the next time, he fails to make a strike. The final panel reveals that the miner has committed suicide when he hits "The Bed Rock." The caption reads: "How life passes with the mine prospector of the Great West—A career which has many mutations and generally ends in a tragedy." The bottle in the illustration links alcohol to the prospector's ruination, where he dies alone in the mountains, birds hovering overhead to pick at his carcass.

The story tried to offer the moral that mining offered fleeting joy and nothing permanent, a rising and falling arc akin to that of a gangster. The sketch places the vice of celebration as a centerpiece, with the eventual hard reckoning, "the bed rock," almost an afterthought. As a public warning, it unintentionally glamorized this lifestyle and fed into wider societal concerns

Figure 2.2. *The Slave of Gold, National Police Gazette*, November 11, 1882. Library of Congress, Print and Photographs Division, Washington, DC.

about the large number of people willing to uproot and risk everything to become wealthy in this distant land. The desperation of the '49ers was treated in terms of needless risk rather than framed in their frustration over eking out a living.[61] These tales certainly heightened the fears of relatives left behind and echoed in the minds of the many thousands of Irish who set off on their journeys to California.

In both their entertainment and their occupation, miners faced dangers on the frontier. Alcohol acted as a ubiquitous social lubricant, and its absence was more noted than its presence at gatherings. The use and abuse of alcohol were widespread and a point of pride in the frontier mining towns. Successful Irish miners would learn that alcohol was a contributing cause to the accidental deaths of many Irish in and around the mines. Newspapers reported tragedies linked to alcohol in understated tones, such as this article in the *Sacramento Daily Union* in April 1859:

> Barnard Bradley, Irishman, mining laborer, was found dead by the road side near the Empire Ranch . . . death had been caused by cerebral apoplexy, induced by excessive use of alcoholic drinks. His age was about forty. He was an habitual inebriate for the last nine years, and single. Barnard Cullen, Irishman, miner, married, age about thirty, fell off a low bridge into a flume in the night, at Sucker Flat. He is supposed to have drank too freely in the evening. Verdict death by falling or drowning.[62]

The paper continued by vividly describing the end for a third Irishman, although in that case the cause of death was a mining accident and, unlike the previous two examples, it did not mention alcohol as a factor.[63] Alcohol was widely available in the many saloons that invariably clustered in mining camps and towns. Even remote locations were served by "travelling ginshops," similar to the traveling whiskey peddlers in coal mining camps in Pennsylvania. They plied their wares in the smaller remote outposts that did not have a saloon or a store. John McTurk Gibson observed "four women dressed in men's clothes" in one of these caravans, leading to the possibility of a traveling prostitution outfit. The use of men's clothes may have been a purely practical decision, given the rugged terrain.[64]

Regardless of whether these portable purveyors of liquor dabbled in sex work, in the eyes of other Americans, any woman who associated with either alcohol or the Irish had compromised morals because of their social degradation. Irish displays of frivolity often perturbed dour American observers. One traveler who spent a night in a so-called Irish groggery, run by an Irish woman, sarcastically referred to the female owner and patrons in quotations as "ladies." The American writer wished for a knife or a gun to quiet the "drunkenness, singing, fighting, and the usual noise of Irish sprees [that] were kept up through the night."[65] The cause of the spree? Irish generosity. A miner who had earned $200 on a recent dig was "freely treating" others to drinks. It was his third day celebrating, perhaps indicating that even among the heavy drinkers of the American West, the Irish leaned toward the more extreme side of the continuum.

Robert Williamson received a letter from his relative Bill in California, advising him to come over: "I am satisfied that you could come here and with the same industry be a great deal better off and place your family in more comfortable positions than you can there." These cheery prospects were followed with grim news about a friend who had migrated to California but had come to a very unfortunate end: "You ask about young May . . . he was murdered in a saloon."[66] This was a sad and violent end, but Bill provided more context for the murder in his letter in an effort to assure everyone at home that law and order did exist in the West: "he was partly intoxicated and his murderer will have his trial this month."[67] Bill himself later decided to quit mining and turn his hand to farming. He wrote, "Artemisia says [to] tell you since I moved on the farm I am a much better man . . . I am not growing around the paunch any more."[68] The miner's lifestyle was not just dangerous but also unhealthy, with treacherous conditions and hard drinking commonly accepted parts of the miner's life. Drunkenness was also blamed for troubles whose roots might have hinted at a deeper cause. One intoxicated man was beaten to death in a Nevada City saloon after he shouted that he could lick any Irishman in the house.[69] This represented a rather extreme example of the assertive attitude of the Irish as an ethnic group in the American West.

Nativist "Law"

The Irish were frequently the targets and adversaries of nativists and vigilantes in the mining country. It was rare for the Irish to participate in the vigilante movement, since the group could target anyone who was not a white, Anglo-Saxon Protestant. The unique cases when the Irish participated are understandable due to a class dimension, which was often able to override targeted ethnic violence. That was the case with the Irishman Peter Mehen, detailed below. This targeted violence almost certainly backfired, as it encouraged the solidification of the Irish as a defensive and cohesive ethnic group, encouraging them to form fraternities and cliques for their protection.

Desperation and greed also led to violent encounters, and one Irish newspaper printed an account of banditry on the California goldfields in 1852.[70] A "young gentleman" from Belfast described a journey in 1852, where he accompanied a friend visiting a possible investment opportunity—a quartz mine in the Grass Valley region. Bandits ambushed his party. The gang included a Mexican, a Belgian, and one man with "the unmistakable accent of Munster or Connaught" he referred to as a "Patlander." The desperados bungled the robbery and a few bandits were captured. Some of the party wanted to administer the "Lynch Law," but cooler heads prevailed, and they dropped the bandits off in nearby Sacramento where the "vigilance committee" offered a reward of $3,000 for the outlaws who remained at-large.[71]

A more serious incident occurred in 1850 in Sonora when four Americans came into town with three "Mexican Indians" and a Mexican in custody and charged them with the murder of two American miners at Green Flat Diggings, eight miles from the town.[72] As they were being processed by the sheriff, a lynch mob formed and began shouting "string them up" and "hang 'em." The crowd then elected a local Irish merchant, Peter Mehen, as a "judge."[73] They took the prisoners outside the town and held a faux trial, in which the three "uncouth" Indians and the Mexican "of gentle and pleasing appearance" were found guilty.[74] Only the aggressive intervention of court officials and lawmen stopped the lynching.[75] A few days later, miners poured into Sonora from the entire southern goldfield for the beginning of the legitimate trial. A crowd of nearly 2,000 miners, armed with pistols and rifles, waited on the streets for the judgment of the courts.[76] Fearing an ethnic conflict between the miners and the Mexicans in the region, Sheriff Work traveled to the nearby Mexican camp with a posse, arrested all 110 men, and confined them in a corral in Sonora.[77] During the trial itself, an accidental firearm discharge almost started a riot. Eventually, the men were acquitted without further incident, largely because most of the miners had returned to work their claims by the time the judgment was handed down a week later.[78]

After this event, false reports of the murder of American citizens circulated, and a vigilante group held a meeting in Sonora on July 21, 1850, declaring that Americans were in danger from "lawless marauders of every clime, class and creed under the canopy of heaven"—including "the peons of Mexico, the renegades of South America, and the convicts of the British Empire."[79] The "convicts of the British Empire" was an obvious allusion to the Irish and Australians. The vigilante group adopted a legalistic tone and passed seven "resolutions." The first stated that "all foreigners in Tuolumne county (except those engaged in permanent business, and of respectable characters) be required to leave the limits of this county within fifteen days from [this] date, unless they obtain a permit to remain from the authorities hereinafter named."[80] These authorities were to be three American citizens elected by US citizens in each camp or diggings. It further resolved that all foreigners were to be disarmed and that 500 copies of the resolutions be distributed in Spanish and English, directing the local so-called authorities to collect money from each camp to defray the costs of printing incurred by the *Sonora Herald*—a tidy racket for whoever was tasked with collecting the money.[81]

Predictably, the paper soon announced the success of these measures: "[they have] had the effect of teaching certain hombres a lesson they will not soon forget."[82] An explicit goal of the measure was to create intimidation across the region and "to silence the tongues of Sydney convicts and boisterous inebriates . . . who palm themselves off as American citizens"—in effect, a denial of the rights of those who were American citizens but were viewed as undesirable by the nativists.[83] The phrase *boisterous inebriates* was a coded reference to the Irish and as a slur was perhaps referring to the broad Irish disrespect of these extra-judicial measures. Newspapers such as the *Sonora Herald* played a central role in whipping up vigilante fervor among certain sections of the American-born population who directed their fury at vulnerable minorities and helped solidify Anglo-American political and social domination. They also had the convenient secondary effect of increasing their own sales by scaring the public with rumors, reports of murders, and the printing of threatening resolutions. The *Sonora Herald*'s publication of these notices served as both advertisement for and advocacy of the nativist message. Unsurprisingly, the paper gleefully announced the introduction of a yearly tax of twenty dollars on foreign miners.[84] After seeing the tax collectors in action a few months later, one observer noted, "I believe the American of California loves the country just in proportion to the amount of dollars and cents he can gain from it; the public officers prove this annually."[85] This tax revealed the covetous core at the heart of vigilantism, although those involved defended the measures based on public safety. The writer in question responded to this argument by noting shrewdly that the taxes themselves were not necessarily objectionable: "no harm in that, if the said foreign miners received any protection from the government, which they do not."[86] More to the point, by not so subtly co-opting the powers of taxation, they implicitly threatened imitating other state powers, the most important of which for them was capital punishment. Finding oneself on the wrong side of nativist American fervor could be harmful, both physically and economically.

Nativism was obsessed with "foreigners," but who exactly was a foreigner? Lines blurred for the second generation who could imitate Anglo-Americans, as in the case of James Mullany who sought to hide his Catholic heritage in McMinnville (see chapter 5, this volume). The definition itself varied based on context, with greater importance placed on class, race, religion, and ethnicity than on any citizenship granted from being born in the territory of the United States. Another Irishman who earned the ire of the vigilantes was "Notorious" Jack Powers, who mananged to earn the title "Notorious" later in his life; his biographer asks whether he was classed as a villain because he opposed the vigilantes, sided against the powerful land interests, or socialized with Indians, Mexicans, and other undesirables.[87] In reality, it was all three coupled with another important factor: he was Irish. On its own merits that made him suspicious, of dubious American pedigree or allegiance, and a perfect archetype in the American Western canon as a troublemaker. Everything else was used as ample fodder to portray his supposed villainy.

One final piece that completed the puzzle of nativism was its rejection of any American principle of egalitarianism and its effort to stymie any possibility of advancement, rejecting the idea of the frontier from its outset. As one writer explained bitterly, "California proved to be a leveler of pride, and [in] everything like [an] aristocracy of employment . . . the tables seemed to be turned, for those who labored hard in a business . . . fared worse than the Irish laborer . . . who made the most money in mining."[88] For a time, the rigid class and social hierarchy was upset. If an Irish laborer was making more than an American, he might start considering himself an equal, or the situation could lead to an even more appalling prospect: believing he was the American's superior. Therefore, the determination to push other groups down to their supposedly proper place was intertwined with an attempt to restore the dented pride of nativist Anglo-Americans:

> No inferior race of men can exist in these United States without becoming subordinate to the will of the Anglo-Americans, or foregoing many of the necessaries and comforts of life. They must either be our equals or our dependents. It is so with the negroes in the South; it is so with the Irish in the North; it was so with the Indians in New England; and it will be so with the Chinese

> in California. The Indians, it is true, would not submit to be enslaved; but they had to suffer exile, hunger and death as a consequence of their intractability.[89]

The nativist author Hinton R. Helper was born in North Carolina in 1829. His father died when he was young, leaving the family destitute. Helper left to seek his fortune, first in New York and, after gold fever spread to the East Coast, then in California. He was a failure both times, with only $160 to show for his efforts after three years of prospecting. Seeing the wealth of others, many of whom were immigrant Irish, he found reality a bitter pill to swallow and lied to his family about his situation: "I'll not admit that I have made so little; for, if I do, they'll accuse me of having been indolent, of gambling, of drinking, or some other disreputable thing that I have never been guilty of."[90] Whereas in their poetry Ruiséal and Ó Muircheartaigh blamed their misfortune on the evils of management and the systemic failures of the United States as a whole, Helper could not bring himself to criticize his failures or the economic systems that perpetuated inequality. His southern upbringing reminded him of his meritocratic hegemony as a white man; with his dreams shattered, he could only find succor by doubling down on his racist ideology.

The adversarialism demonstrated by Helper's words lived on in the intimidating targeting of certain groups that were undeterred by the establishment of official state structures of governance, including law enforcement and courts. We know of several notable murders of Irishmen in California, perhaps the most egregious of which was a plot to destroy Irish Catholic political opposition to nativism. This solidified into a plan to fix a duel by tampering with the pistols and thereby murder notable Irish politician David C. Broderick. Nativists demonstrated their unassailable position to the public by entrusting the task to a fellow nativist who was a judge. Californians were horrified by the merciless killing, and newspapers reported that there was "no justifiable ground" for the unfair contest.[91] Another murder of an Irishman was detailed in a different account in the *Los Angeles Star*. The courts acquitted a southerner for the murder of headwaiter Thomas Keating, and the newspaper wrote that it was a disgrace that someone should be "dragged before the courts simply for killing an Irishman."[92] The writer, Henry Hamilton, was Scots Irish, again highlighting the venomous distinction between that group and Irish Catholics.

The case of Peter Mehen stands in marked contrast with the killings of Broderick and Keating. Mehen was an Irishman who traveled from Australia to California; therefore, he should have been twice condemned by the vigilantes' printed resolution. However, instead of being persecuted, Mehen was elected lynch judge during the fake trial discussed earlier. The cause? Mehen's change in occupation from miner to businessman in California.[93] Contemporary accounts refer to him as a wealthy individual with many properties, thus elevating him to the vaunted "respectable character" exception. Perhaps he was responsible for including this clause in the nativist manifesto, which again reveals the hypocritical standards of nativism. There is no mention of Mehen's time in Australia as a miner in Lang's *History of Tuolumne County*, a volume very sympathetic to the vigilantes' version of events; this omission was no doubt an effort to distance him from the "Sydney convicts" and possibly draw attention away from the double standards of their "prominent men."[94]

Foreigners were more vulnerable to running afoul of these violent groups than were Americans. An American writing to his sister on the East Coast described how a man named "Holt" was found shot to death near the town of Weaverville: "Suspicion immediately fell on an Irishman" who had been seen speaking to the dead man earlier that day.[95] An official judge antagonized the crowd by informing them that it would take months to put the man, Michael Grant, on official trial. Several impatient people began to argue that Grant needed to be executed, very soon and in public. The author of the account sympathized with the mob and did not dwell on questions of basic legality, such as whether the trial should take place. So great was the mob's bloodlust that they almost turned on one another, arguing about whether Grant should be strung up immediately after an impromptu "trial" or on an execution date ten days in the future. The latter proposal won out, and a Catholic priest attended the young man before he was hanged, "which he said was a great consolation and died perfectly resigned to his lot."[96] The account reaffirmed the writer's belief that mob lynching was a wholly justifiable course of action, almost as if he were reassuring himself after having witnessed a harrowing spectacle. He did express certain reservations: "Although no one has the slightest doubt as to his guilt and I think he suffered justly yet I say Heaven preserve me from falling into the hands of an excited people. It is a hard tribunal and if circumstances are against you, however innocent you may be, you stand no chance."[97] These events were not a consequence of the benign excitement of the crowd; they were the well-defined parameters of privilege.

An individual's safety might well depend on whether compatriots were with them or nearby—indeed, such presence was a matter of life or death. In Grass Valley, a drunken argument between Patrick Doyle and Andrew Byrne escalated to Doyle challenging the other man to a fight. Byrne replied that he would fight him in the morning, but when Byrne turned to leave, Doyle shot him in the abdomen. A crowd seized Doyle, but his appeals to the bystanders—including some other Irish who opposed the lynching—convinced the crowd to hand him over to the sheriff.[98] Thus, ethnic solidarity played an important role in protecting the safety of the Irish in the face of such dangers as summary lynching. Ethnic solidarity could also be abused, and the networks of trust based on cultural nationalism meant that Irish communities were vulnerable to treachery within the ranks. An Austrian priest, Father Florian, wrote home in 1859 about an incident that had taken place a month before when an Irish merchant promised to buy cheap flour on behalf of some compatriot miners: "The poor people gathered together 1,300 dollars and gave it to him. The merchant disappeared with the money and up to date has not been caught."[99] The wide expanses enabled men to abscond from their families, thieves to escape retribution, and new identities to be realized, if so desired.

In danger of debt or death in California, Irish Catholics faced more metaphysical dangers, jeopardizing their eternal souls. Catholics priests saw it as their responsibility to safeguard the Irish, viewing them as exiles, not economic migrants: "The emigrant is indeed an object of pity and commiseration . . . at home he has been persecuted and pursued."[100] The All Hallows annual report tried to inspire the increased numbers applying for missions to the American West with intense pleas referencing the most famous Irish saint: "Are there no Priests in Ireland with the spirit of St. Patrick, to come and snatch the souls from the fangs of the destroyer?" The destroyer in this case was the twin dangers of the remote wilderness beyond the reach of the church and the temptations of the "phalanx of errors," Protestantism.[101] Fr. Andrew Twomey, a Cork-born priest who served the Irish mining town of Smartsville, California, wrote in the Record of Internment book for the Diocese of Sacramento of an Irishman, Patrick Galligan, who died on October 10, 1888, age "almost 60," of heart disease: "[He] Was found dead alone in his cabin. Seemed to have been dead a couple of days and consequently did not receive the last rites of the Church. *It was a clear example of how one should be always prepared and how dangerous [it is] for [a] person to live alone and remote from assistance*."[102] Both the preparations and the dangers were spiritual in nature.

This Irish Catholic worldview was shared by the congregation, which approached life with a long-term perspective and one eye fixated on the hereafter. "The life to Come is the important part," wrote Denis Hurley as he contemplated his declining health and old age.[103] In their letters, the Irish bemoaned the absence of priests and their inability to attend mass.[104] The clergy went so far as to debate organized Irish settlement in the West, with those who supported such schemes arguing in favor of larger numbers clustered together to protect individuals and guarantee the long-term spiritual survival of these communities.[105] The most pointed assessment was written by Fr. John O'Hanlon, in his guide for Irish immigrants, who advised that "in all cases in which Catholics meditate removal to a remote and partially settled country, we would advise the formation of companies and colonies; not so much for the purposes of mutual protection and assistance in encountering the dangers of wilderness and its colonization, as to obtain for themselves the spiritual succor they require, and which, under other circumstances, cannot often be procured."[106] Here, Protestant ideas of independence and reinvention on the frontier clashed with Irish notions of community and religion.[107]

The Irish frequently traveled to the American West with other Irish and remained in the same network of friends and family that helped, through funds or information, enable their journey. This trend established the Irish clusters seen in urban centers in the US.[108] Other denominations picked up on the clerical tendency to dissuade movement westward, and one Irish Protestant charged that the Catholic Church was "vehemently opposed to emigration to the West, since they [the Irish] are more difficult for the Church to control when so scattered."[109] If the Catholic Church was seeking to "control" its flock, Irish communities happily demanded the instruments of that control—priests. The all-important caveat was that the priests be Irish, and, perhaps more important, they often succeeded in getting what they wanted as Irish seminaries sent dozens of Irish priests who acted as extensions of the Irish community in these mining towns. Many Americans abhorred the Irish loyalty to the Catholic Church, equating it with slavery, but for Irish Catholics it was not a mark of "slavery" but a reaffirmation of their identity as well as their religion.

In California, the Irish were one among many different nationalities of Catholics.[110] Some, such as Father Florian, found the diversity energizing, a manifestation of the *katholikos* namesake, meaning universal. He felt like "a true Catholic Christian" because of the diversity of his congregation.[111] However, the deflating realities of negotiating inter-ethnic relations proved difficult to navigate. He defensively excused the poor attendance of his fellow German Catholics (numbering fifty out of a thousand of his parishioners) at mass because "the rude Irish did not leave any places for the German in the Church benches," revealing the limitations of his ecumenical enthusiasm and the failure of denominational solidarity to trump ethnic, cultural, or class differences.[112] Other accounts detail that Latino Catholics did not mix with the Irish, and class divided Irish Catholics into the respectable, middle-class "lace-curtain" Irish who set themselves apart from the working-class, "shanty" Irish.[113]

Religion remained important for Irish Catholics, often imperceptibly existing in the background of their lives. Father Florian wrote, "If I do not assert that all these are practical Catholics yet the majority are decent Catholics, and, in fact, many can be counted among the good and best people. Even the bad ones still want to be counted as Catholics and die as such."[114] Setting aside the fraction of Catholics who would remain if Catholic membership depended wholly on either meeting clerical standards or never being bad, Father Florian and Father Twomey, discussed below, misunderstood two important points for Catholics who lived throughout the American West. Isolation from the structures and rituals of the church did not diminish their faith; they remained Catholic. Furthermore, their Catholicism was intertwined with their identity, and most of these Irish Catholics found it no easier to walk away from their religion than they would have to walk away from their Irishness.[115]

As priests warned of the spiritual dangers of isolation, newspapers printed articles with titles such as "Horrible Story from the Far West," which originally appeared in Helena's *Montana Republican* in 1866 but was reprinted in papers as far away as Ireland:

> In May a man, representing himself to be a minister of the Gospel, stopped at an Irishman's ranch at Tobacco Plains, on the road from Pend Oreille Lake to Kootenay mines, and that some time afterwards the Irishman was miss-

> ing, and the quondam preacher stepped into his shoes, informing all who inquired after the previous proprietor of the ranch that he had bought out the Irishman, and that the latter had gone to the lower country. This seemed a plausible story; but a discovery of human bones led to an investigation, which has proved that the supposed lay-preacher not only murdered the Irishman, but hacked the body to pieces, and served it up in variously-prepared dishes to the travelers stopping at his place.[116]

What must have titillated many Irish readers imagining the savage wilds of Montana must equally have terrified the relatives of those who had gone so far beyond the boundaries of their reach.

Since resources were finite, miners themselves were by their nature a transient workforce. Accounts of destructive fires mention in passing that the fires were started by tramps, increasing suspicion and fear of migrant workers and setting the tone for later paranoia.[117] Placer miners had to be mobile, not only to find a claim or to travel to the next boomtown but also because harsh winter weather and hardened earth forced them to quit their claims as transport and work in heavy snowfall became impossible. As James Gamble wrote to his brother: "Times is very dull here at the present time, it's very cold here and a hard frost and some snow on the ground. There is very little doing here only those that are working in tunnels. As for my part I have done nothing this month."[118] In the 1850s John Boyd from Upper Holywood, County Down, came to California seeking adventure. During the off-seasons (from November to April) he, like most other miners, made his way to the coast working at road building or as a lumberjack. He commented on the intolerable loneliness working for the Hudson Bay Company for fifty dollars a month and claimed that the place "was inhabited by Kanakas, French Canadians and mosquitoes." He seemed unsure which was worse.[119]

The greatest danger early migrants faced was disease. One particularly deadly outbreak, the 1850–1852 cholera epidemic, accounted for half of all deaths on the overland trail between 1840 and 1860.[120] The journey over the Isthmus of Panama was equally deadly, and James Gamble wrote to his brother Abel in Belfast concerning their brother's death in 1858.[121] Andrew had worked as a stone cutter in New Jersey, but increasingly irregular work led him to try his luck in California.[122] He contracted Panama Fever, quit his trip west, and survived just long enough to return to his wife in New Jersey before he died.[123] Falling sick while traveling or alone could be particularly dangerous, and an Austrian priest was forced to spend several days "forsaken in a lonely hut" in the Sacramento Valley tended to by two poor Irish families who ran an "Irish Inn" and who, according to the priest's diplomatic language, could not support him.[124] Some German Protestants took pity on the priest and paid for better lodgings and a doctor's visit.[125]

The Great Land of Gold

Letters formed a global communication network linked throughout the Irish diaspora, with Irish relationships between family and friends acting as delicate strands routing information and details to one another. All peoples of the nineteenth century used letters, though few as effectively as the Irish. Even those who did not like the Irish understood that their migration patterns were distinct from others, especially Chinese migration patterns, and statistics further draw out the differences between groups in different counties in California. Letters obviously notified others of dangers and difficulties in California, but some focused almost entirely on economic fluctuation, with letters to Ireland recommending against going to the eastern cities: "Times are said to be very dul [*sic*] away down East, I don't expect that wages will ever be so good again around Eastern cities as it was [*sic*] for a number of years after the war" wrote Charles Canning in Riverton, California, to Lizzie McSparron in County Derry.[126] Instead, Charles suggested that the West offered a more promising alternative for Irish emigrants: "I also think that the West is by far the better place for us wild Irish from 'the Island of the Saints.'"[127] Lizzie knew well the lure of the American West. Her brother William had left the Pennsylvanian coal mines to go to California.

The opening passage from the following letter from Irish-American James Mullany, in Walla Walla, Idaho Territory, to his sister in Philadelphia keenly details the isolation many gold-hunting miners felt as they scoured the Rocky Mountains for riches:[128]

> Once more do I engage in the pleasing task of conferring with dear and absent friends in the hopes to revive in them a kindred spirit. I have looked with no little anxiety for each successive mail in hopes it would bring me some glad tidings from you but the arrival of each one only doomed me to

> more disappointment. But still I have not dispaired [*sic*] that you have given me as yet up for lost. I still shall hope on ever.[129]

He ended the letter pleading "Please write soon."[130] Others requested that letters be as detailed and as long as possible. Denis Hurley mildly admonished his brother when he read in a letter that he had married: "Why did you not say more about the affair? You know it would interest me . . . I will expect full and minute answers to all these queries and anything further you may think of."[131] Others were more direct. Paschal L. Mack heaped equal parts disillusionment and guilt on his sister when he arrived to San Francisco from his small mine camp to find no letters waiting for him: "what was my disappointment not to find one . . . I have arrived to the conclusion that you as well as the rest of the folks have forgotten that there is in being such a chap as Pick Mack—and this is the last time I will put you in mind of the fact that there is such a fellow living."[132] Miners' mobility from one mining claim to the next, following various rushes and living on speculation, often hampered communications with home, as did their regular isolation from post offices.[133] "I have wrote [*sic*] no letters these past four years" James Gamble wrote to his brother after going from California to Oregon, adding "nor I have got none from any of my friends."[134] For some, detachment seemed to be a state of existence one could slip into without consciously deciding on it as a course of action.

The absence of women among these early hordes of young prospectors may have added to this sense of isolation. In 1850, 93 percent of the population of California was male, and in one town, Grass Valley, 90 percent of the population was made up of men ages twenty to forty.[135] John Orr wrote, "In a city of some 10,000 Inhabitants, you will not see more than twelve or twenty women in a day there are only about 300 in [the] whole city."[136] Observers blamed the lack of women and families as the source of general immorality, yet life for the few women in California was difficult. Paschal Mack contrasted the extreme wealth with the shocking poverty he saw in San Francisco: "there are poor women in this city who have to beg for a living or starve and little girls around the wharfs barefooted getting wood—and men about the street picking up old rags and crods etc. etc. Such is California, the great land of gold."[137] Sex work existed throughout California "not only in cities but also in villages," and Fr. Florian noted five "houses of ill repute where low women publicly prostitute themselves" in the small town of Weaverville.[138] The priest viewed sex work as a secretive urban phenomenon, and it disturbed him to see it so proudly advertised in California's Eden-like frontier. Other moral outrages included "California marriages," informal relationships between men and women that Fr. Florian labeled "concubinage as long as it please both parties."[139] One Irish emigrant believed the American West was no place for a young married couple, given the harsh living conditions: "It is a ridiculous thing for persons to be looking for fortunes when getting married to give a girl lief [life] to work and to eat stirabout."[140] Women were seen as too weak for the frontier, either in the moral sense (corruptible) by some or physically (delicate) by others.

Even if many men and women "follow[ed] an evil life" in California, it did not mean an absence of others who did follow strict moral codes.[141] Patrick Manogue was an early Irish gold miner at Moore's Flat. Born in Ireland in 1831, he immigrated with his family to Chicago in 1846. In his younger years he worked to support the large family and attended the college seminary Saint Mary's of the Lake. He abandoned his studies to travel to the goldfields of California in 1853.[142] At the diggings at Moore's Flat, he became known as a hard worker and an honest figure in the camp: "Whenever a dispute arose Manogue was always the arbitrator. Both sides knew that he was perfectly fair and were willing to abide by his decision. He put down every quarrel that arose and prevented many a fight among the miners." His imposing 1.2-meter (6'4") stature undoubtedly helped.[143] After amassing enough money to comfortably provide for his relatives in California, he used his fortune to travel to Paris and attend seminary there. He was ordained and served as a priest in the silver mining town of Virginia City and later became bishop for the Diocese of Sacramento. Manogue was known as a towering figure of leadership for the Irish community in the city and, later, the state.[144] He was portrayed as more authentically Californian for having participated in the early gold rushes and more morally pure for having withstood the easily accessible temptations of the time for men with money. In a certain sense, Manogue embodied the ideal of Irish Catholic masculinity in the American West—pure and faithful to the end.

All Irish emigrants noted the bustling newness of the mushrooming mining towns. When one Irishman, Robert Williamson, visited his relatives in California, they took him on a tour of the surrounding hills. He later wrote,

Table 2.2. Number of Irish-born and as percent of total population, California counties, 1870–1920[a]

	1870		*1880*		*1890*		*1900*		*1910*		*1920*	
County	*No.*	*%*	*No.*	*%*	*No.*	*%*	*No.*	*%*	*No.*	*%*	*No.*	*%*
Yuba	927	9.6	830	7.4	574	4.2	304	3.8	205	2.0	83	0.8
Kern	96	3.3	197	3.5	258	2.6	289	1.8	540	1.4	424	0.8
Mono	22	5.1	858	11.4	109	5.5	76	3.5	32	1.6	13	1.4
Nevada	1,806	9.4	1,287	6.2	895	5.2	598	3.4	332	2.2	138	1.3
Placer	816	7.2	709	5.0	578	3.8	390	2.5	349	1.9	140	0.8
Sierra	496	8.8	345	5.2	208	4.1	126	3.1	58	1.4	28	1.6
California total	54,421	9.7	62,962	7.3	63,138	5.2	44,476	3.0	52,475	2.2	45,308	1.3

[a] We have only county-level data on ethnicity from 1870 on. Data compiled from US Census, 1870–1920.

"This country is full of mines which will be better known to the next generation."[145] Some gold-strike mines did develop into mining towns, but most of these few strikes boomed and faded in a few short years, leaving little trace of their existence in the wilderness apart from some deep holes in the ground, some wooden shacks, and high piles of tailings that appeared as small hills to the untrained eye, indistinguishable from the surrounding landscape. Williamson's relatives bought their farm after making their "pile" (miner parlance for fortune). The movement from mining to agriculture occurred for the Irish too, although it remains difficult to track exactly how frequently, primarily because miners' geographic mobility was sometimes matched by occupational transience.[146]

Table 2.2 details the distinct differences in the patterns of growth among various California counties, such as Mono, Nevada, Placer, and Sierra Counties. The latter three experienced their high point in the 1870s when shallow quartz mines were easily worked and profitable. As miners dug deeper, many petered out while others required increasingly expensive machinery. In 1880, a gold rush to Bodie in Mono County drew many Irish miners to the richer, easier pickings there. Yet the Mono goldfields proved fickle and were soon exhausted, with only one-eighth of the Irish who lived in Mono in 1880 still there a decade later. During its brief boom and decline, the town of Bodie had a significant Irish dimension. Little more than a year after the Land League was founded in County Mayo, there was a Land

Table 2.3. Irish-born and Irish-born with Irish parentage, California mining counties, 1880[a]

	Irish-Born		*Irish-Born with Irish Parentage*	
County	*Number*	*% of total pop.*	*Number*	*% of total pop.*
Yuba	830	7.4	1,689	15.0
Kern	197	3.5	272	4.9
Mono	858	11.4	1,166	15.6
Nevada	1,287	6.2	2,794	13.4
Placer	709	5.0	1,388	9.8
Sierra	345	5.2	690	10.4

League in Bodie. Judge Thomas Ryan opened the chapter, declaring, "For seven centuries Ireland has been fighting for liberty [and] those who have gathered here tonight should not respond as Irishmen, merely, but as citizens of the leading republic of the earth and aid in liberating the oppressed people from English rule."[147] It would have been easy for Ryan to simply state that all loyal Irishmen should respond to English oppression in Ireland, but he broadly implied a global dimension in challenging the British Empire. Reading Ryan's words, it is hard to believe that they were written for an audience of Irishmen assembled in a dusty Miners' Hall at the foothills of the Sierra Nevada. Ryan emphasized the dual identity of Irish-America, and rather than imply that its loyalty was divided, he declared that its twin identities strengthened each other.

Since Bodie had a short life span as a mining town, the Irish had less time to establish deep roots. We can see from table 2.3 that the increase in Irish parentage in Mono County is less than half the increase in the other counties. This shows that there were fewer children in the overall group, which is further confirmed by reading the full census returns. This made the migration from Bodie even swifter from 1880 to 1890, when the gold mines produced less and more attractive strikes drew people elsewhere—for example, to Butte, Montana, which by then was demonstrating surprising longevity and continual prosperity for a mining town.

In contrast to Mono County, Yuba, Nevada, Placer, and Sierra Counties all had larger and more established communities, with married couples and children, as seen by the fact that the parentage figure more than doubled the size of the Irish community.[148] These figures also show that while the

Table 2.4. Nativity and nativity with parentage figures, Yuba County, 1880[a]

Country	*Nativity*	*Nativity and Parentage*	*Difference*
Ireland	831	1,689	+100%
England	202	309	+53%
Scotland	61	114	+87%
Wales	31	54	+43%
China	2,142	2,153	0.0%
Canada	202	142	-30%
Germany	365	738	+102%

[a] The parentage (second-generation) figure does not include mixed marriages. IPUMS.

Irish had a significant presence in Mono County, comprising 11.4 percent of its population, a comparatively smaller percentage had Irish parentage. So, when a massive drop of Irish-born residents occurred, from 858 in 1880 to 109 in 1890, it happened quickly because of the greater number of individuals than of families with children.

A comparison of the different ethnic groups in Yuba County in 1880 (table 2.4) reveals one stark fact. Both the Irish and Germans had large numbers of second-generation residents, indicating a degree of stability in both groups. The opposite is true for the largest ethnic group, the Chinese, whose figures reveal a lack of women or children in the group, as seen in the minuscule increase in "Nativity" and "Parentage" numbers. Why was this the case? The Chinese imperative was to see Gaam Saam (Gold Mountain) and return home with their fortunes, not to stay in the US.

Puck Magazine noted this tendency among the Chinese in a picture titled *The Difference between Them* (figure 2.3). Beneath lines of Chinese entering the port of San Francisco is written "We are going home," contrasting with the lines of Irish who are seen as permanent immigrants, with "We have come to stay" written under them for emphasis.[149] The cartoon mocks both groups' accented English in captions above their heads. The Chinese caption reads "Wantee plentee monee to takee home," whereas the Irish caption reads "Wanted immajitly a chance to work as alderman or coroner." It is also noteworthy that the advertisements for departing ships only list Chinese ports. The cartoon alludes to the Irish propensity for political activism but uses it primarily to highlight the corresponding lack of Chinese influence in the political sphere and the different overall experiences of the groups.

Another cartoon from *Puck Magazine* repeats the point more explicitly, with political figures courting Chinese voters in a room called the "Naturalization Bureau" with a sign under it that reads "Voters Made Here." The caption beneath the picture reads "The Chinese Question would be settled if the

Figure 2.3. The front page of the popular nineteenth-century magazine *Puck* shows a line of Irish and Chinese, drawn as racialized caricatures, entering the US. The Irish indicate that they want to stay, whereas the Chinese carry signs saying they want to return home after they have made enough money to do so. *The Difference between Them*, *Puck Magazine*, May 15, 1878, Library of Congress, Print and Photographs Division, Washington, DC.

Chinee, Chinee would votee! Votee!! Votee!!!"[150] The temporary nature of the Chinese migration to the American West meant they could not effectively oppose legal efforts in California that excluded them from certain spheres of society, aided by Denis Kearney's populist nativist vitriol in San Francisco. The passage of the Chinese Exclusion Act in 1882, four years after the picture in *Puck*, legally formalized the fleeting Chinese presence in America by barring them from citizenship, enshrining in law the fewer options the Chinese had vis-à-vis the Irish. These legal barriers prohibited the emergence of permanent Chinese communities, thereby stifling any possibility of balanced demographics.

Nativist cartoons mocked the Chinese failure to engage in politics and also insulted the apparent excess of Irish political participation. This Irish activism was not unopposed, and it repeatedly drew the ire of nativist elements in American society. These sentiments existed in California during the gold rush, and written accounts reveal examples of discrimination against the Irish that run the gamut from subtle employment discrimination to outright bigotry. An example of the former was noted in one American's diary: "I have made an arrangement with . . . an agent for an English Mining Company here, to take his four Irishmen into my pay, as he finds some difficulty about employing them at the Maripa."[151] The latter is exemplified in a different account from a WASP from Maine, Franklin Buck, who wrote home from California after he learned of his father's election to the state House of Representatives on the Know Nothing platform: "I hope he will carry out the Know Nothing principle until the *Irish race* is utterly annihilated from these United States."[152] Although he urged exterminating the Irish, his view of other groups was more tempered. He explained his reasoning: "The Chinese we can get along with . . . They don't want to be made citizens," whereas the problem with the Irish was that they were too involved in politics: "The Irish have a great deal more to say about the election than we do." Buck sought to rectify this and explained that he was a "judge" at the previous election where he and other native-born Americans forced every voter with "an Irish name" to swear their vote on "the Holy Cross."[153] Irish accents or appearance did not offer a wide enough net in detecting Irishness by their standards, so they attempted to find any Irish background by using names. Buck boasted sarcastically that their efforts stopped approximately 150 "good American citizens" from voting. His assumption that these Irish were not citizens was based solely on his prejudiced stereotyping of them as a threat, and he was untroubled by the possibility that he might have perverted the democratic process by denying citizens the right to vote through intimidation or the addition of a sacrilegious oath before they voted.[154]

Americans had a range of views on the Irish, but suffusing their perception was a latent suspicion of Catholicism permeating society. One Sunday, American Stephen Davis could find no "protestant [*sic*] worship" and went to the nearby cathedral, "which consisted of the ceremonies, formalities and rites of the Catholic Church in all their bigotry and superstition."[155] He did not explain why the Catholic mass was bigoted, but his use of the phrase *ceremonies, formalities and rites* draws a curious parallel between Catholicism and the rites of Freemasonry. After mass he decided to follow the priests to a cockfight that was being held for their "special benefit," but he did not enjoy the "bloody scene" and soon left. Social boundaries could be crossed and prejudice set aside, if only out of boredom. Bigotry was common, and a travel writer observed the actions of two other Americans in a small hotel who referred to an Irishman staying there as "the meanest thing in all creation." They also referred to the "Scotchman" as a half-bull, half-donkey "critter" and said they enjoyed giving "the Britishers . . . particular h-ll."[156] Sometimes even those who disliked the Irish were capable of empathy, and the writer who noted Davis's account thought of them as strangers in a strange land, stating "how sad it must be to die in a foreign land, among strangers—without father, mother, brother, or sister to offer consolation, sympathy, and the many little offices that affection would devise."[157] This separation from family and home struck the Irish particularly hard and encouraged their perception of themselves as exiles.

Smartsville, Yuba County

For a group that considered themselves exiles, putting down roots in a foreign country would always present a unique set of problems. The permanency of life in the more stable Irish-American communities, such as Marysville and Smartsville in Yuba County, created different frictions with other groups and fractures within the group than occurred in the more unstable mine camps. Difficulties with nativism continued, as towns developed that were dominated by American-born (Timbuctoo) and other ethnic groups such as the Cornish (Grass Valley), and class distinctions within the Irish community slowly began to show themselves. Perhaps most important, the rooted-ness of the Irish facilitated the establishment of a distinctly Irish Catholic Church served by Irish priests. Through the descriptions of one of these priests, Fr. Andrew Twomey, the circumstances and challenges of Irish-American life in Smartsville become more intimate and clear.[158]

The rapidly growing population in the 1850s quickly exhausted the surface deposits of gold, drawing a curtain on California's famed placer mining phase in a few short years and necessitating the use of other gold mining techniques. Quartz mining and hydraulic mining gained widespread popularity, but these methods required infrastructure and a larger workforce, which, in turn, gave rise to permanent settlements. Some overly excited miners in the early gold-rush towns took the title "mining town" literally: the minutes of the Marysville City Council for August 12, 1851, disclosed that the town's first mayor, S. M. Miles, raised objections to miners sinking shafts at the intersection of E and Front Streets. The mayor issued a proclamation that read: "It having been represented to me that sundry persons have laid out and staked claims on the bar in front of the steamer landing for mining purposes, now, therefore, I, S.M. Miles, Mayor of the town of Marysville, do hereby caution all persons against trespassing on or injuring the public grounds within the limits of the City of Marysville in any manner whatsoever."[159] The warning succeeded in putting an end to further tunneling and prospecting under buildings and in streets throughout the town.

Irish miners often named their mines and claims after things that reminded them of home, for example, the Hibernia and the Maid of Ulster. Sometimes they named mines after themselves, such as the Kennedy Mine in Jackson County, California, or after a particular person they knew, as was the case with Marysville in California and Montana. The Irish in California gravitated toward towns and mining camps that had other Irish, as had happened in earlier mining regions across the United States.[160] Names held a deeper meaning and could reveal prejudice, as seen when one man sneered at an Irish surname in the name of a mine company: "Smith O'Brien Quartz-crushing Company . . . what a title for a company! Quite enough for a man to bear it!"[161] For example, in Yuba County, it was common knowledge that Grass Valley and the smaller Timbuctoo were largely American and British Protestant mining towns, whereas Irish Catholics dominated Marysville and the smaller town of Smartsville.[162] Names had ethnic attachment, and this ethnic magnetism helps explain the differences in the numbers of Irish in Kern County and Yuba County (see table 2.2)—two regions that had very different reputations.

Yet it would be incorrect to interpret this preference as constituting isolationism from other ethnic groups. Focusing on the area around Smartsville, the *Daily Alta California* wrote these revealing lines: "Closely adjoining Timbuctoo are the smaller villages of Sucker Flat, Smartsville, Temperance Hill and Mooney Flat. On Temperance Hill are several handsome cottage residences, surrounded by gardens, the brick office of the Tri-Union Canal Company, and a Masonic Hall. At Smartsville there is a Catholic church, for the accommodation of the large Irish population, which predominates in all the localities named except Timbuctoo."[163] The more open-minded American miners who lived near Irish miners often freely mingled with them. For example, an American miner named Twogood was camped near Smartsville and took a break in the evening from his part-time farming and gold mining in the hills of Yuba County to socialize and drink with the Irish nearby, writing in his diary, "After a night with the Irish County at Moore's diggings, we return home."[164] Similarly, the census reveals interesting interactions between the ethnic groups. The 1860 census details the miner settlement Colfax in Placer County within which we find four households nestled between homes whose residents are Chinese, with every occupation listed as miner.[165] In the first house lived James White from Tennessee and Peter Donnelly from Scotland; in the second lived Lawrence Austin and Michael Graham, both Irish-born; in the third lived John Larkin and Philip Mulligan, both Irish-born; while the head of the last household was Charles Rice, a forty-one-year-old miner born in Ireland who lived with Al Song and Al Lehing, two twenty-three-year-old Chinese miners.[166] These daily interactions between ethnic groups should temper any general portrayal of a segregated and fractious population, perhaps pointing to a broader working-class solidarity seen more easily in granular census returns.

Despite the population's generally rapid adoption of territorial and federal law, the informal retributive justice of the vigilantes occasionally reared its head, even in towns near the more established and secure Irish communities such as Smartsville.[167] In 1856, a Vigilance Committee in Timbuctoo lashed an "idiotic and crippled" Frenchman for allegedly stealing $250 from a miner.[168] A man named Lyman Ackley, supported by three Irishmen, put a stop to the public flogging:[169]

> Had he not interfered with the Committee while doing their duty, no doubt is entertained but that a full confession would have been obtained from the prisoner before he would have been borne many more lashes; but the sympathies he received from the "elderly gentleman," [Lyman Ackley] made doubly stubborn the guilty thief, well schooled in his profession, and the Committee were obliged to give it up and let him go with as little punishment as is often meritedly received by school boys. Here let me say the number of lashes he received is meanly exaggerated, and not a drop of blood was drawn so lightly were they applied.[170]

The above extract is part of the response from "Vindicator" printed in the *Marysville Daily Herald* detailing the incident and defending the actions of the vigilantes against a damning letter written by Lyman Ackley and printed in the *Marysville Daily Herald* on August 22, 1856. Ackley's letter stated that the vigilantes were "sympathizers with the Vigilance Committee of San Francisco," an accusation the writer in *Marysville Daily Herald* denied, adding that these men were the "strongest advocates of 'Law and Order'" and "the first citizens of Timbuctoo."[171] This statement implied that the vigilantes were high-standing members of pioneer and Masonic organizations. The glorification of the vigilante movement in pioneer literature reflected the huge numbers of former vigilantes who established and led western pioneer organizations and historical societies.[172] "Vindicator" went further, stating: "Who is Lymen Ackley? I answer he is an Irishman, who is well known here and elsewhere, and if in him are to be found the qualities that adorn a gentleman, Heaven send different decorations to be used by all others."[173]

As in other towns with large Irish populations throughout America, the Irish in Yuba County formed several fraternal organizations that supported causes important to them. A letter, probably written before the 1870s, drafted a proposal for the formation of a fraternity of Irishmen:

> To the Irishmen of Marysville *and vicinity*,
>
> We the undersigned citizens of Marysville in assembling together expressly for the purpose of forming a society for the formation of sociability, [illegible], Industry, Charity and Good morals most sincerely hope and beseech our countrymen at large *without distinction of creed or polliticks* to come forward on this

> occassion and take pattern by other organising societies which have been and are a benefit and honour both to themselves and their country . . . We therefore in hopes this will resound to the benefit and honour of ourselves and children do solemnly confess, agree and promise to form a society of purely Irish both in feelings manners and respect for fatherland . . . As this is purely for the advancement of the Irish character at large it is to be hoped that *no wolfes in sheeps clothing* or any person with estranged feelings will attempt to enrole themselves as a member of such society as unfortunately hithertofore the feelings and interests of said citizens have been estranged devided and kept aloof from each other.[174]

From the start, the writers stress the importance of including all Irish in the area around Marysville, regardless of religion or political persuasion, discarding the exclusive nature of the Ancient Order of Hibernians (Catholics with Irish ancestry only). Rather strangely, they refer to Ireland as their "fatherland," perhaps an influence of the German societies, as Ireland was generally referred to as female.[175] However, the most interesting portion of the letter is the express hope that "no wolfes in sheeps clothing" join the fraternity to subvert its inclusive mission, suggesting that the authors knew about the infiltration of other organizations—whether by "estranged" Irishmen or detective agencies.[176] The organization's goals are somewhat hazy and only the draft of its mission exists, but it remains another example of Irish organizational skill and cultural pride, their hopes and fears in this new land.

Irish miners in Yuba County became entrepreneurs and businessmen, reinvesting earlier mine earnings into the more capital-intensive ventures required for hydraulic mining. These projects offered employment for miners and laborers during economic downturns, seasonal shifts, or sudden financial difficulties. During the winter of 1871, James Gamble worked on the construction of a "large stream for mining purposes," nearly 100 miles long.[177] To ensure the stability of the workforce (largely fearing the miners would join the next gold rush), the company paid them in early spring for their work over the previous season; although pay was low by mining standards, fifty dollars a month, it included board and there were few amenities or distractions to entice the men to spend what little money they had.[178] The *Sacramento Daily Union* noted that "the 'Shamrock Company' are digging away, and have as high hopes of its richness as they have of the ultimate triumph of the shamrock in 'Ould Ireland.' Creary's claim is being worked energetically—labor going on day and night—and is said to be giving evidences of great richness."[179] News reports saw the devastation wrought by this type of mining on the environment as largely inconsequential: "The site was formerly a high hill, but is now down deep in the earth many feet below the surface."[180] Geographic shifts and changes were largely an uncommented-on consequence of industrial processes.

Labor agitation in this area consisted largely of protesting workers "shaming" local owners and managers by marching around the area advertising their dissatisfaction and seeking support from the local community. On February 5, 1877, the *Sacramento Daily Union* reprinted the story from the *Marysville Appeal*, which the day before had reported that miners in Sucker Flat, Timbuctoo, and Smartsville had united to protest a general reduction of wages and paraded peacefully through the towns of Yuba County, starting and ending at Mooney Flat.[181] Local miner James Hanley led the procession, accompanied by a brass band. The 175 miners in the march were also raising funds to aid "the poorest families" in Marysville.[182] The incident demonstrated the broad-based alliance of miners in the region and their support for the local community.

Hydraulic works became more extensive over the years, requiring greater investment for longer aqueducts to supply the increasingly distant sites. The Excelsior Mining Company largely controlled hydraulic mining in Yuba County through the Excelsior Water Company, thanks to its thirty-five-mile canal connected to the Yuba River (figure 2.4). The trustees of the Excelsior Mining Company in the 1880s were James O'Brien, Daniel McGanney, James Pierce, William Ashburner, and Charles Webb Howard.[183] The capital stock was $5 million, divided into 50,000 shares of $100 each. The actual amount of the subscribed stock was $3 million, of which D. McGanney took $90,000, James O'Brien $100,000, James Pierce $1.8 million, Wm. Ashburner $100,000, and C. W. Howard $100,000.[184] As the *Sacramento Daily Union* noted, "The stockholders in this ditch are all wealthy men, and they are determined to make this enterprise among the most successful of all the enterprises of Northern California."[185]

An 1886 article in the *Daily Alta California*, reprinted from the *Grass Valley Union*, attests to the influence of these very wealthy individuals:

> The *Union* has received from James O'Brien, of Smartsville, a box of oranges, of beautiful color and of delicate flavor, raised on his residence grounds.

Figure 2.4. Hydraulic mining near Timbuctoo. Library of Congress Print and Photographs Division, Washington, DC

> Smartsville is but 700 feet above the ocean level, and is in a warm belt, which makes the conditions very favorable for citrus culture. The locality has been famous in the past for the richness of the production of its hydraulic mines, which are yet far from being exhausted, but the senseless crusade of the anti-miners has for the present ruined that industry, but this cannot rob the locality of its ability to raise fruits that are golden, and which will yet be a product of important value.[186]

While the article prattles on about O'Brien's agricultural successes, the newspaper acted as O'Brien's mouthpiece in his effort to curry favor with another newspaper, the *Grass Valley Union*, so as to dominate the local media and demonize those who opposed the mine company's actions, the so-called anti-miners. A little over three-and-a-half years later, the *Daily Alta California* published an article on one of the business partners, Daniel McGanney, noting that O'Brien's business fortunes had taken a turn for the worse: "The causes leading to the assignment are not known, but are strongly hinted as being due to a depreciation in landed and mining interests . . . His business lately has been chiefly in cattle and outside speculations, and to the latter are attributed his present difficulties. His private residence, [a large colonial-style mansion] at Smartsville is valued at $20,000."[187] He had lost the vast sum of $120,000 on mining ventures.

O'Brien hired Chinese workers extensively, both for the Excelsior Mining Company to build ditches and tunnels and in the construction of his large home.[188] As a symbol of gratitude for their employment, they planted Zelcova trees in his front garden.[189] While the wealthier figures in Smartsville willingly hired cheap Chinese labor, local miners and laborers saw them as a threat to their livelihoods. On the morning of May 4, 1866, 100 workers in Smartsville assembled and walked to R. L. Creary's claim: "quietly loading the Johns [Chinese] with their traps, they placed them at the head of the procession and marched them over the hill out of town."[190] After the group evicted the Chinese workers, they gathered in the town hall, appointed a president and secretary, and named their organization the "Union League of White Men."[191] It remains unknown how successful this organization was at attracting Irish support, but it disappears from the records after this incident. Although the numbers of Chinese in the region recovered in the 1870s and 1880s, in 1884 "Chinatown in Timbuctoo was totally destroyed by fire this evening. Loss not known."[192] This may have represented an aggressive form of eviction and the lingering strain of anti-Chinese prejudice from the time of the earlier "procession."

Not everyone who suffered economic setbacks at this time projected their fears onto the Chinese. In a letter to his sister commiserating with her over her recent loss of a child, Lawrence Austin wrote: "My dear sister it is as you say one trouble never comes alone. I have not had good luck lately for the mine that I have been working in has shut down and throws lots of men out of work but I am working for myself and hope to do pretty well after a while but it takes time."[193] Austin's optimistic tone might have been for the benefit of his heartbroken sister, but the absence of any reference to the Chinese was contrary to much of the popular anti-Chinese sentiment in California at the time. Perhaps his experience of living next door to Chinese miners in the past led him to be more open-minded.

Meanwhile, in March 1882, citizens held a public meeting on the Chinese question at which they enthusiastically advocated a ban on hiring all Chinese workers.[194] Two weeks after the Chinese Exclusion Act was passed in May, locals met again to establish a branch of the Pacific Coast League of Deliverance, which advocated vigilante violence to expel the Chinese. Patrick Reddy, once a miner but now a one-armed lawyer, was the organization's leading opponent.[195] He questioned the reasons for "doing violence to helpless slaves, who are not here at their own bidding." As he shifted the blame from the Chinese, he redirected the crowd's attention elsewhere: "If we are to do violence, why be cowardly about it? Why not attack those who bring them here? If there is any boycotting to be done, do it in the right direction and strike like men."[196] Reddy's unusual deflection resonated with the townspeople, and the town did not establish a branch of the league or experience any anti-Chinese violence. The example of Reddy and the events in Smartsville shows that many Irish viewed and interacted with the Chinese differently than did other Anglo-Americans.

Other organizations, like the Father Mathew Society, were more popular with the local Irish Catholic community. The Irish temperance fraternity held "well-attended" meetings that were an indication of the other activities of the community.[197] A unique primary source, the internment book for St. Mary's Catholic graveyard in Smartsville, offers a closer look at life in the town. It lists all those buried throughout the parish between 1878 and 1918. Smartsville at this point had a population of approximately 800 people, and the local priest, Daniel O'Sullivan from Listowel, County Kerry, had been the founding pastor of the Catholic Church there.[198] In 1882, Fr. Andrew Twomey took over the duties of parish priest, and his comments in the internment book during the waning days of the Smartsville mining era provide a unique insight into the priest's perception of his congregation and the makeup of the overwhelmingly Irish Catholic community.

Twomey commented in the margins of each person's entry, including little thoughts or notes about the person and their funeral. For example, he reveals an important reason why Irish fraternities proved popular among emigrants. In the entry for J. J. Sullivan, a miner who worked seasonally between San Francisco and Rose's Bar near Smartsville, the internment book states: "Died in San Francisco. He belonged to the A.O.H. [Ancient Order of Hibernians] who marched at Funeral. Lot. O in cemetery."[199] Without family or relatives, these groups were the only way these men could ensure that someone attended their funeral and gave them a proper sendoff. Knowing they would be missed and given proper burial rites far from their homelands gave the living some peace of mind.

Although the Catholic Church had firmly brought the Ancient Order of Hibernians under its control in the 1870s, the church maintained a relative distrust of fraternities through the rest of the nineteenth century out of fear of affiliation with or influence by anti-Catholic Masons.[200] These fears revealed themselves in Virginia City when Father Lynch forbade members of the Order of Chosen Friends from dropping sprigs of evergreen, the symbol of their group, into the grave of Colonel T. Brady in that city. When the fraternity ignored his order, he announced at the funeral that no member of the fraternity was allowed to enter the consecrated grounds of a Catholic cemetery as a representative body; "he also spoke disparagingly of the custom of placing costly emblematic offerings on the graves of deceased persons."[201] The solemnity of the occasion did not get in the way of the priest making his point.

Observing the dangers of working the mines, Twomey commented on the death of Gareth Hoare, a sixty-two-year-old Irishman who died in a mining accident in 1892: "This poor man was killed with a mishap from a powder blast. It shows how careful one must be when his duty is explosives. Also shows the necessity of having always [to be] prepared for death as this plays on the mind."[202] Commenting on the death of eighty-one-year-old Michael Powell in another part of the internment book, he reiterated his earlier warnings: "The death of this poor man is a clear lesson that when old age and decripited [*sic*] set in it is wise always had ministered the Last Sacrament even though death may not be imminent. Also it shows how dangerous it is to live alone."[203] Here, the spiritual dangers of isolation were added to those of old age. The priest took these dangers seriously and made every effort to visit his sick and elderly parishioners scattered throughout Yuba County.[204] Twomey wrote about James Gleeson, who died of Bright's disease at age fifty-five: "This poor man through the mercy of God received the last rites of the Church sometime before. Always administer sacraments when sickness is serious. Oh holy St. Joseph I thank you."[205] It is difficult to quantify the spiritual importance of such work to the faithful who were offered some measure of comfort as they lay sick or dying.

Other comments highlight the onward migration of the Irish community. For example, Timothy Early died of consumption at age seventy-five in Butte, Montana, but was interred in St. Mary's, Smartsville. Another Irishman named John Daly (listed in the internment book as "Daily") was born in the mining camps of Nevada County, California, to Irish parents and died in Dublin Gulch, Butte, on September 11, 1898, when he "was shot by a masked man or robber in a polling booth."[206] That vote was dominated by the efforts of one of the copper kings, William A. Clark, to win a senate seat. Another Butte copper king, Marcus Daly (not a relative), was sufficiently disturbed by the murder of this local election judge in the heavily Irish district that he hired the Pinkerton Detective Agency to investigate the case. Although the *Anaconda Standard* speculated that John Daly was killed in an attempt to steal ballots, the Pinkertons never discovered the identity of the murderers.[207] Clark's circumspect involvement only sharpened the mutual loathing between him and the Irish Catholic community in Montana.

Unlike other locations, the Irish seemed to consider Smartsville a second home in America even as they followed other Irish to find work. Despite their time in places such as Butte or San Francisco, which had larger Irish populations, they still sought to be buried in Smartsville. The graveyard for St. Mary's Church also testifies to the almost unique Irish tendency to detail their homeland on their headstones, with many including their native parish or village. The design of nineteenth-century Irish headstones remained plain and straightforward. For example, Peter Dillon, who died on the cusp of his twentieth birthday, had a cross emblazoned on his, along with Requiescat and "native of Ireland." Samuel O. Gunning's headstone recorded that he was a native of County Sligo, Ireland, and Richard Walsh's notes that he was a native of County Waterford. Others specifically stated the parish and sometimes the village: on Micheal Sweeney's headstone is chiseled "native of the parish of Easkey, County Sligo, Ireland." These details are seen in greater number and detail for Irish emigrants and emphasize their deep attachment to their local area. Also of particular interest is the headstone of Daniel Creedon, a native of Macroom, County Cork, who died on December 7, 1877. He hailed from the same town in Ireland as Father Twomey (figure 2.5), indicating that there may have been some degree of direct migration between the barony of Muskerry in Cork and Smartsville. These relics from the past—the internment book and the graveyard headstones—attest to the

Figure 2.5. Fr. Andrew Twomey, the Cork-born priest of Smartsville. Diocese of Sacramento Archives, Sacramento, CA.

exceptional localism of the Irish migrant even beyond death, as well as to their intense cultural and religious loyalty.

Father Twomey primarily attended to his flock's spiritual needs, but he was not focused solely on that aspect of their lives. He commented on John McGovern's death by paralysis: "Another great example of the Mercy of God. For this poor man was at his duty a short time before. Also it is my belief that it would be better for men of his class who are alone to be in the hospital."[208] Father Twomey believed that McGovern, a single miner, should have received appropriate comfort and care in addition to the added peace of mind by being near those who could offer him his last rites. Twomey noted the pecuniary troubles of his deceased parishioners in sympathetic tones, but he could not help but write about the death of Nicholas Petit, age thirty-four, in 1881, that "his friends cried a great deal but did not settle with the Monsignor."[209] He still believed the church and himself as its representative were offering a vital service to the community and deserved to be supported by them through appropriate financial donations.

However, Father Twomey's main focus was on the spiritual well-being of his parishioners, and he repeatedly demonstrated his intense devotion to his community's religious needs, as exemplified by the entry for Michael Beatty who died in 1891: "That it is well always to go on a sick call when summoned no matter how far the distance may be and to go at once."[210] The internment book entry for March 11, 1902, gives lasting testimony to Father Twomey's efforts to reach distant parishioners; it lists the name "Reverend Andrew Twomey" with the comment "Drowned while trying to ford Dry creek, on his way to say mass at Parochial House March 8th 1902. R.I.P."[211] A swell in the river overturned his tram and he drowned, trapped under the vehicle, while striving to serve his parishioners. His death at age thirty-six was a terrible spiritual and psychological blow to the Irish Catholics of Smartsville, and his replacement, Father Hynes, an American priest, could not sustain the Irish community as Father Twomey had. The decline in the mining industry throughout the region further hastened the demise of Smartsville as an Irish town, even as local memory struggled with the loss of Father Twomey. Smartsville's faithful built an imposing ten-foot monument to mark the grave of their beloved priest.

The decline of the local economy left little hope for young people and families, who increasingly returned merely to visit their parents or to be buried in the family plots. Newspapers recounted the visits of relatives. In 1905, the *Marysville Daily Appeal* wrote that "Timothy Lenihan of the Southern Pacific Bridge Company, is up from the southern part of the State to visit his parents" and that "P. Callihan and son, Thomas, Robert Byrne and Richard Barrett left last week to seek a fortune in San Francisco."[212] In 1907, the paper reported that "Miss Mary Lenehan, who has been absent for several months at the bay preparing herself for a trained nurse, is up on a visit with her parents in this neighborhood."[213] The departure of the younger generation indicated that a fortune could no longer be found in Yuba County. Smartsville had merely become a place where elderly Irish parents grew old and were buried; it failed to sustain its Irish population thereafter.

Randsburg, California

In the 1900s, the mining town of Randsburg in Southern California entered a second mining boom, and the smaller Irish population negotiated very different circumstances than they had in earlier Californian gold ventures or in towns known to be Irish-dominated, as was the case with Smartsville. When Joe Meehan approached the Yellow Aster Mining and Milling Company in Randsburg, California, on May 7, 1915, looking for work, the application process included a comprehensive screening of workers, including details such as name, age (28), nationality (Irish), length of residence in the US (6 years), number of years mining (2), where employed (G. Consul Mine Company, Goldfield, Nevada), kind of work (mucker), marital status (married), wife or children (no children), persons dependent on you (wife), have life or accident insurance (no), member of lodge or fraternal society (no), next of kin (Mrs. Joe Meehan, 1515 Carro Gordo St., Los Angeles, CA.), can you read (yes), what language (English), have you read the rules of the company (yes), have you had them explained by an interpreter (blank), and do you understand the rules (yes); the form also required a signature.[214] The application contained two additional sections, one for the company surgeon, who examined Meehan and noted that he had a "weak abdominal wall" but was otherwise healthy, and one for the superintendent who also thought he was "fit" for labor.[215] This information gave mine companies tremendous power over workers, and they expended significant resources on administration to develop and maintain this detailed paperwork as a source of information and control over their workforce. The records disclose that Meehan was young and married, but because Randsburg lacked a substantial Irish community or sympathetic mine management who might offer a more enticing place to live, Meehan's wife lived in the Irish community in nearby Los Angeles while Joe worked in the mines in Nevada and California for several months at a time. Instead of building and rebuilding Irish communities in each mining town, the Irish were settling in established hubs in larger urban areas and temporarily working in the mining towns for weeks at a time.

The possession of detailed personal information was only part of the changing relationship between workers and business during this period. Businesses increasingly viewed any labor activism as an imminent threat and a challenge to their corporate power. In this atmosphere of distrust, when the Yellow Aster Mining and Milling Company (YA) heard rumors of mine workers' dissatisfaction, it decided to hire the Thiel Detective Agency (TDA) in response to workers' growing disquiet and the possibility of a strike after the YA decreased wages from $3.00 to $2.50 a day. The TDA had an office in Los Angeles, and the company wanted it to infiltrate the Randsburg Miners' Union—a branch of the Western Federation of Miners (WFM)—and report on the workers' feelings. When the TDA operative arrived, he found that the strike had already begun.[216] He spoke to Phil Conley, a miner who believed a strike could be won "inside of 60 days," but he also knew that the "Randsburg union is in very bad shape and the Federation wants to win the strikes around home in Colorado . . . as the Federation has not enough money to fight all the small strikes and has already dropped one or two in order to keep the Cripple Creek strike going."[217] Conley was eager to return to his home in Los Angeles, the union headquarters, but with the threat of the WFM calling off the strike in Randsburg he wanted to wait until "Lewis," the union man, arrived to clarify the union situation.[218]

Both effective unionization and the formation of ethnic communities proved increasingly elusive in these isolated mines. Ironically, this resulted in part from the growth of regional unions such as the WFM. As heavily unionized regions such as Colorado faced powerful eastern business interests' determined efforts to destroy unions, these efforts were felt in the WFM across the American West—even in smaller mine camps such as Randsburg, where workers also faced company blacklists and private detectives. Unions in these small towns were hard-pressed for funds as attention turned toward larger strikes in Colorado.

Instead of calling off the strike, mine workers in the region concentrated their efforts against the YA and decided to end the strike against the nearby Butte Lode Mining Company by reaching a settlement whereby the supervisor of the mine, McMahon, would agree to hire four or five union men.[219] While this gave the union at Randsburg breathing space, its morale was shaken. Consequently, the men showed signs of uncertainty: "the union men are very changeable. One day three or four will want to call the strike off and the next day they will be all out to win."[220] This shifting mood gradually gave way to despair, and the TDA reported that "most of the union men would be willing to call off the strike but the officers keep banging on, as they feel that to call it off now would wipe the Randsburg union out of existence."[221] The strike sputtered on, and in October a union man nicknamed "Frenchie" arrived in Randsburg from the strikes in Telluride, Colorado. He stated to the undercover detective that the Yellow Aster Mine was "the first mine he ever scabbed at."[222] Frenchie was not proud of the admission, but the fact that he was unafraid to admit the truth to a stranger shows the weakness of the union in the region.

The operative also reported on other miners migrating from Kernville and added that a man, "Burk," had arrived to work drunk, but as he and his supervisor belonged to the Elks (fraternity) he knew Burk would not be discharged.[223] The account points to the importance of other non-union organizations such as fraternities and connections, a point that was noted by many other observers. As one Catholic clergyman wrote, "Here in the United States societies are the order of the day, and the young man who is not attached to some organization can't get on."[224] The writer called the societies as "all-pervading as evil" and emphasized the solution posited by the contradictory papal encyclical on labor, *Rerum Novarum*, that each parish should enable Catholics to band together for both social and spiritual purposes.[225] This supposed solution collapsed entirely when faced with the mélange of cultures and religious backgrounds in the US workforce.

The union's disintegration continued with the local election in November 1904, when the union rallied behind the Democratic Party candidate, Thomas McCarthy. Once a miner but now a stationary store owner in Randsburg, McCarthy had remained a member of the WFM and attended its meetings, although the operative complained that WFM rules specified that only miners be allowed attend meetings.[226] McCarthy unwittingly spoke to the undercover operative after the vote and stated his conviction that he carried Randsburg by a 50-person and perhaps a 75-person majority, which would allow him to win the election by about 18 votes.[227] The next day the Thiel detective wrote that the YA-backed candidate, a Mr. Peterson, had won.[228] How could McCarthy have been so wrong?

Later, the Thiel detective overheard the local Democratic leader, Judge Manning, telling workers that they should be proud, as "the Democratic Party had one of the strongest combinations to buck against on the Desert as there is anywhere," but the detective added a note at the end of this story that he "thinks that Judge Manning voted for Peterson."[229] The union, unaware of this betrayal in the party's highest echelons, still knew well the implications of the loss: "the union now realizes that it can not elect anyone of their men in Randsburg at all."[230] Men such as the scab "Frenchie" were part of a larger influx of former or non-union men, and the union members seemed aware that the organization's strength was seeping away. One member stated angrily at the meeting that "if the election had been held three months later McCarthy would have lost his own town by 50 votes."[231]

"His own town" did not just mean his political control of the town. To solidify his support within the town, McCarthy allied with the social reformist tendencies that were gaining popularity throughout the West in the late nineteenth and early twentieth centuries with the solidification of progressive ideas into a movement. In 1904, as Christmas approached, McCarthy rallied the citizens of Randsburg to issue a warrant against a woman named Marguerite "for running a disorderly house in the business portion of the town."[232] A "disorderly house" was code for a brothel. Saloons often acted as fronts for brothels, quite literally, with the saloon in the front and the sex workers' private rooms in the rear of the building. Marguerite responded to this threat to her business by making a list of men who could act as witnesses for her in the case. She had eighty-four men on the list, seventy-two from the mines and twelve from the town, proving that the mobile mining workforce provided most of her business.[233] The townspeople had fifty men willing to act as witnesses, "among them a number from the hill [i.e., miners]."[234] The YA decided to stay out of the dispute, though the operative noted that the company possessed the ability to have the trial postponed or thrown out on the "Q.T." (quiet) if it wanted—an indication of its considerable local influence in legal circles.[235]

In the new year, a miner named Frank Lane was demoted to mucker and quit after his pay was subsequently reduced, from $3.00 to $2.50 a day.[236] As he spoke to others about his pay cut, a nearby operative overheard one remark: "the Supt. would not last long if he kept on cutting down wages."[237] Most of these idle threats were a vain effort by the men to recover some respect against the imbalanced power dichotomy. The reports should also be contextualized as those of persons working for a business whose trade relies on a management afraid of its workforce. Vague threats were not the only way workers exhibited animosity toward the company. Direct violence was risky and could bring severe legal and financial repercussions and so, with no stake in the welfare of the company on one hand and a decrease in wages on the other, workers engaged in nonviolent resistance.

Now, in addition to hearing worker dissatisfaction, the operative also noted an increase in those who bunked off on the clock or drank on the job. "On Friday Dunning wasted fully two hours talking," wrote the exasperated Thiel detective.[238] He further reported the Irishman's thrifty and very dangerous habit of cutting the fuse short, forcing others to run for cover when it was lit. Dunning stubbornly replied, "I am doing this and what the h_l have you fellows got to say about it? If the Y. A. don't like it they can give me my time."[239] Such behavior irritated the detective and he kept a close eye on Dunning, detailing how he broke a pick through "pure carelessness" and had taken three-and-a-half days off from work after a drinking and gambling spree.[240] The operative added that Dunning had no one to blame but himself: "The fact is that Dunning is a very poor miner and drills his holes only in what the miners call 'muck pile' and then never gets more than two holes about 18 inches deep in half a shift; in fact, when he gets that done he will do nothing else until shooting time."[241] Whether Dunning's work was due to incompetence or an enmity toward YA is not revealed in the documents.

The operative also noted what Dunning said in conversations about the recent influx of immigrant workers: "All Italians, Swedes and other foreigners should be run out of town as no white man can compete with them as the[y] work like h_l and the men on the hill are working hard enough without foreigners."[242] Here, Dunning's attempt to label Swedes as anything other than white betrays his limited understanding of race and whiteness and reveals his true intention—to ally the Irish with German, Cornish, and Anglo-American miners against the newer immigrants.[243]

Unlike the situation in the Coeur d'Alene mining district discussed in chapter 5, the townspeople expressed sympathy for the workers' plight, especially the muckers who they believed should earn at least $2.75 a day.[244] Low pay caused friction in the town between workers and the businesses to which they owed money, but businesses knew that if the workers were paid better, it would lead to more spending (and an increase in payment of dues). A more immediate problem for the YA sprang from the critical shortage of muckers. The YA asked the TDA to recruit laborers for it. Inquiries in San Diego and advertisements in Santa Anna's papers yielded the response that "plenty of Dagos [Italians] could be obtained, but there was no white men who could be employed for that class of work as they themselves had corralled all of that class of men for the Water Work Company."[245] The terms of employment for the TDA must have been extremely broad, which points to the detective agencies of the period acting less like crime prevention groups and more like corporate troubleshooters for any job—legal or criminal—companies wanted to carry out.

The Thiel detective also spread misinformation among workers. Nicholas Martin, "a union man from Colorado," told the operative he was going to apply for work at the Yellow Aster Mine and "asked [the] operative if it would not be well for him to use an assumed name when applying. Operative replied that it would not be necessary if he did not tell them where he was from, but Martin did not seem to be entirely reassured."[246] While the Thiel detective's primary task was to infiltrate and note union activities and membership, the priorities of the YA changed over time. By 1907, the theft of ore by miners had become the manager's primary concern and the detective's sole focus. The operative wrote on several occasions about workers keeping particularly rich "specimens" and of "shovellers" (muckers) going home with "their pockets full of rocks."[247] This "high-grading" was not simply a matter of theft to the miners; "there was more to it than money, more than picking up a few extra bucks, it was a matter of dignity."[248] In a system in which they were demeaned and distrusted by their employer, theft was a way they responded in-kind, repossessed a fragment of dignity for themselves and their families, and returned the disrespect.

Union workers had noticed the detective by the end of 1905, and the operative wrote the following account on November 21, explaining the omission of reports for November 17 and 18:

> In the evening operative went down town until 10.30 p.m. And wound up in Andy Nixon's saloon and Mr. Nixon was tending bar and asked operative to take a drink. Operative called for a drink of water and in less than two minutes flat became violently ill. There were several men in the place at the time and by making unusual effort operative managed to get out of the place without notice and reached home. He will always believe the water was doped as he had drunk very little liquor on this evening. Sat. Nov. 18: Operative unable to leave home.[249]

It is possible that the operative invented the story to excuse a spree, but the story gains more credibility by setting the scene of the event at Andy Nixon's saloon, a well-known union bar hostile to company spies if their true purpose in the camp was discovered. His use of the phrase *wound up* was a nice way of avoiding responsibility for possibly getting drunk. Equally unlikely is the idea that the operative, after hanging around saloons all evening, asked for a glass of water late in the evening; after all, it is much easier to spike a drink of whiskey or beer than one of water.

The next reports date from a year later, but due to their anonymous nature, it is not known if this is the same operative: "[Operative] reports several new miners at work in the Shoshone Mine from Butte, Mont., and Goldfield Nev. All are radical union men with no love for the [YA]. They are in Nixon's saloon almost every night."[250] Less than a month after the newcomers were initiated into the local branch of the WFM, the operative reported: "a boy of 20 named Zern, from Prescott, Mich. Also an Italian of 19 named Negros, who could not speak English, both working in Tungston, were rated and put through. Also an Irishman [John O'Leary], whom the operative has seen around Pat Byrnes' saloon was initiated."[251] The mélange of workers seemed to have successfully united under the WFM umbrella.

By 1907, the detective's identity had become common knowledge within the community, and a man named Frank Higgins approached the detective while he was drinking at a saloon: "Higgins assured the operative that he would say nothing about him to any one, but he did not keep his word."[252] Strangely, most townspeople did not shun the detective, probably for fear of antagonizing the company, but they refused to give up any names of union men or high-grade thieves. When the detective asked about specific people, one woman said "she never knocked on anyone and never would."[253] A barman named Jim O'Donnell told the operative "he had never yet played stool pigeon on any one," but he went on to clarify that his loyalty was negotiable and that "under the circumstances he could not do it now unless there was some object in it for him."[254] O'Donnell's willingness to negotiate proved too tempting an offer for the operative to refuse.

O'Donnell, who owned a saloon and dance hall in Randsburg, had heard that the operative was a Thiel detective from a strange source, a Mr. Bocha and a Mr. Benson, two Pinkerton men also hired by the YA to investigate the thefts. The Thiel detective tried to charm O'Donnell: "Operative made a 'good fellow' of himself by dancing and treating O'Donnell and his wife, etc., but could not get O'Donnell to divulge any names."[255] The operative's dancing and munificence did not have the desired effect, and at the end of the night O'Donnell threatened him, telling him that if he mentioned anything he had said, the operative "would go over the hill in a wooden box."[256]

Perhaps O'Donnell was playing the detective agencies against one another for his own gain.

The diverse groups the Irish interacted with across California during different phases of its gold rush illustrate a wide diversity of experiences. San Francisco and the nativist fervor of Denis Kearney proved that the Irish could deploy nativist and racist language to appeal to wider American society, and the other examples that litter this chapter point to a much more varied range of interactions. While the Irish in Randsburg negotiated the varied ethnic and labor difficulties, the town possessed limited opportunities for them occupationally or socially. Irish experiences in Smartsville and Randsburg proved very different; in the latter, despite holding positions as union leaders and saloon keepers, their influence remained restricted and tenuous without a larger community to support them. The Irish presence in Randsburg was therefore more limited and less enduring than the one in Smartsville, and Randsburg never became known as a particularly "Irish" place to live or be buried. California, with its large, excited influx of people, acted as a prototype for how the Irish would find their place in the mining American West. Their successes and failures highlighted the importance of Irish community formation during this period as well as its limitations. Gold rushes continued across the West in the following decades, and, like others, Irish migrants earned and lost fortunes. However, the emergence of larger mining towns also led to the development of more established Irish-American communities such as Virginia City, Nevada, and later, Butte, Montana.

3

Mirages in the Desert

Irish fortunes brightened with the emergence of a large mining city in the American West, in the form of Virginia City in Storey County, Nevada. The city surfaced from the wealth of the Comstock Lode, and the most powerful mine owners were Irishmen—the Bonanza Kings. John Mackay, James Fair, James C. Flood, and William S. O'Brien, also known as the Irish Big Four, controlled the Consolidated Virginia—the most profitable mine—and managed to wrest control from a group of investors known as the Bank Crowd. Virginia City and its nearby partner town, Gold Hill, set the tone for other emerging industrial urban landscapes the Irish entered and illustrate the challenges they faced in establishing the necessary structures to support their burgeoning communities. For the Irish leaving California, it must have been a shock to find themselves in this mining city or the remote outposts of Nevada or Utah. Here, unlike the lush green of Ireland, they were greeted by the dusty winding foothills of the Sierra Nevada and the vast deserts of the Great Basin. In Virginia City the social and cultural threads of a distinctive

https://doi.org/10.5876/9781646422517.c003

Irish Catholic community appeared, challenging the difficult conditions and the expectations of American society.

Matching the geographic hostility of the region was the reemergence of some of the same adversaries the Irish faced in California. As surface deposits of gold disappeared, they were replaced by quartz mining, which required that miners drive shafts in their efforts to uncover seams of silver and gold. A miner nicknamed "Old Virginny," James Finney, accompanied by Peter O'Riley and Patrick McLaughlin, accidentally uncovered the rich silver veins of the Comstock Lode in Nevada while placer mining for gold at the head of Six Mile Canyon.[1] The origin story, recounted by historian William Hickman Dolman, is an interesting one and details how vigilantes' eagerness to hang claim jumpers led them to inadvertently dupe themselves out of a fortune:

> The entire camp except two men—Peter O'Riley and Patrick McLaughlin—formed a Committee of Vigilance and started with Sides for the Valley, as they said, to try and hang him. These two men left in Six Mile Canyon were not the owners of their claim then. They commenced work on other claims to make a little money, and no one was there to say "nay." The first day they struck the rich croppings of the Ophir Chimney and took out three hundred dollars. They covered up the rich deposit, and with this money they purchased an interest therein.[2]

This account sought to highlight the Irishmen's duplicitous nature but inadvertently provides another example of the social division between the Irish and those who supported vigilantism.[3] The distinction between the choices of the Irish vis-à-vis other ethnic groups became a recurring theme for the mining towns Virginia City and Gold Hill, built from this strike. From the earliest discovery of the famous silver mines, the Irish were a major part of the history of Storey County, Nevada.

Mines sprung up around the area, though development stalled until the Irish Bonanza Kings struck the Comstock Lode and its fabulously rich seam. These excavations yielded over half a billion dollars in silver during their operations. The deep mines necessitated reliable, skilled workers and a commensurately higher wage to attract them. More important than that was the presence of Irish mine owners; this sympathetic management allowed the Irish community to flourish. Ethnic bias in favor of the Irish on a larger scale was an unusual reversal from the discrimination in other mining ventures run by Anglo-American businessmen or skilled Cornish management, and they took full advantage of the circumstances, imprinting a strong Irish-American character on the region.

These strong influences were expressed through Irish miners' social interactions. They had several spheres of such interaction: the mine, their lodgings (be it boardinghouse or home), the Union hall, bars, the Catholic Church, and the fraternal hall. Irish women had fewer social outlets, centered on the home and the church. Women were vital for the formation of ethnic communities; their arrival in mining towns, coupled with endogamous marriage patterns (Irish marrying Irish), allowed for the survival and expansion of a community with a strong Irish identity. While limited sources survive from Irish women in mining communities, their importance was not simply that they gave birth to the next generation; they were key agents in the transmission of Irish identity, especially Catholicism. Irish women, heavily involved in church activities and organizations, were largely responsible for the loyal attendance of Irish congregations, which was an important facet in supporting the Irish clergy and allowing them to hold complete sway within the Catholic Church in nineteenth-century Nevada. Research on women in Virginia City confirms that Catholic "ladies fairs" in the city were almost exclusively Irish affairs, but this was not true of the entire Catholic Church in Virginia City.[4]

The Irish welcomed those who shared their goals and did not challenge their control, but when threatened, Irish congregations and clergy flexed their power in slight and almost imperceptible ways. In May 1864, a second Irish parish, St. Patrick's, was established in Virginia City, but jurisdictional disputes arose between the Irish parish and the newly arrived Italian Passionist clergy. These disputes led the Passionists to criticize Irish bishop Eugene O'Connell, who responded quickly and quietly. An influential letter to the Passionists' Provincial resulted in the transfer of the Italians' mission to Mexico in September 1865, which forced the closure of their church and its absorption into the Irish parish.[5] This event sent a strong message to other ethnic groups vying for influence within the American Catholic Church and Irish-heavy towns. Any attempt to oppose the Irish had significant risks. This power stemmed not only from the large proportion of Irish Catholics in the West but also from their ability to request and keep Irish-born clergy.

By the time of the 1880 US Census, the Irish-born in Storey County numbered 2,501—by far the largest foreign-born population—in a region that

Table 3.1. Birthplace and birthplace with parentage breakdown, Storey County, Nevada, 1880[a]

	Birthplace	*Parentage*	*Difference*
Ireland	2,501	5,060	+102%
England	1,395	2,017	+46%
Scotland	195	341	+75%
Wales	88	132	+50%
China	642	642	0%
Canada	711	465	-35%
Germany	585	1,011	+73%
US	9,217	3,165	-66%

[a] For the US parentage figure of 3,165, 22 were born in Canada to US-born parents, 3 were born in the West Indies, 1 in Austria, 1 in Mexico, and 1 at sea. IPUMS.

encompassed both the Comstock Lode mines and Virginia City.[6] With a total population of 16,168 in Storey County, the parentage figures reveal that the Irish community composed roughly one-third of the total population of Virginia City in 1880. Using the parentage figures for the US-born population (table 3.1), we see that 66 percent of the 9,217 US-born were children of the foreign-born population.[7] Virginia City was cosmopolitan, an example of the immigrant American West, where more of the population was foreign-born or the children of the foreign-born than were Anglo-Americans.

This fact holds true for the skilled working population. Indeed, the 727 Irish-born men employed as miners in Storey County provided the economic backbone of the Irish community in Virginia City, generating a large pool of income from which social supports could be developed and religious, fraternal, and social organizations could thrive. In contrast to the narrative of the Irish only existing as laborers, they constituted the largest ethnic block of skilled miners in the county (table 3.2), followed by the English, a category largely composed of Cornish miners.

This population, concentrated around the mines, lived side by side with the industry that sustained it. One Irishman wrote, "If you were in Gold Hill near Virginia City you could scarcely stand the noise of the mills crushing the rocks that comes [*sic*] out of the mine[, which] never stops Sunday nor holidays night nor day"; in the nineteenth century the industrial and family spheres overlapped.[8] Contradicting a misconception of the city as a feckless center of nothing but bars and brothels, a higher percentage of Irish-born men and those of Irish parentage in this urban space were married compared to the general population, which had a more equal ratio of single to married men. This signaled that the industrial center supported a larger Irish community for workers and their families.[9] Delving deeper into the statistics, they also show that the Irish-born population of Storey County was more likely to be married and to have a spouse present than was the general population. The Irish had come to stay. The presence of a large number of married couples explains the large footprint of the Irish community and the creation of social services for families, such as Catholic schools, churches, and hospitals.

Table 3.2. Birthplace and birthplace with parentage breakdown of miners, Storey County, Nevada, 1880[a]

	Birthplace	*% Total*	*Parentage*	*% Total*
Ireland	727	33.2	843	38.4
England	594	27.1	610	27.8
Wales	31	1.4	29	1.3
Scotland	50	2.3	74	3.4
Canada	165	7.5	87	4.0
Germany	50	2.3	63	2.9
US	458	20.9	262	12.0
Others	117	5.3	224	10.2
Total	2,192	100	2,192	100

[a] IPUMS. There was only one Chinese miner in Storey County, Nevada, in 1880. 54.5 percent of all Irish-born males were miners, and 15.3 percent were laborers. 32.7 percent of all those with Irish parentage were miners, 11.1 percent were laborers. Within the "others" birthplace category, eight were born in Canada and one was born at sea. See chapter 1, this volume.

The married figures explain the high number of Irish-American children in Virginia City, as suggested by table 3.1. Children are represented by blank entries in the census, which we can use to deduce that there were only 128 Irish-born children and 1,574 children of Irish parentage in Storey County in 1880—indicating that there were few families migrating directly from Ireland to Nevada. The figure for Irish-American children grows to over 2,000 when we include those under the category of students, further contrasting the popular image of a mining town as a place of rough men, loose women, and rampant drinking. Within the space of two decades, Virginia City had become a large and thriving mature settlement filled with young families.

The machinery of industry did take a toll, not just in the noise generated by the machinery but in its effect on people's health and lives. A significant number of Irish-born people lost their spouse, and the 10 percent figure for Irish-born widowed is noticeably higher than the 6.3 percent for the general population (table 3.3).[10] Although the majority of those who died were men, women suffered directly and indirectly from the dangerous work, as seen in the hospital records detailed below. Women were listed in the census mostly under the title "keeping house," a category that obscured the daily work of purchasing

Table 3.3. Marital status of Irish-born, Irish-born and Irish parentage, and general population, age 18 plus, Storey County, Nevada, 1880[a]

	Irish-Born		*Irish-Born and Irish Parentage*		*General Population*	
	Number	*Percent*	*Number*	*Percent*	*Number*	*Percent*
Married, spouse present	1,432	57.6	1,735	53.5	4,937	48.1
Married, spouse absent	126	5.1	166	5.1	881	8.6
Divorced	17	0.7	20	0.6	77	0.7
Widowed	249	10.0	268	8.3	643	6.3
Never married/single	661	26.6	1,055	32.5	3,728	36.3
Total	2,485	100.0	3,244	100.0	10,266	100.0

[a] IPUMS.

food, cooking, cleaning, minding children, and performing a multitude of other tasks that made a house a home. Women, married or otherwise, were vital to the functioning economy of the Irish community in Storey County.

Fraternal and Social Life

A plethora of fraternities existed in the US in the nineteenth century. The overwhelming majority were supposedly secret societies, which often printed the time and location of their meetings in city directories—somewhat helpfully under the title "Secret Societies." Unfortunately, the records of the organizations themselves rarely survived, and the lack of this documentary evidence has led to a void in the historiography because the organizations offered an important, hidden series of networks for many. Those who could not rely on ethnic or familial connections used fraternal groups to provide them with the support vital to economic stability, often through employment. Membership depended on a range of economic, cultural, racial, and religious criteria. Many were linked to nativist-leaning Masonic lodges, which excluded Irish Catholics from a range of groups; others such as the Benevolent and Protective Order of Elks were more inclusive, and they acted as insurance companies in case of workplace injury.

The Irish were well aware of the utility of these organizations, as seen in the popularity and diversity of Irish fraternities and clubs in Virginia City. Active branches of the Emmet Guard, Sarsfield Guard, Montgomery Guard, Knights of the Red Branch, Ancient Order of Hibernians (AOH), Fenian

Brotherhood, and the Land League were all supported by the attendance and funds of the Irish community.[11] Primarily men's social clubs, these fraternities helped fashion a sense of identity and belonging among Irishmen in Virginia City. Although the clubs only had male members, they organized many events for the entertainment of the whole community, such as picnics, shooting contests, and marches. While Irish fraternity members were usually Irish—the AOH enforced a rule that its members must be Irish-born or of Irish ancestry—they were not always exclusively so; one non-Irish man, Wells Drury, editor of the *Territorial Enterprise*, rose to the rank of lieutenant in the Sarsfield Guard.[12]

Debate sometimes swirled around these associations' activities. One dispute in 1871 illustrates interesting facets of Nevada politics:

> WILL STAY AT HOME—The Emmet Guard of Virginia (City), which had contemplated visiting San Francisco to assist in celebrating St. Patrick's Day, since they learned that the colored military had been assigned a place in the procession have concluded to stay at home and allow San Francisco to do its own celebrating. Their judgment and good taste is to be commended by all right-thinking people.[13]

At first glance, this newspaper piece appears to sustain the notorious perception that the Irish were racist and to congratulate them on their discrimination. However, the article was a reprint of an earlier one in the nativist, pro-Masonic *Carson Register*. Alfred Doten wrote a follow-up article in the *Territorial Enterprise* and clarified the situation: "Capt. Arnold of the Emmet Guard called upon us last evening and requested us to state that a single word was never spoken in the company in regard [to] the 'colored military' and that the only reason why the company did not conclude to go down [to San Francisco] on this occasion was that many members were miners, and the foremen and superintendents of the mines in which they are working do not wish to spare them for so long a time as they would be absent."[14] The article in the *Carson Register* represents a subtle attempt to manipulate details of events related to the Irish community, which Doten corrected in his piece.

Alfred Doten keenly observed and commented on social life in Virginia City while he lived there. In his extensive daily diaries, kept with impressive regularity from March 18, 1849, until his death on November 11, 1903, he detailed the comings and goings in the city. On July 8, 1871, he watched as the "Irish exiles,

Luby and Burke arrived at 8½ oclock [*sic*] this evening and were escorted into town from the Geiger Grade by the Emmet and Sarsfield Guards . . . Band of music & quite a procession."[15] They had come on a speaking tour, a popular means of generating income for political and religious commentators as well as many Irish nationalists. The men would have been skilled speakers, having honed their oratory at dozens, if not hundreds, of engagements. Dr. Thomas C. Luby gave a speech on July 10 in the opera house titled "The Prospects of the Irish National Cause," earning $250 by charging a $1 attendance fee, while Col. Thomas F. Burke gave a lecture on July 11 at the Miners' Union Hall in nearby Gold Hill titled "The Necessity of Irish Organization."[16] The turnout for these speeches in mining towns reveals an audience both sympathetic to Irish nationalism and possessing a disposable income.

The Irish readily organized political and religious societies, and such events reinforced Irish immigrants' nationalist fervor by keeping the situation in Ireland fresh and relevant in their minds. They also acted as excellent recruitment vehicles for local fraternities by stressing the importance of Ireland's status as oppressed and exploited by Britain, thereby encouraging immigrants' pride in their ethnic identity and emphasizing the importance of unity. Doten also mentioned hearing a talk by Fr. Eugene Sheehy, who was known as the "Land League priest," titled "The Genius of Irish Liberty."[17] The priest's trip to America was ostensibly to collect funds for the construction of the local church in Rathkeale, County Limerick, but the topic of his talk and the background of strong church opposition to Irish nationalism in the US and Ireland suggest that he gathered funds for more than the cause of building a church.[18] The popular Irish-American newspaper, the *Irish World*, described him as a "young and gifted Irishman, not of the West-British type . . . a good priest and a true patriot."[19] The definition was interesting because of its implication that there was a range of qualifiers for a "good priest," most important, that he was not "West-British," distinguishing him from the emerging class of "respectable" Anglicized Irish clergy who were largely uninterested in Irish culture or nationalism—typified by Father Mathew and his reformist social temperance crusade. In contrast, Father Sheehy spoke Irish fluently, and he and his relatives were targeted for persecution by the British authorities. The radical priest was more interested in forceful political and economic reform for the Irish than in limited "respectable" reform of the Irish, and it was this that impressed the American observer Doten.

Nationalism was a rallying cause for the Irish in the US, a powerful symbol of their support for the homeland, and an expression of communal solidarity in a distant land from Ireland. These speeches, however, contained multifacted critiques of other aspects of society, both in Ireland and in the US. Indeed, when Michael Davitt spoke to the Virginia City branch of the Land League in a packed opera house, he told the audience, "We can afford to put away the harp until we have abolished poverty, mud cabins, and social degradation."[20] Thus, he upended the nationalist claim that independence alone would relieve Ireland of all social ills.[21] Davitt himself wrote about his visit to the US in his book *The Fall of Feudalism*, where he detailed his visit to Virginia City and an encounter after his talk:

> When in Virginia City, Nevada, Mr. J. W. Mackay, "the Silver King," made me his guest in the hotel. He attended our Land League meeting, but could not be prevailed upon to make a speech. He did not believe either in the agrarian or any other Irish movement. It was all a waste of individual and national energy and means. "Why not leave the whole island to England, bring your people all over here, settle them down in Nebraska or Colorado, and call the State 'New Ireland' or 'Home Rule,' or whatever you like, and end the whole trouble?" "And give American millionaires the chance of buying up the land of 'New Ireland' in advance, I suppose?" He laughed at the retort, but he believed that nearly all the Irish people would ultimately find their way across the Atlantic.[22]

The language of the two men reveals their understanding of and approach to political and ethnic questions. Mackay, fully embodying Irish-America, argued that living in Ireland was no longer an obligatory part of the definition of being Irish. Davitt himself was born in England to Irish parents and understood his meaning. Mackay's attendance at the event and his desire to meet with Davitt betrayed an interest in Irish affairs and a concern for the Irish people, though outwardly he denied this, believing the causes Davitt supported were "a waste of individual and national energy." Davitt, in contrast, tied Irish misfortunes to economic systems of oppression; when Mackay claimed that America represented opportunity, Davitt's retort about "millionaires . . . buying up the land" underlined what he saw as the exploitative failings of the US capitalist system.

After his meeting with Mackay, "the boys" of the Land League, as Davitt called them, eagerly asked him how much Mackay had donated to the cause.

Mackay was one of the wealthiest Irish-Americans but notoriously tight-fisted, and the Land Leaguers hoped the meeting might loosen his purse strings. Davitt replied that Mackay gave nothing: "'Not a cent.' 'Did you ask him?' 'No.' 'But what is the blank, blank use of the league sending over a man to beg money who does not ask for it?' And, I confess, I left a very small reputation for obtaining funds behind me in the city of the bonanza mines."[23] The Fenian Brotherhood held a fundraiser for the Land League simultaneous to the one in the opera house but chose a public house called Maguire's for the venue.[24] The occasion demonstrated how both the theater and the bar, which collected the same amount of money for the cause, acted as venues for a wide variety of events, including political ones.[25] The large turnout also reaffirmed the Irish tendency toward political engagement, a trait much maligned by nativist observers.

Throughout the nineteenth century, plays were another common social function in larger mining towns, and Virginia City hosted a variety of Irish-themed plays including *The Shaughrawn*, *Arrah-Na-Pogue*, and *Rory O'More*. The crowds showed their appreciation by showering the actors with silver coins or, on one memorable occasion, presenting them with a brick of solid silver.[26] A series of interviews Duncan Emrich conducted with old miners in the famed Delta Saloon in Virginia City detail events throughout the mining towns of the American West. One old-timer named Edwards recounted a performance of the play *Othello* in Butte, Montana. Though they were well-paid, the actors often had to deal with crowds that contained what Edwards called "a rough mining element":

> And we come to the bedroom scene. Othello has just strangled Desdemona, and he turns to the audience and he says, "She's dead! She's dead! And what shall I do?" Just then a voice from the back yells, "Fuck her before she gets cold" . . . The curtain is run down. Southern and Marlowe/Baldwin sit and they say they will not the following night. But the mayor and officials of Butte, Silver Bow County, Montana, cajole, and promise that no such contretemp [*sic*] will occur again . . . At the following night's notice there's a seat vacant in the center aisle. Suddenly down the aisle strides the typical western sheriff . . . A six shooter on each hip and Bill Durham is sticking out of his pocket and he's wearing a ten gallon hat . . . He says "Folks, I guess you know who I am. I'm Jerry Donovan. I'm the sheriff of this here Butte, Silver Bow County,

> Montana and I'm here to preserve law and order. And the first one of you bozos that makes a wisecrack like last night, I'm going to shoot between the eyes. And I can hit the ace of spades at fifty feet." He sits down . . . The play proceeds. We come to the balcony, the most tender scene ever written by the Bard of Avon . . . Romeo is strumming his guitar and Juliet is above listening . . . And suddenly unable to withstand the throes of passion . . . He throws her a kiss. As she catches the kiss she says, "Oh, a kiss—a kiss—what's more wonderful, more beautiful? What is more sublime?" Just then, the sheriff leaps to his feet, draws his gun; he says, "The first one of you sons of a bitches that says 'fucking', I'm going to shoot in his tracks!"[27]

Such events served western audiences quite well as public entertainment. The way Edwards recounts the Butte tale of rowdy miners reprimanded for their lack of social graces by the even more obscene, "western," and Irish-American sheriff Donovan indicates an obvious awareness, indeed pride, among its populace of its identity as western, boisterous, and partly Irish. The story shows the miners taking ownership of Shakespeare in their own way, and the location in which it was recorded—a bar in Virginia City—illustrates the mobile nature of jokes and stories among the mining population.

Doten remained a keen observer of the Irish community in his diaries and was close enough to the community to note divisions within Irish ranks. On July 14, 1871, the town received telegrams from New York about riots between "Catholic Irish" and "Orangemen." Doten wrote that this news "creates quite an excitement here also," but he did not mention any outbreaks of violence in Virginia City.[28] The Irish Catholic community rallied around the St. Patrick's Day parade (figure 3.1), and this remained the undisputed date for the manifest expression of strength and unity of Irish identity in Virginia City. Even non-Irish Catholic clergy understood its importance and expressly endorsed the holiday. A French priest serving the mining town of Truckee in Nevada told his congregation that "St. Patrick's Day is a holiday and Irishmen at least ought to observe it as a great holiday for the love he bore to that country who have so many saints to God and so many prophets to the world[.] Irishmen are scattered all over the world carrying the word of god with them wherever they goe [*sic*] he said."[29] The Irish holiday was a representation of their culture and an intertwining of the religious and nationalist aspects of their identity.

Figure 3.1. Emmet Guard march in military regalia in Virginia City, Nevada, on St. Patrick's Day, circa 1870s. Department of Archives and Special Collections, William H. Hannon Library, Loyola Marymount University, Los Angeles, CA.

In 1873, at nearby Gold Hill, Doten witnessed the widespread popularity of St. Patrick's Day, noting that it was the "most celebrated I have ever seen it."[30] This was a consequence of the Irish community reveling in its success in establishing roots and an example of defiance toward its adversaries and triumphalism to its competitors. The Emmet and Montgomery Guard of Virginia City and the Sarsfield Guard of Gold Hill marched along with Divisions No. 1 and No. 2 of the Ancient Order of Hibernians and another fraternal organization called the "Irish Confederation." Altogether, approximately 300 armed men marched at Gold Hill in full military regalia with three bands, while a smaller gathering of Irish companies marched in Virginia City. In a display of ecumenical interaction, the big event of that evening was a ball in the Odd Fellows Hall, held for the benefit of the Daughters of Charity Orphans Asylum, illustrating the widespread support for the nuns' vital community work. Of the ball itself, Doten wrote that it was "fearfully crowed—No show to sit down and hardly stand up—I took it out in drinking whiskey with a lot of Micks with celebrations eventually concluding at 5 in the morning."[31] One estimate states that the typical amount of money collected at an event such as this would be approximately $10,000–$12,000, a vast and vital source of funds for the Catholic charities serving the community.[32]

Thirteen years later, in 1886, the holiday had changed significantly, and Doten noted that there was "no observance of the day beyond a few green neckties."[33] The pomp of the parades had been replaced by a grand ball in the evening, a sign of the overall decline in Virginia City's population but also perhaps an indication that the Irish community was more secure in its prosperity. The Irish had risen in social status in the town, and they felt less need to assert their presence with overt public marches. This further indicated a decline in militarism among the Irish-American community, possibly because of the older age of the members or possibly because of the lull in pro-independence Irish nationalism across the US. The proceeds of this ball, hosted by the Emmet Guard, went to the Irish Parliamentary Fund to assist "Parnell et al in securing Home Rule for old Ireland" and not to the more militant Clan na Gael.[34] The switch from parade to ball proved popular, and Doten ended his day's observations by writing that the ball was a huge success and one of the biggest ever given in the city.[35]

On March 17, 1900, Doten was in Carson City, Nevada, where he attended a ball in Armory Hall: "Charley Bray pinned a sprig of imitation shamrock on the lapel of my coat this morning, the Queen having proclaimed that the Irish may wear it hereafter."[36] He continued bitterly, stating that he "didn't wear it long—'Nothing too good for the Irish' now, since the good fighting they have done in the British army in South Africa."[37] For Doten, the actions of the pro-Boer Irish Traansvaal Brigade, totaling 300 men, did nothing to counterbalance the 28,000 Irish fighting on the British side against the Boers.[38] Although he somewhat unfairly blamed the Irish for what he perceived as a betrayal of nationalist principles, his assessment of the conflict in South Africa shows a well-informed and nuanced understanding of the chasm between the rhetoric and reality across the Irish diaspora. Doten's radical friend Drury joined the Sarsfield Guard in the 1860s despite not being Irish. He felt he had found kindred spirits among the rebel Irish, yet, as Doten noted sadly in his later accounts, time muted some of their ealier fiery passion as many rose to middle-class respectability.

The Daughters of Charity

In 1864, three members of the Daughters of Charity of St. Vincent de Paul arrived in Virginia City to establish a school and a hospital as the bustling town mushroomed into existence from the dusty scrublands of Nevada into an industrial city. Thirty-year-old Sister Frederica, born in Ireland and raised in Philadelphia, was an experienced administrator, and she led the group west to establish its charitable projects.[39] The Daughters' Mission in Virginia City consisted of a school, a hospital, and an orphanage serving the needs of the sick and the young.[40] Though they were members of a religious order, the sisters' approach to their vocation represented the opposite of a cloistered convent, and they offered vital social services to the mining population. The sisters found little more than a boisterous camp when they arrived, but newspaper editor Wells Drury commented that Virginia City had the "marks of civilization," although "the rough element of society was never entirely dispersed."[41] Wearing distinctive cornettes, these religious women might have seemed out of place in a mining town, but, as Anne M. Butler suggests, they actually fit the profile of the general population: "Either immigrants or first generation Americans, they were young adults, supporting themselves in work created by and dependent on the mining boom."[42] Mine accidents meant dead or dying miners, and the dust that gathered in miners' lungs as they toiled in the mines left them feeble and susceptible to ailments such as consumption. Shootings and faction fights led to injuries. Orphans needed shelter. Children needed to learn to read and write. The Daughters of Charity thus provided reliable, affordable, and accessible services for the Catholic community and were open to others in the region. These services also offered a Catholic alternative to potentially exploitative or prejudiced subpar healthcare.[43]

The Daughters of Charity provided a comprehensive medical service. For example, they offered a health insurance plan that miners could subscribe to for a dollar each month, compared to the county hospital near the town, which charged two dollars a day for treatment.[44] The sisters noted miners' special need for aid due to the danger of their work. Responding to the collapse of an embankment at the mines, one sister wrote that the memory was seared into her mind, especially the sounds of some trapped miners: "Their lamentable cries are heart rending—no one can reach them."[45] They cared for and comforted the sick, the crippled, and the dying and offered medical or spiritual relief from their suffering, granting them a degree of dignity during and after their deaths.

The litany of the sisters' deeds was recorded in the Hospital Logbook of St. Mary's.[46] Its pages tracked all those admitted to the hospital's care between 1874 and 1904; these thirty years span the high point of Virginia City as the premier mining city in the American West and the richest silver mine in the world, as well as its decline. Each page lists thirty patients, giving their name, age, occupation, birthplace, date of admission, ailment, and date of release or death. Furthermore, the logbook shows that while the Daughters of Charity accepted anyone who needed medical aid, it became an "Irish hospital" of sorts, as indicated by the hospital's shortened name, St. Mary Louise's, to St. Mary's.[47] In total, the Irish number 59 percent of those listed the logbook, almost double their population in Storey County, again reinforcing the hospital's Irishness.

The logbook reveals details of some of the unusual everyday dangers faced by people who lived in the mining city. Irish housekeeper Margaret Baker received a gunshot wound on June 20, 1878, while an explosion under a house killed Irish domestic servant Johanna Brenna.[48] Mrs. Jane Sarah Bloom, age forty-two, entered the hospital on January 12, 1878, with "frozen feet" from walking outdoors in the subzero conditions. She died two days later.[49] Common causes of death for men included gastritis, pleurisy, pneumonia, consumption, typhoid fever, rheumatism, debility (old age), and "broken bones" or "general injury." One sister wrote, "The poor miners, who live in the bowels of the earth, frequently suffer from ill health and sometimes encounter serious accidents . . . [but] never fail to be consoled by the care and sympathies of the Daughters of Charity who carry to their poor 'shanties' little delicacies, and words of consolation, which are so valued by these Men of Faith."[50] The sisters were deeply respected in the Irish community, and from this nun's description, the feeling was mutual.

A change occurred in the entries between the years 1880 and 1887 when the sister responsible for logging patients started listing the county of birth for the 90 Irish. They were: Cork 26, Kerry 17, Galway 16, Waterford 4, Donegal 3, Limerick 3, Leitrim 2, Longford 2, Meath 2, Roscommon 2, Tipperary 2, Tyrone 2, Wicklow 2, Westmeath 1, Offaly (King's County) 1, Cavan 1, Clare 1, Down 1, Dublin 1, Kilkenny 1. Regardless of whether the sister or the patients began this practice, both seemed pleased to continue entering the details in the logbook and offering them when asked. The birthplace distribution reveals the dominant contribution from Ireland's southern and western counties. The chain migration pattern was most obvious between the Counties of Cork, Kerry, and Galway and the far distant mining town of Virginia City. The notable absence of County Mayo is partially explained by the preference of people from there to settle in Pennsylvania, although that tendency began to change in the 1880s following company persecution in the anthracite region, coinciding with Butte's rise as the major mining city in that later period. With the exception of Cork, none of the counties listed had a significant mining population.[51] Many of the birthplace entries that list "Nevada" also have the word "Ireland" in brackets afterward, suggesting that the Irish nuns serving as nurses were sharply aware of the importance of being identified as Irish-American rather than just labeled American.[52]

As in the case of Smartsville in California, the headstones of Irish graves in the Catholic cemetery in Virginia City remain unique among all other graves in mentioning specific localities. The overwhelming majority of Irish headstones list the Irish person's county of origin, and many also list the home parish and often the specific village. Putting these details on their headstones indicates the importance of such ties for the Irish who came to Virginia City. Rarely did the Irish use the obelisk, popular as a Masonic headstone, and most have simple crosses or Celtic crosses on their headstones. Many used common American headstone symbols, such as Patrick Fahey's, which has a relief of a weeping willow, symbolizing sorrow (and often death from a sudden disease), and a hand pointed upward, symbolizing that the person has gone to heaven (figure 3.2).

Wells Drury scoffed at the disposable income of the working class, noting that "the community gloried in crude opulence," but it also supported the many charitable organizations, including the Daughters of Charity. Mine workers donated a dollar a month toward the running of the hospital in addition to making other contributions for management of the churches, schools, and orphanage.[53] The Daughters of Charity welcomed donations from a range of fraternal organization, Irish and otherwise.[54] The success of these collection drives reflected the admiration held for the sisters' work in wider society and among other Christian denominations. For example, a Mrs. Sunderland (an Episcopalian) and a Mrs. Theall (unspecified Christian

Figure 3.2. Some surviving Irish headstones in the Catholic cemetery in Virginia City, Nevada. *Top left*: Thomas Stanton. Parish of Carabane, County Galway. Died August 25, 1870. *Top right*: Patrick Fahey. Parish of Kinvarna, County Galway, September 6, 1873. *Lower left*: Thomas O'Sullivan. Parish of Aglish, County Cork. Died September 16, 1865. *Lower right*: Julia O'Connell. Caherciveen, Parish of Filmore, Kerry County, Died August 17, 1864. Author photos.

denomination) traveled to different mines to collect money for the sisters' hospital, going to "the most dangerous places" to "get the last dollar." It was perhaps this subtle intimidation that encouraged miners to donate one or two days' wages to these charities.[55] The Daughters of Charity's ecumenical outreach did not end with Christians; they maintained friendly relations with the small Jewish community in Virginia City, teaching Jewish girls in their school next to the Church of St. Mary's in the Mountains. Such openness and generosity did not go unappreciated, and it shielded the sisters from some anti-Catholic hostility.

At the state level, a wider and more general distrust of Catholic institutions revealed itself in a funding dispute over the sisters' orphanage and school. In early 1867, some members of the state legislature in Carson City objected to the orphanage receiving $2,500 on grounds that legislation for the funding explicitly stated that the asylum could only care for white children. When the sisters agreed to this stipulation for funding, they inadvertently drew further controversy as opponents argued that funds drawn from the common taxation of citizens could not go to a public agency with an exclusionary policy. The Daughters of Charity were trapped. Their effort to refuse education to "colored" children did not result from a principle of exclusion; rather, it was a political stipulation foisted upon them. As Anne M. Butler assessed, their Vincentenian traditions defined them as "social ministers rather than as political reformers," and it was this that constrained them.[56] Following these precepts, Sister Frederica refused to defend or attack the policy and tried to avoid being drawn into a political spat. She refused to continue the debate by pointing out that in 1870 there was a solitary "colored" person of school age in Storey County, Nevada, perhaps realizing that making this point could have been seen as an indirect endorsement of the funding provision.[57] The *Territorial Enterprise* came to the nuns' defense by offering a backhanded insult, stating "they are not expected to understand either constitutional law or party politics," thereby denigrating their intelligence and their educational mission.[58]

Feeling that her identity and vocation had been disrespected, Sister Frederica wanted the last word and let loose her final volley in the dispute in the newspaper on February 21, 1873:

> The Sisters established the asylum in 1867, when the state allowed but a heartless indifference for parentless children. All know, though some reluctantly admit, they were Pioneers in the state for Orphans. When they threw their doors open to the public, no questions asked there was no distinction of creed—distress and poverty directed [the Daughters of Charity's] actions. The

> State not from charity, but a sense of justice rewarded [them]. But of late, a hostile feeling has risen against them . . . If we are not entitled to the appropriation in justice, we do not look for it in charity.[59]

Sister Frederica framed the spat as one instance in a pattern of growing anti-Catholicism, further exemplified by the formation of organizations such as the Order of Caucasians and the Vigilance Committee in Carson City. She laid the charge at the feet of elected officials that "anonymous persons" approached officials running for state positions and informed them that their election rested on their opposition to the Daughters of Charity.

If the Irish were the dominant force in Virginia City, that was not the case in the state capital of Carson City. A letter titled *"La Feile Naomh Padraig"* (Feast of St. Patrick) by Denis Hurley to his parents in Clonakilty, County Cork, on March 16, 1874, further suggested the relative weakness of the Irish community in Carson City—the political heart of Nevada—noting the strength of the Anglo-American nativist-leaning population. He added disdainfully that the Irish holiday would see "lively times throughout this country tomorrow. Won't be much in Carson."[60] While Hurley supported the Democratic Party in the nineteenth century, by the twentieth century he had decided to join the Republican Party, even running for political office several times but losing by a single vote each time due to his nativist opponents' use of bribery and powerful connections (which explained why he left the Democratic Party).[61] He continued his involvement despite these setbacks, writing with a measure of pride to his brother in Clonakilty, "I like to show that an Irishman is able to take a hand in affairs of government."[62] Obviously, his presence and activities were his way of trying to prove a point against the bigotry he encountered in Carson City.

The political pressure on Catholic associations that emanated from Nevada's state capital extended to Irish fraternities, which were required to demonstrate their loyalty to the state through an Official Oath of the State of Nevada. In the oath, the person solemnly swore that they and their organization would "support, protect and defend the Constitution and Government of the United States, and the Constitution and Government of the State of Nevada, against all enemies, whether domestic or foreign" and that they "will bear true faith, allegiance and loyalty to the same, any ordinance, resolution or law of any State Convention or Legislature to the contrary notwithstanding."[63] Strangely, the entire second half of the oath is preoccupied with forbidding duels.[64] If the oath showed the weakness of the Irish community in Nevada state politics, the legislature still managed to pass some populist measures favoring the Irish. A bill introduced by James Phelan on January 31, 1873, incorporated both the Miners' Union and the Ancient Order of Hibernians in the state of Nevada.[65] Two years later a bill contained a provision that the Miners' Union would receive an exemption from taxes. Proposed by a Mr. McDonnell on January 28, 1875, Assembly Bill 54 passed by a vote of 38 to 6 and in part read "all Real Estate and Personal Property, belonging to the Miners' Unions, throughout the State of Nevada, is hereby made exempt from all taxation whatsoever, so long as such Real Estate, and Personal Property, claimed by each separate Miners' Union shall not exceed the appraised value of more than $15,000."[66] Apparently, the Miners' Union, as a broad-based organization, remained more acceptable to the state government than did certain ethnic-based fraternities like the AOH, but the high exemption rate showed the effective negotiating ability of pro-union politicians when compromise was required.

Many others vied for a different sort of public affection. Sex workers and the brothels in which they lived and worked were a common feature in larger mining towns and operated with varying degrees of public acrimony and legal and political oversight. Although for the most part the wider community tolerated these businesses in the nineteenth century, society directed its shame toward the women who worked there rather than toward the men who purchased sex.[67] The division between respectable society and the stigma attached to these "others" was often a thin one. Irishwoman Kitty Shea owned a lodging house on A Street in Virginia City and felt her reputation was valuable enough to warrant an advertisement in the *Territorial Enterprise* stating publicly that she was not a prostitute.[68] Perhaps some townspeople chose to misinterpret her friendly demeanor and necessary close proximity to single, migratory men as enough reason to suspect her of impropriety, or she may have been the target of idle gossip. The fact that she was a working woman who owned property probably provided enough fertile soil for jealousy to mature into rumors.

The presence of sex workers in the heart of these towns proved endlessly titillating to the many newspaper publications in the American West. The papers graced them with a multitude of titles: soiled doves, sisters of sin, bawds, chippies, sporting women, ladies of commerce, women of easy virtue, the fair but the frail, fairies, and tarts.[69] Whatever the numbers of Irish miners who patronized these women, the Irish appear to have had limited involvement in operating or working in brothels. Even though figures presented in Marion S. Goldman's research on Virginia City sex work err on the high side, her work reaffirms that the number of Irish sex workers and pimps was tiny.[70] In 1880, only 3 of the 137 sex workers were Irish, and of the 21 male pimps, only 1 was Irish-born while there were 6 German-born pimps.[71] Perhaps the familial support structures and religious devotion of the Irish community limited the numbers of Irish willing to associate with such businesses in Virginia City, but given the numbers of Irish in the city, it is remarkable that so few are detected in the historical records.[72]

Protestant Americans women who sought to "rescue" immigrant sex workers failed to understand how they were separating the women from their communities and, worse, their cultures, since conversion was a stated goal of their religious mission.[73] As such, the heavily Irish dimension of Catholic religious orders may have been another cause for the low Irish numbers, as their outreach efforts were more likely to succeed because of their shared cultural background.

Despite the numerically low representation, some of the most famous sex workers and madams across the American West were Irish. They, too, followed the fortunes of the mines, as seen in the case of Rosa May, nicknamed "the hooker with a heart of gold."[74] As a young woman, she fled an abusive home in Pennsylvania. A daughter of Irish immigrants, she decided to go west and arrived in Virginia City in 1874, early in the boom time. For twenty years she lived there as a parlor girl for Caroline "Cad" Thompson, the Irish madam of the famed Brick House brothel—a property Thompson bought for $3,200 in 1871. Thompson stayed in Virginia City for a long time, well into its decline, and was offered a mere $20 in 1892 for the same property. May was not forced to move on to survive but had few options as a sex worker in her thirties. She bought a small cottage in Bodie, California, where there was a sizable clientele and little competition. The apocryphal tale that she died after contracting a fever tending to miners graced her with the famous nickname, but the tale, real or not, did not earn her a burial spot on consecrated ground. She was buried outside the church graveyard because of her unrepentant attitude toward her profession.[75]

Virginia City's rapid economic decline caused the population to migrate outward across the state or onward to the next mining boom towns, the most notable of which would soon become the largest industrial city west of the Mississippi: Butte, Montana. Virginia City lingered on, but in 1883 a reporter for the *Chicago Herald* wrote: "The 35,000 people have dwindled to 5,000. The banks have retired from business. The merchants have closed up and left. The hotel is abandoned; the gas company is bankrupt and scores of costly residence have either been moved away or given over to bats."[76] The town's boom period had lasted twenty years. The town had flickered and flamed into existence and yet, despite the community's firm footing and the creation of a vast range of modern infrastructure, it died away. A few buildings on two streets are all that survive to the present day. Many other mining towns in Nevada had an even shorter life span and entirely ceased to exist after their mines failed, leaving a few rotting wooden shacks and deep holes in the ground.

Scattered Towns

Ward was one such town in Lincoln County, Nevada—named after Irishman Thomas F. Ward, who discovered promising veins of silver, lead, and copper there in March 1872. The mining camp grew rapidly into a town, with two smelters, a twenty-stamp mill with three furnaces, a tramway, two breweries, fraternal orders (including a branch of the Land League), stores, saloons, a hook and ladder company, a school, a post office, a city hall, and two newspapers (one called the *Ward Reflex*). By 1877, the population had reached 1,500, and Irish-American newspapers noted that Ward's many Irish citizens made generous donations to nationalist causes. However, the lead content in ores from the mines began to decline; with no improvement in sight, the population began to drift away. A fire in 1883, which started at Roach's blacksmith shop, destroyed one-third of Ward, including the schoolhouse and city hall. Between 1883 and 1885, most of the remaining buildings in Ward were salvaged for building supplies for the town of Taylor, situated across the valley. The post office was disbanded in 1887, and today half a dozen stone walls are the sole remnants of the town of Ward.[77] Although its existence was brief, the Irish who lived in Ward tried to establish an Irish community. Their swift arrival and the subsequent funding of Irish groups point to a degree of Irish prosperity, enabling further mobility in the face of such rapid changes in the economy. Their sense of place could obviously shift rapidly, accommodating this movement.

The 1850s were a slow period of mining development in Nevada; however, its location just beyond the border of California allowed it to act as the main gateway for the rest of the Southwest. In the 1860s, central Nevada was more fully explored by prospectors who formed the Reese River mining district in May 1862.[78] By Christmas, the discovery of some rich ore samples caused a rush to the region, and in early 1863, Lander County was established.[79] Jacobsville became the first administrative center but within a year was superseded by the Irish-sounding Clifton, situated five miles closer to the rich silver mines, which the town of Austin supplanted in turn to become the main staging and supply town for prospectors in central and eastern Nevada. Austin's economic cornerstone continued to be mining, and the population reached 1,800 by 1864. A severe stock market downturn in gold mining caused hundreds of fraudulent or unprofitable mine companies to collapse as they lost access to capital investment for almost two years.[80] By 1866, 133 producers of ore had consolidated the hundreds of rich claims.[81] Three years later, only two mines were worked regularly, the Oregon and North Star and the Bhuel North Star, but Austin had solidified its position as the economic and administrative center for the surrounding twenty-three mining districts up to 145 miles from the town.[82]

The Irish-born of Lander County made up 10.1 percent of the population in 1870 but declined to 9.1 percent in 1880.[83] The 1880 census shows that those of Irish birth and Irish parentage formed 15.6 percent of the population of Lander. The Irish in the region organized early and met in August 1865 in Judge Logan's courtroom in Austin to form a chapter of the Fenian Brotherhood.[84] In Austin, the Fenians became the major Irish fraternal organization; while they advocated the overthrow of British rule in Ireland, in effect they acted just like any other Irish fraternity. They met regularly during the year, raised funds for Irish nationalist causes, and organized the large St. Patrick's Day parade every year. These and other events, such as the open ball in Austin in 1867, were advertised in the local newspaper, the *Reese River Reveille*.[85]

As in other locations, the Irish in Austin were not solely defined by the activities of the fraternity and remained part of the wider community while maintaining their own distinctive identity. For example, baseball became hugely popular among the Irish in the area, and various local teams sprang up in the late nineteenth century. Other inclusive events such as the Fourth of July celebration in Battle Mountain, an outlying town near Austin, were advertised widely. "Goin' to Battle Mt? Sure Mike" was emblazoned on handwritten posters for the Independence Day gathering, which included "Horseshoes, Contests, Games, Sports" and promised a "Grand Ball in the evening." The special train service scheduled for the event left Austin at 6 a.m. and returned the following morning at 7:30 a.m. The fares were advertised as two dollars for a roundtrip and half price for children and Indians.[86] Lander County's small and scattered population encouraged inclusive events such as these that transcended ethnicity, and they became important elements in drawing the varied communities together when the mine companies tried to lower wages.

The district's Catholic infrastructure grew significantly during this period, and the pastor of the entire state of Nevada since 1862, Father Manogue, sent Fr. Edward Kelly to Austin in 1865, where he began constructing a Catholic Church. Fr. Dominick Monteverde replaced Kelly and completed St. Augustine's in 1867 with help from the local Fenian Brotherhood, which had donated $500 toward the building of a church from its successful St. Patrick's Day Ball in 1866.[87] A. B. O'Dougherty ran a boys' school from the church basement, and the Sisters of Mercy opened a school for girls in 1871.[88] Father Monteverde served the Irish Catholic community until 1876, when an Irish priest, Fr. Joseph Phelan, replaced him. Phelan, like Manogue, was an Irish gold miner turned priest but was much less successful: "I did not realize any fortune. Then I felt I had a vocation for the priesthood, but having little means I was aided by my brother Martin—God Bless him."[89] He traveled back to Ireland where he entered Mount Melleray for three years, thereafter joining the seminary at All Hallows.

Phelan returned to America in 1874 after his ordination and was appointed pastor to the Irish flock of Austin, Nevada, where he served for eighteen years.[90] His Lander County parish was a vast desert 400 miles long and 100 miles wide as well as a prosperous mining region. Echoing the earlier refrain of Father Twomey, he, too, decried the dangers of isolation and attached a providential element to his duties to his dying parishioners: "I never arrived at a house and found a person dead before me, in all those years, though many died just after my arrival. God is good, and He was especially good to me. Many who were nearly dead came back to receive the Sacraments, and three who were said to be dead came back for a time."[91]

The Savings of a Lifetime

Lander County's mining economy stalled for two years, 1881–1883, suffering from the effects of unregulated mine stock speculation.[92] A significant bubble could prompt a flurry of activity, whereas a dip could grind all work to a halt and empty a mine camp. Stock speculation became a popular form of gambling for many, regardless of occupation or status. As with each gold rush from one location to the next, vast sums of money swelled a rush for the next stock that would magically multiply investors' fortunes, big and small. Appealing names and alluring stock tokens tried to coax people to spend their money, and the choice to invest was often based on a rumor of a great find that perked the ears of the populace, who soon emptied their wallets in the frenzy to get in before everyone else had the same idea. In 1884, James Stewart wrote to a friend in Belfast recalling 1874, when he got caught up in one of the speculative crazes "which have so long cursed this community." His savings were wiped out, but fortunately he held a stable job that sheltered him from destitution: "Since then I have plodded steadily along in the daily routine of business life, trying to regain by hard work what I foolishly lost, I recognize the fact, however, that my life has been a failure. I remain a bachelor, and that word is sufficient to express as much as I could write in a whole page."[93] Stewart felt the pangs of his solitary life far more intensely than he did any financial disaster, believing the outcome of his financial disaster led to personal disaster and that his dreary hard work became an attempt to atone for his sinful gambling.

Others never learned from repeated harsh lessons of the vagaries of the stock market. Despite the frequent disappearance of their investments, they continued to gamble in the hope of striking it rich. Two brothers, Michael and Denis Hurley, born in the Tawnies near Clonakilty, County Cork, immigrated to Carson City, Nevada, in the 1870s and worked various jobs, mostly supervising railroads. Denis had followed his brother out west and got his job in Carson City thanks to Michael, who had moved from there to work in Spokane Falls, Washington. There he was fired due to his involvement in the Pullman strike, after which he returned to San Francisco to work odd jobs.[94]

Denis Hurley settled in Carson City and, noting the extreme distances for travel and communication in the American West, wrote: "I have not heard from Michael recently although he traveled 1700 miles last summer he was still several hundred miles from me, and as he had but a month's lay off he could not make much of a visit."[95]

Carson City had a small Irish community compared to Virginia City, but it was more affluent. As proof of this wealth and organization, Denis wrote to his cousin in 1873 about the condition of the Catholic portion of the town: "We have got a nice Catholic Church, attended by a very eloquent Irish priest, in this town. All persons are provided with good comfortable seats. The appearance of the congregation is very respectable on account of the rich clothes which all the members, especially, the female portion wear."[96] Denis also noted the tremendous diversity of the community:

> Men from every nation under heaven can be found in Carson. The long-tailed, sombre looking Chinaman; the black haired red skinned Indian, with chalk lines drawn across their face on each side of the nose, with their women called Mahalies, having their infants called Pappooses wrapped on their backs. The swarthy copper-coloured Mexicans; the slow-thoughtful looking German; the more lively Frenchman. With Swedes, Swiss, Italians, Spanish, Portugese [*sic*] . . . .[97]

He noted the diverse multi-ethnic nature of this new society but did not celebrate it. His next line ends with a sober observation that "in conclusion I have wandered away from home and domestic relations," hinting at his loneliness in this alien land.[98]

He wrote about the food in detail, giving his parents his satisfied assessment of what was a rich diet:

> I must say that I don't know the name of many of the smaller novelties supplied; you will be asked what you will have, whether ham, beef-steak, or mutton chops, well you will get some loaf bread, cakes of many kinds, some soft bread steeped in milk and some other mixture, your tea, or coffee as you like, some highly flavoured soups excetera [*sic*]. This is repeated three time a day with some variety a few peeled potatoes and some mashed ones are also placed by.[99]

Denis assured his parents that the brothers were not wasting their money on drink, which he noted was cheaper and more widely available than at home: "Where ever else our money may go much of it does not go into saloons for whiskey. It is the cause of a great deal of misery more so here than the old country owing to the greater opportunities working men have to indulge their brutish appetite. The inferior kinds which indeed is a very poor article can be had very cheaply, it is the devils [*sic*] own stuff."[100] These "inferior kinds" of drink included so-called rotgut whiskey and could be made from pure alcohol or be watered-down and adulterated whiskey. They might contain ingredients such as creosote or strychnine to increase the potency, with chewing tobacco or molasses helping give them a darker color.

Denis scattered Irish phrases and words within his letters home and titled a letter on January 6, 1876, *La Nollaig beag*, Little Christmas, the Irish term for the Feast of the Epiphany.[101] This letter contained extensive information on mining stocks he bought and profits on some of the leading mines. It ends with the postscript "3.30 pm, stocks away up to-day."[102] Later, in May 1877, the brothers met in Carson City and made a request to their parents that would "undoubtedly appear very odd."[103] Normally, money flowed from the US to Ireland, but instead they asked for a loan of 100 pounds ($500), which they intended to pay back with interest plus the cost of transmission within six months:

> Father, don't hesitate, if you knew how things were, you would not. We have several thousand dollars at stake, and with your loan *we are almost morally certain that next Fall would render us independent of works and Christmas find us home together with a handsome fortune.* If you wish to hasten that consumation [*sic*], dont [*sic*] deny us our request. You may be sure that it is not [a] trifling affair that has induced us to take this step, and we trust we wont [*sic*] be disappointed. We know with how much pains a little money is put together at home, and we would never think of asking it for any risky or doubtful venture. If we are as successful as we expect we will return your money doubled at least two-fold. Father we dont [*sic*] ask this in pity, or for charity, as even if you refuse it (as we cannot believe you will) we will still have means thank God, but your refusing will cause us to a loss of at least £1000. If you intend to send, lose not a day, no excuse is worth a cent. We are neither drinking nor rawdying nor squandering but a rare chance has presented itself and we want to avail of it working all a man's life is played out.[104]

Caught up in speculation fever, the letter contains a strange combination of pleading, reassurance, and gentle emotional blackmail. After assessing all the surviving letters, there is no reason to believe the brothers lied and spent the money on drink or other pursuits, particularly given their comments on the ills of alcohol abuse and their assurances that they themselves did not imbibe. After they succeeded in making a "handsome fortune," they said they would return home, a statement that may well have been born from a genuine desire to return home or one that was possibly enticing bait for their parents to send the money. Their true motive is unknown but not their burning enthusiasm for the chance that they might become wealthy.

Their parents dutifully sent the money, and the brothers invested it in mining stocks, writing back that it would be "2 or 3 months" before there might be "any important move."[105] Denis spent most of the rest of the letter detailing the vast earnings Flood and O'Brien made from the California Mine and Fair and Mackay made from the Con Virginia, "controlled by those Irishmen who were themselves miners . . . These 4 men are now about the richest firm in America. They control other mines adjacent to those rich ones, on which you [*sic*] humble friend is keeping an eye."[106] When Michael next wrote to his parents in December, he commented forlornly that "if we had it [the money] six weeks sooner it would have been four times better, however it was good when it arrived." While the letter displayed gratitude to their parents—"We will never forget your kindness in sending it"—it also attempted to temper the earlier boasts of tremendous profit: "I expect the time won't be far distant when we will sell stocks at a big profit and send back that amount or double. It would take too long a time to explain here how that money was so much needed by us it was I told Denis sent after it and did well by doing so."[107] The surviving letters do not mention whether the request for funds from home ever paid off or whether the brothers paid back the loan, though it seems possible that they did, as in a later letter Denis mentioned his estranged sister with the lines "I suppose our dear amiable sister and her henpecked charge would not recognise [*sic*] any of us. Her wrongs are too deep (£100) to be condoned."[108] Denis also sent a week's wages to his mother in 1891, returning to the more general pattern of remittance flowing from America to Ireland.[109]

Later letters from the brothers show that they indeed hoped to return to Ireland. Both appeared ashamed over the way their investments had turned out. Michael revealed his abiding fear of debt and a deep insecurity (likely stemming from his own investment losses) when he chastised his parents for buying land:

> I was astonished to see that you gave 750 for that place I think you must have been out of your mind. No wonder times would be hard in Ireland when people are that foolish to pay so much for such a little place and such rent after. I have got that much money after all I have lost but I don't want to give it all for that place and then going into debt for stocking it. I can do better here.[110]

Based on his letters, Michael seemed to have suffered a series of financial setbacks before the request to his parents for a loan, and further financial misfortune plagued him later in life. Denis wrote of Michael: "His means are not large, always laboring with the delusion that he would get rich by speculating in mining stocks. He never wasted or dissipated, was not intemperate or immoral."[111] These failures weighed heavily: "He was never more discontented with things than he is at present. He has spent the savings of a lifetime in mining speculation expecting to strike the crock of gold someday. Now he has not much hope—and with age and infirmities coming on he is becoming somewhat despondent."[112] It is likely that Michael shared Denis's desire to return home, writing "after some time we expect our income to be much larger we may if God spares us, see old Ireland some day."[113]

They set their desire on the hope that they could return to Ireland with a sufficiently sizable amount of money to buy land and retire. However, Denis's later letters to his brother John in Clonakilty exhibited disappointment at how things had turned out: "I may some day possible [*sic*] see the land of my birth, but I doubt it as the world has been treating me."[114] In another letter, Denis stated his hope that "if we had a comfortable home in Ireland we should appreciate it and be happy and contented. Foreign fortune seeking may occasionally meet with big rewards but the disappointed ones far outnumber them and their hardships are not much written about."[115] These Hurley family letters trace the hardships and disappointments inflicted upon Irishmen of modest means after losing their savings and investments because of the false hope of making a profit from mining shares.

A disillusioned Denis Hurley summarized his advice for Irishmen seeking to immigrate to America in the early years of the twentieth century:

> I have no encouragement for emigrants. Most of them would be better off at home than here by putting forth the same industry and energy. The Irishman must compete with the hardy sons of all nations here. We have a relative—my wife's brother, and we see no reason to encourage him to come here. Of course

> he, and any others are free to come, take their chances and see for themselves. Our principal sources of employment are mining and railroading, both dangerous to life and limb. *The pay is good if a person gets a job and can keep it.*[116]

While Denis warned of the physical dangers of mining and railroad work, this final piece of advice noted the vital caveat that emigrants who were seeking to take advantage of good pay needed permanent employment. Both Denis and his brother Michael suffered from uncertainty brought about by speculation rather than a lack of steady jobs in the American West. If his reference to other "hardy sons of all nations" might be construed as Irish bigotry, Denis wrote in another letter during the recession of the 1890s that "the fewer that leave Ireland for this country under the present conditions the better all round," indicating that he believed the cause of employment troubles was rooted in a surplus of labor.[117] His sentiments may have stemmed in part from a longing to return home or a reading of labor literature, but they were hewn from a lifetime of hard labor and the relentless chipping away and erosion of dreams of wealth embodied in the fortunate few Irish mine-owning millionaires.

While many made a living and some became wealthy, these letters reveal the failure and isolation of many Irish miners on the frontier. Failure was a spectrum of experiences; even with their Irish networks, some found themselves trapped by the shame of poverty or by the length of time they had spent in America. W. L. Kennedy, a miner in Kern County, California, wrote to his cousin James Gilmore in County Down in 1879 and asked, "I have not had a letter from home now—from anyone of our family for more than a year—Isabella used to correspond with me regular until lately and I fear very much either she has been sick or is dead, the latter I fear . . . I would like to know if Cousin John Gilmore is living and his address I presume if living he is a very rich man in Australia."[118] He concluded on a resigned note, saying he would prospect in Arizona and that "I have mined so long now I am hardly good for anything else—and I have lived for so long on the frontier I can hardly live any where else."[119] He also mentioned that he had a brother in Australia but had not communicated with him for decades and thought he might be dead. Despite the lack of contact, his relatives were never far from his thoughts.[120] Such a lack of contact over a long period was not uncommon, and the cause might have been neglecting to write or a family dispute. Patrick

Dunny laid out one spat that had gone on for months: "I am very much surprised at Christy not writeing [*sic*] to me before this time I expect he is angry at me for what I told him about Richard but I congratulate myself for what I told him nothing but what is true and what he requested of me to do he told me to send him the true account of him good or bad and that only I done."[121] These communications were personal and human, not just consisting of well wishes but also describing news, arguments, and petty grudges. Letters ran the full gamut of the family experience transformed into a communication system across vast distances.

Sitting down to write a letter gave the migrant precious time for reflection. Reviewing his life in one letter, Thomas Higgins asked himself "'was it worth the price?' 'Would I do it over again?' Knowing the future as I now know the past I certainly would hesitate before embarking on that field of adventure a second time."[122] Higgins was one of the miners who had struck it rich, and coming from him the meaning of loss changes from a financial one to a loss of opportunity—the opportunity to return home if they wanted, to earn the respect of friends and family they left back home in Ireland, to have something more than a deep sense of regret and loss after all their work and pain. The refashioning of these Irish communities in mining towns throughout the West was not just an attempt to consolidate jobs, power, or respect within American society. These communities offered the Irish an intimate familiarity without which they were often sorely detached, not just from home or parish but from their *Logos*. Most of them would never return home or make their fortune, but these communities gave meaning to many during their lives in the dusty realm of the Great Basin.

When a fire engulfed Virginia City in 1875, Irish mine owner John Mackay shouted at a crowd trying to save the Catholic Church, "D—n the church, we can build another if we can keep the fire from going down these shafts!"[123] Historians quote this line as evidence of the relative worth of religious versus economic infrastructure on the frontier, but this is only part of the story. Mackay's statement was a reinforcement of a social contract among the Irish mine owner, his Irish workforce, and the Irish community. The church was rebuilt, and Father Manogue took the opportunity to make the new building even more magnificent than the earlier one. It still stands there, on the rugged slopes of the Sierra Nevadas, a statement of endurance by the Irish of Virginia City and the communal bonds that supported and sustained them.

Without this mutual respect and shared affinity to come together, business leaders reaped different consequences for their actions—namely, fractiousness and violence as the Irish tried to carve out a piece of the American flag for themselves.

4

Mollies in the Mountains

Irish-American solidarity was expressed in many different ways, through songs, stories, friendships, faith, and organizations. Religion was one physical manifestation, as seen by the prominent Catholic structures in towns such as Virginia City and Smartsville. Labor unions were another, and while they had a physical presence in union halls, libraries, and events, their presence is most apparent in the notable labor conflicts that exploded across the American West during this period. From the early days of mining in the US, the Irish were immersed in organizing, directing, and controlling the shape of unions. Since business owners as a class were overwhelmingly wealthy Anglo-Americans, they perceived unions as bastions of Irishness and, by association, centers of subversive resistance. They repeatedly exhibited a distrust and hostility toward working-class organizations, and the experiences of the Irish and labor in Colorado more broadly reveal the roots of what became the typical friction and violence linked to business-labor conflict in the American West.

The Colorado gold rush was as uncertain as those in California and Nevada and similarly held the ever-present and unlikely possibility of a miner

https://doi.org/10.5876/9781646422517.c004

obtaining great wealth from striking the mother lode. The idea of "gold fever" proved just as contagious ten years after those earlier rushes when a gold rush to Colorado Territory drew thousands, the so-called fifty-niners, to the Continental Divide. The rush proved little more than a mirage, sparked by exaggerated reports from newspapers and eager outfitting merchants in Missouri, but the arrival of some skilled prospectors within the hordes of hopefuls led to the discovery of rich outcroppings of gold-bearing quartz on the north and south sides of Clear Creek.[1] Limited placer mining deposits coupled with unsuitable conditions for hydraulic mining left miners with little choice but to drive deep shafts to get at the veins of ore. However, Colorado gold was combined with sulfides, making it resistant to traditional processing, and nearby ore mills were only able to recover about a third of what their California counterparts had.[2] Propped up by eager investors from the eastern states and Europe, the hard-rock mines that were not scams began excavating in earnest. Importing machinery from factories in the East as the Civil War raged was difficult and expensive. This, coupled with difficulties in transportation due to weather and renewed conflict with Indians, led the majority of the population to make the easy decision to move on to placer diggings in Idaho or Montana.

By 1870, the population had stabilized somewhat, with 1,685 Irish-born included in the territory's 39,864 residents. Most of the Irish-born were concentrated in two counties: Arapahoe (545) where the trade town and later ore-refining center of Denver was founded, and Gilpin (511) where Central City, Blackhawk, Nevadaville, and many smaller mining camps dotted the landscape. The arrival of the railroad to Denver in 1870 and its expansion to Clear Creek Canyon in 1872 opened up the Gilpin County mines for more workers, cheaper goods, and—most important—cheap transportation of ores and fuel.[3] As it did throughout the rest of the West, the arrival of the railroad marked a new era for Colorado.

Given the huge numbers of Irish mining coal on the East Coast, in particular in Pennsylvania, it is logical to expect that many would have migrated to the new Colorado coalfields; instead, they decided that the opportunity for higher wages in the deep and dangerous hard-rock mines was more enticing. Pockets of Irish miners worked in the coal mines, mostly in the southern Colorado coalfields that provided an important fuel source for the new railroads, the mining industry, and the wider public. Here, the Welsh, unlike the Irish, preserved a heritage proudly tied to coal; and they came to Colorado specifically to continue these traditions, forming a significant percentage (11.2) of the skilled coal mining population in Fremont County.[4] Fremont County, separate from the main southern coalfields, also contained the Cañon City coalfield and became home to a slightly larger proportion of Irish coal miners in the region. Twenty-two of the 329 miners had Irish parentage and an unusually small number were multi-generational migrants, with 16 born in Ireland. The growing Irish community was concentrated in Leadville instead of the scattered smaller coal towns so that by 1880, there were more miners with Irish parentage who were born in Pennsylvania working in the hard-rock mines of Leadville (158) than in the three Colorado coalfield counties combined (50).[5]

Smaller numbers of coal miners also worked the bituminous pits surrounding Crested Butte in Gunnison County. In January 1884, more than fifty miners died in an explosion at the Jokerville Mine. The *Denver Tribune* reported the carnage: "Hands were raised as if to protect the face, the skin and flesh hanging in burned and blackened shreds, arms broken, and in some cases boots torn off by the force of the blast."[6] Rumors later abounded that the coal miners were seeking revenge on the company for the disaster. In response, the Colorado Coal and Iron Company, which controlled the railroad, refused transportation to any workers traveling to the site. Miners snowshoed across the mountains, arrived at the mine, and began digging out the bodies of their comrades. Although several almost succumbed to the gas, they dug out all the remains and returned home. Meanwhile, newspaper reports redirected responsibility for the disaster from the company to a strange source. One breathless account detailed: "It seems there is here an organization of Mollie Maguires," laying the blame at the feet of an Irish secret society.[7]

Names such as O'Neil, McGregor, and Donegan among the dead did not dampen the growing popularity of the story that the Mollies or Molly Maguires, an organization supposedly eradicated five years before with the execution of ten Irishmen in Pennsylvania, were culpable for the incident. During the rescue, a miner rushed in and handed Gibson, the mine boss, a gun, telling him to protect himself because "a gang of Molly Maguires were coming to lynch him."[8] The newspapers capitalized on the labor tensions between Gibson and his workforce (or possible ethnic tensions between

Gibson and the Irish miners under him) to sensationalize the story by adding the Mollies to the Jokerville disaster. Historian David A. Wolff argues that the Denver newspapers were happy to ascribe the label Molly Maguire to possible violence, as they were "unwilling to accept the notion that local miners could do such damage." More than that alone, they sought to demonize labor unrest and organization in any form; with the Irish at the heart of mining union membership and leadership, the label Molly Maguire was a natural fit, even if the use of the phrase had no supporting evidence.[9] Guards patrolled the streets of Crested Butte and guarded Gibson while a quickly convened coroner's court found Peterson, a "green" Swedish miner, responsible for the blast after he entered the work room with an open flame despite a warning from a fire boss.[10] Whether Peterson or the company was to blame for what happened, the court succeeded in calming the tense atmosphere and those grieving took their revenge on the deceased by excluding Peterson's name from the plaque on the mass grave containing some of the dead; other remains were sent to relatives throughout the state and to Illinois, Ohio, and Pennsylvania.[11]

In spite of the dangers, Irish continued to filter into the coalfields over the years, and the Irish-born population in Colorado's coalfields doubled between 1880 and 1890, in line with the total population.[12] In 1903, the Colorado Fuel and Iron Company still recorded a significant Irish presence in its workforce on the eve of the Colorado Labor Wars, with 900 Irish, 3,700 Americans, 3,500 Italians, 2,000 Austrians, 1,000 Mexicans, 800 English, 600 Slavs, 600 "Colored," and smaller numbers of Hungarians, Welsh, and Scots, to name a few of the many nationalities that made up the workforce.[13] Many mining stories mention the Irish working with other ethnic groups in the coalfields, and one such story details the interaction between an Irishman and his Slovakian coworker:

> In the heyday of Hudson's coal mining industry work about the mines went on by night as well as by day. In this night work two men worked together at driving a new slope. One of them was a Slovakiaan [*sic*] by the name of Mike Malaski, the other an Irish lad whose name was Bill Flynn. They always ate their midnight lunch together and Flynn would give Mike a piece of pie or a piece of cake from his lunch, for which in kindly reciprocation Mike would give Flynn a piece of rabbit. "Mike," said Flynn to the Slovakian one night, "You don't

> have time to go hunting. Where do you get all the rabbits?" "Oh," replied Mike blandly, "the wife she kill 'em when they come around the house at night and cry out." "Cry out?" echoed Flynn in consternation. "Why, Mike, rabbits don't cry out." "Yes, oh yes," Mike defended stoutly. "They go 'Meow, meow.' "[14]

In the joke the Irish are depicted as the more civilized and familiar of the two mining ethnic groups, revealing that in the popular mind the Irish "foreignness" had been tempered and coupled with a warmer naïveté, in particular when compared to the strange and barbaric customs of the newer eastern European immigrant stock. In spite of jokes, the Irish still had a threatening reputation among many Anglo-Americans; whether the association was the Molly Maguires, the labor unions, or the Catholic Church, any organizational ability was seen as an integral part of the perfidious Irish character in coal or hard-rock mining over the course of the nineteenth and early twentieth centuries.

For the hard-rock mine industry, the 1870s were a period of rapid transformation. Once difficulties in refining gold and silver were overcome through the use of coke, thanks to its ready availability near the mines, many skilled Irish and Cornish hard-rock miners entered the vacuum left by less skilled workers who abandoned the harsh mining conditions.[15] One Irish miner recalled the change in the population: "Some of the prospectors who had failed to find mines left as early as 1878 to search elsewhere, rather than take jobs with the producing classes."[16] This "producing class" referred to the wage-earning, hard-rock miners. Whether it was prospectors' disinterest in overseeing mines as managers, a desire to retain their sense of independence and carry on with their footloose existence, or perhaps just a desire to stick to what they knew, most prospectors sold their promising claims and moved on to the next promising digs. The subsequent expansion of these profitable mines, coupled with the available high-paying employment, drew large numbers of Irish to the booming Colorado towns.

By 1880, Irish-born miners made up 7.9 percent of all miners in Colorado, with the Irish constituting 13.6 percent of the total when those with Irish parentage on both sides are calculated (table 4.1). English miners appeared to have traveled more recently and more directly because of the smaller increase in their parentage figures, indicating that fewer were born in the US and were part of a multi-generational migration, as in the Irish case. The US and Canadian

Table 4.1. Miners in Colorado, 1880[a]

	Birthplace	*Percent*	*Parentage*	*Percent*
Ireland	2,252	7.9	3,872	13.6
England	3,096	10.9	3,590	12.6
Wales	417	1.5	483	1.7
Scotland	486	1.7	690	2.4
Canada	1,355	4.8	976	3.4
Germany	940	3.3	1,374	4.8
US	18,501	64.9	14,292	50.1
Sweden	564	2.0	576	2.0
China	163	0.6	163	0.6
Other	746	2.4	2,504	8.8
Total	28,520	100.0	28,520	100.0

[a] IPUMS. There were 4,466 miners in Colorado who had at least one Irish parent. Within the "others" birthplace column category, 8 were born in Canada, 1 at sea. The listing "others" in the parentage column includes persons not included in the birthplace category for the countries listed, as well as those whose parents did not have their country listed. Because of this and the aforementioned question about determining the ethnicity of persons of mixed parents, this figure is inflated. A longer synthesis with a more granular analysis could clarify these edge cases. See chapter 1, this volume.

Table 4.2. Miners in Lake County, Colorado, 1880[a]

	Birthplace	*Percent*	*Parentage*	*Percent*
Ireland	1,049	14.8	1,741	24.6
England	518	7.3	614	8.6
Wales	108	1.6	119	1.7
Scotland	158	2.2	225	3.2
Canada	466	6.6	315	4.5
Germany	240	3.4	347	4.9
US	4,243	60.0	2,872	40.6
Sweden	150	2.1	156	2.2
Other	142	2.0	685	9.7
Total	7,074	100.0	7,074	100.0

[a] IPUMS. There were no Chinese miners in Lake County.

Table 4.3. Ethnic breakdown, Lake County, Colorado, 1880[a]

	Birthplace	*Parentage*	*Difference (%)*
Ireland	2,093	3,969	+90
England	1,050	1,388	+32
Scotland	363	536	+48
Wales	195	253	+30
Canada	1,272	913	-28
Germany	1,075	1,764	+64
US	16,470	11,533	-30
Total	23,570	23,570	–

[a] IPUMS.

figures see a predictable drop in their numbers with the more accurate use of parentage to measure the size of the ethnic group, while the Swedish and Chinese parentage figures are stable, indicating the former's recent arrival in the US and the latter's male-dominated population. Leadville drew most of the Irish in the American West in the 1880s, with its mines running day and night perched high in the Rockies.

A more careful look at the mining population in Lake County in 1880 (table 4.2) reveals that Irish-born miners account for one person in seven, but parentage figures reveal that one in four miners were Irish-American. This 10 percent jump is the largest of any ethnic group and is reflected in the broader population figures. The inclusion of those whose parents were Irish-born increases the size of the Irish community in Lake County by almost 90 percent.

The ethnic breakdown table (table 4.3) shows that native-born Americans made up only 49 percent of the total population of Lake County, Colorado, in 1880 rather than 70 percent of the total, as it appears when the birthplace figures are used.[17] Both Canada and the US had similar decreases of close to a third, while other nationalities had increases of about a third except for Germany, which increased by 64 percent when we account for the entire German-American population. These figures reveal other important distinctions between ethnic groups; for example, the Irish-American community was deeply involved in the mining industry—with almost two-thirds of the working Irish-American population as miners or laborers—whereas the German-American community had a more diverse workforce, with only 20 percent mining.[18] Also impressive, though somewhat unsurprising considering the size of the Irish community, were the 9 clergy who had Irish parentage in a population of fewer than 4,000. The Irish were not the only Catholics, but their ability to command this number of priests who shared their identity was one more indication of the strength of their solidarity and the transnational threads of the Irish community that enabled Irish-Americans in Leadville to imprint a deeper hue of green among the spires of the Rocky Mountains.

Hard Rock, Hard Lives

Leadville's population changed rapidly during its early mining years, and some changes were felt more keenly by certain ethnic groups than by others. The town boomed from 1,500 people in May 1879 to 18,000 by the end of the year, reflecting a vast influx of people to the new state. The town's rapidly expanding industrial output boasted fourteen smelters that collectively cast a pall in the thin air. Thirty-odd working mines running day and night gave the town a boisterous quality, and it remained lively late into the morning with drinking, gambling, and vaudeville acts. One contemporary recounted the popular couplet, "It's day all day in the day-time. And no night in Leadville."[19] Like many other mining towns, Leadville had a wild reputation; however, as a visiting preacher observed, "Leadville is the most lied about city in the west, sure it has its wickedness on the surface, but there are more murders in Brooklyn than in Leadville. Their churches are crowded on Sunday." Many of the churches were Catholic and were filled with the Irish.[20]

The Irish-born in Lake County constituted 9 percent of the total population in 1880. By the time of the next census, in 1890, the number of Irish-born in Lake County had declined by 25 percent.[21] However, that same year the total population declined even more sharply, by almost 40 percent, thus leaving the Irish-born percentage of the total population at 11 percent—its highest historical point and a full percent more than the number in Silver Bow County, Montana, with its bastion of Irish-America: Butte.[22] The Irish had become more entrenched in Leadville, and the mining community's decline took another decade. Between 1900 and 1910, the sharpest drop in numbers was seen for the Irish-born in Lake County, a decline of almost 60 percent.[23] Many of the older miners died, and most of the others had moved on to greener pastures in places such as Butte.

The Irish path that led to Leadville and other mines in Colorado usually consisted of a series of hubs rather than a direct route. At these stopovers—large cities or other mining towns—the Irish found work and company. Rumors of higher wages in Colorado lured them onward. Stories about the mining boom and "big wages" led Thomas Quillen from County Monaghan to Leadville.[24] Monthly remittances of ten pounds acted as confirmation to neighbors and relations that traveling this migration path was worthwhile. For many, Colorado did not mark the end of their travels; it was simply another hub from which they could move on to the next frontier of opportunity.

The story of three brothers, Michael, Dan, and Dennis McGee, illustrates how the lure of high wages brought Irish immigrants to mining towns and shows how they described Leadville to their relatives in Ireland. For the Donegal-born Michael McGee, his meandering route was typical of other Irish who came there. His indirect journey began in 1873 when he and four friends went to Redington, Pennsylvania. Soon after, they continued their travels, moving 200 miles west to work on driving a tunnel underground for three dollars a day.[25] At the time, he seemed pleased by the US and regretted not having come earlier.[26] Five years later he joined relatives who had settled in Marshall, Illinois, where he worked intermittently in different Illinois towns over the next few months. The recession, the poor pay, and the sporadic nature of his employment blighted his formerly positive perception of America, and he warned his parents in a letter that "men are working for a dollar a day . . . and glad to get a dollar a day . . . I wouldnt advice [*sic*] you to come out here at present."[27]

Two years later, his location and his perspective had changed again. In lockstep with his improved wages he wrote:

> I am writing up on top of the Rocky Mountains where they are digging gold and silver. This is a good place to work wages is from 12 to 16 shillings a day [approximately four dollars a day] I like this country only for one thing its [*sic*] too cold. We have snow here 8 months in the year and in some places it never leaves the people are dieng [*sic*] fast here when they get sick they don't live anymore 2 or 3 day[s].[28]

Despite its geographic isolation in the Rockies, newspapers and letters kept Leadville's sizable Irish community up to date on the latest news, detailing the starvation and hardship in Ireland and prompting the community to organize a collection for relief back home. Fearing for his parents' well-being, Michael sent twenty pounds home and added that despite the dangerously unhealthy conditions in Leadville, he believed "some of them lads will do better in this country then [*sic*] they will there."[29]

Michael often mentioned visiting home in his letters—"If I am living and well and in good health I will be there with you before this time"—but he repeatedly postponed the trip to look after his mining claim.[30] His parents spoke to him about land they were thinking of purchasing, perhaps in the hope that they might entice him to return to Ireland and take up farming. Michael replied, "About that piece of land that you was talking about yous can do as yous [*sic*] please about it I don't want no land in Ireland if I want land there is plenty of land [here]."[31]

Michael wrote to his parents concerned about the effect of alcohol on immigrants—"this is no country for a drunkard"—in particular in Leadville's high altitude: "Anybody that drinks hard here won't live no time on account of the lite [*sic*] air."[32] When his younger brothers Dennis and Dan joined him in Leadville and began working in the mines, Michael observed that his brother Dan "seems to like to tak [*sic*] drink of whisky once in a while but told me that he won't drink another drop while he is in this country."[33] In a letter a few months later we see that Dan was only trying to placate Michael, as the older brother wrote that "Dan is well and got the same habits he had there. There is no use in talking to him he only laugh [*sic*] at me."[34] Dennis wrote that while Michael worked his claim sixty miles outside the city, Dan was happy getting steady work twenty miles outside of Leadville in Summit County, adding "it won't mkie [make] know [no] odds to him where will he be working, he must get his beer."[35]

These conditions were tough, and the miner's hard-working, hard-drinking lifestyle contributed to the huge numbers of deaths linked to pneumonia: "The man who, in Leadville, becomes drunk, and throws himself down in the floor, or in a gutter, to sleep off his debauch, is exceedingly likely to be frozen to death, or receive a fatal attack of pneumonia."[36] Dan eventually reformed and "queit [*sic*] all his bad habbits [*sic*] and is working here Study [steady] for [the] last year."[37] Meanwhile, Dennis up and left without telling anyone where he was going; five months later he wrote to his brothers informing them he was working on the Northern Pacific Railroad near Missoula, Montana.[38] Michael chided his parents, believing the reason he left might have been because they did not mention or thank Dennis for the money he sent them the previous March: "He did not pick that money up from the road side nor at the base of a tree he had to work hard for it and I would like to here [hear] what become of it."[39] Ingratitude, or the perception of ingratitude, could have a deep emotional impact over the vast distances.

Irish emigrants, often susceptible to the pleas for remittances, could still be angered by the lack of appreciation those at home showed for how difficult life was in the US. A false perception grew at home, perhaps inadvertently fostered by glowing emigrant letters with money enclosed, that life in the US was both prosperous and easy. Although angry at his parents, Michael did include twenty pounds from Dan and added sternly, "Now I want a correct answer back for this money as soon as possible."[40] This was an attempt to make the recipients pause and think about what they were spending it on and where it came from.

No reply survives, and shortly before Christmas 1883, Michael died in a mining accident after a rock fragment from a blast struck him on the forehead.[41] His body was sent to his relatives in Illinois, and they buried him the day after Christmas.[42] When Dennis heard about his brother's death, he returned to Leadville to settle his estate. Dennis, in his words, "made"—that is, forced—the mining company to give him eighty pounds for his brother's death, which he dutifully sent to his parents. Like vultures, those present at the time of his death divvied up what money Michael had in his possession.[43] The questionable honesty of Michael's work colleagues extended to his business partners, in particular Pat McNellis, who Dennis believed was mismanaging his mine claim. Dennis wrote to his parents asking them to notarize and return a legal form proving they were Michael's parents and the inheritors of his estate and granting Dennis power of attorney, later writing "they want to beat me out of Mick's rights if they can."[44] Adding to the confusion, Dan had disappeared, and the Panic of 1884 led to the collapse of the banks in Leadville.[45]

Over the next year, relatives in Illinois and Donegal repeatedly made inquiries about both Dennis and Dan but received no reply. In 1887, Dennis reappeared when he visited his relatives in Illinois and his brother's grave.[46] In 1888, the brothers wrote to their parents informing them that they had wrested ownership of Michael's mining claim from his former partners but were waiting for better economic times before selling it.[47] In spite of the sluggish economy, both brothers were working, Dan ensconced in a mine five miles outside Leadville and Dennis planning to leave for "some far away camp" in the near future.[48] "Times ar[e] dull," wrote Dennis, but rather than explain the particular skills a person needed for mining, he stressed the importance of connections in getting work: "A man that is aquanted [*sic*] will get all the work he wants."[49] This was the last contact relatives had with the two brothers. A query ran in the *Irish World* from the youngest McGee brother, Patrick, who had stayed in Keeldrum and inherited the family farm. He received the reply from a friend of the brothers letting him know that Dan had died in Leadville at the turn of the century, and Dennis went to St. Louis "to finish out the iron-molding trade and has not been heard of since."[50]

The finality of death marked an abrupt, sad end to communications between family members. These family ties were meaningful to the emigrant brothers, and they said as much in their letters. They were personal, emotional connections treasured by the Irish and the distant branches of the family. Furthermore, they served as vital lines of communication that offered migrants economic, political, and social news through which they could frame their current position against that at home in Ireland or in other places in the US where relatives and friends were located. Dennis understood the importance of these connections in the US and that economic position relied heavily on networks of familiarity. Background and this network made all the difference in the circumstances and opportunities available to an individual, and his comments are not-so-veiled references to the social barriers and networks unavailable to the Irish. These often informal, concealed cultural and class divisions emerged as rifts and were major causes of the difficulties faced by the Irish community in Leadville in the 1880s.

Michael Mooney and the Committee of Safety

The McGee letters do not directly detail industrial unrest, perhaps because the brothers did not want to worry their parents, but Leadville was at the heart of major industrial confrontation during this period—sharpened by an undercurrent of nativist and class hostility toward the Irish. The Irish of the Colorado Mining Belt, where the Leadville mining district was located, were central in organizing the unions that dominated the negotiations between workers and businesses; the disputes revealed the prejudicial and bigoted character of management in the Gilded Age. Irish miners arrived with a cultural understanding of organization and resistance intensified by recent experiences with other workplace hostilities. Their numbers and pro-union tendencies caused them to be accused of being Molly Maguires by the Gentlemen Vigs ("Vigs" was shorthand for vigilantes), a group of nativist Anglo-Americans active in Leadville in 1879 and 1880.[51] The use of "Gentlemen" in the name of the vigilant group made the class dimension unambiguous. In their most infamous action, the Gentlemen Vigs took two men accused of robberies from the city jail and lynched them, pinning the following note to one of the bodies:

> Notice to all lot thieves, bunko steerers, foot-pads, thieves and chronic bondsmen for the same, and sympathisers for the above class of criminals. This is our commencement and this shall be your fates. We mean business, and let this be your last warning. Cooney, Adams, Connors, Collins, Hogan, Ed Burns, Ed Champ, P. A. Kelly. And a great many others known to this organization. Vigilantes Committee. We are 700 strong.[52]

Seven of the eight names are of Irish origin. Interestingly, the list of names is excluded from the primary historical account, replaced by the coded language "Here followed the names of a number of notorious characters."[53] The names expose the Vigs' ethnic and class prejudice, and Dill's sensitive omission of the names indicates his awareness of the links; thus, he sought to sanitize the group in the historical record.

Notwithstanding the hostility toward the Irish, the miners in Leadville quickly organized themselves into unions. On May 26, 1880, miners struck at the Chrysolite Mine and paraded throughout the Leadville district, shutting down mine after mine and bringing more men into their march until thousands assembled to hear Michael Mooney, their impromptu leader, speak to them.[54] The cause of the strike was unknown, and explanations ranged from a recent wage cut in some mines to the growth of the Miners' Union to workers' supposedly secret affiliation with the Knights of Labor.[55] Regardless of the cause, once let loose, the strike could not be recalled. The workers had two key demands: "Our compensation is too small and we demand an increase in our wages from $3 to $4 per day, and a reduction in our working day from ten hours to eight hours . . . Fair play is all we want and that we will have."[56] Mooney presciently warned his men to "avoid drinking whiskey and making themselves in any way offensive to peace and harmony."[57] He was quick to realize the importance of public sympathy and that the influential businessmen opposing the strikers would seek to use their political ties with the governor to bring in the state militia and break the strike under the guise of maintaining the peace if there was violence. Temperance would have the additional benefit of extending the workers' ability to sustain the strike. Mooney's emphasis on fair play reflects a common theme used by contemporary Irish-Americans, and it became a phrase increasingly adopted by trade unions. Conditions in the mines were another source of discontent. Miners complained that "water, foul air, caused by floating poisons, lead, arsenic, change of atmosphere from furnace or mine to the raw outside air, etc." severely damaged their health.[58] It was not just the mining that was injuring them; it was the specific conditions of these particular mines.

The dangerous conditions of the mines are borne out in the records. Historian James Walsh painstakingly compiled the causes of death between 1880 and 1900 for Catholic parishioners in Leadville from church internment records. Lung diseases, lead poisoning, typhoid fever, and mine accidents accounted for two-thirds of all deaths.[59] During Leadville's long winter months, men would leave work sweating from the heat of the deep mines and return home through freezing conditions. Miners realized that the "change in atmosphere" was a deadly problem—after all, how could they fail to notice that pneumonia alone accounted for just under half of all deaths despite an average age of death of thirty-two years.[60] While improved conditions, such as the building of changing rooms at the mines, were requested, pay and reduced hours were the main demands. Perhaps this reflected the miners' outlook; they knew there was a low ceiling on what they could ask for and were willing to risk ill health and death for the pay that had brought them to Leadville in the first place.

In a sizable mining settlement with an Irish population, the Catholic Church played a vital support role in the community. It provided spiritual comfort to the dying and those mourning their departed as well as vital healthcare and education to the community through a school, St. Mary's, and a hospital, St. Vincent's. Similar to institutions in Virginia City, these were staffed by the Daughters of Charity, with thirteen nuns (all Irish-born or with Irish parentage) running the hospital in 1880. With its distinctive Irish tinge, it is unsurprising that a third of patients until 1901 were Irish-born Catholics.[61] If the hospital primarily served the Irish community, it did not do so exclusively; the hospital treated many other ethnic and religious groups, including Jewish and Black people of the town. Sister Mary made a definitive statement on who the hospital would treat the year it opened (1879): "We never inquire about a patient's religion. So long as there is room we take all who come, whatever may be their color, creed, or nationality. Otherwise, it would not be a charity."[62] Likewise, in the silver mining town of Georgetown in Clear Creek County, a local American journalist investigating the local Catholic hospital admitted that "neither religious beliefs nor nationality enters into the matter in any way whatsoever," adding that the Catholic sisters ran the hospital in a "far more efficient manner than a private concern can."[63] One journalist expected his non-Catholic friend to be ignored and wrote of his surprise at the low cost and high level of the care he received.[64] The depth of gratitude can be seen in the actions of one miner, Jeremiah O'Brien, who left his mining claims to the sisters in his 1884 will.[65]

The broader community was supportive of the Leadville hospital, but it was also the target of unwanted attention.[66] When the Denver and Rio Grande Railroad built a depot near the hospital, property values rose dramatically. Soon, the sisters received warnings to move their hospital or be burned out.[67] In April 1880, they received a direct threat stating that the hospital would be fired on that night. The nuns, with Father Robinson, asked the Wolfe Tone Guards to protect the building. Despite their presence, a gang arrived that night and started to tear down the fence surrounding the hospital. The guards fired, wounding one man who later received treatment in the hospital until he healed. The sisters' request for help from an Irish fraternity rather than from local law enforcement indicates their deep distrust in law enforcement's ability or willingness to protect their lives and property. Vandals broke into the church again in 1882 and cut up sacred paintings, broke candlesticks and chalices, and destroyed statues.[68] Another attempt to "lot jump" hospital property was made in May 1883, as several men tried unsuccessfully to steal twenty-five feet of land in front of the hospital.[69]

Another episode highlights the ways ethnic and fraternal identity could be adopted and abandoned by some Irish: "Some years later as a common beggar this Sullivan called upon Mrs. Mary McCarthy asking for food. Mrs. McCarthy after one shrewd glance abused him by saying 'It was you who mutilated the statues in the church!' 'Yes,' he said ashamedly, then bristled a bit as he added, 'But I never touched St. Patrick!' "[70] If Sullivan was of Irish ancestry, which is almost certain given his surname, his thinking demonstrates a remarkable attempt to reconcile anti-Catholic actions with his Irishness. These outrageous attacks on the physical structures of Irish Catholicism in Leadville, their "stakes of permanence in a sea of transience," certainly demonstrate the fierce undercurrent of nativism—in particular, anti-Catholicism—in Leadville at that time and also explain the turbulence surrounding the Irish as they asserted their position in the town.[71] The Irish felt their community was under threat—not without justification—and the nuns, like the wider Irish community, had to forcefully reassert their place in Leadville.

The *Leadville Chronicle* warned striking miners in racial terms that "this whole business . . . will have the effect of bringing the Chinese here, and then goodbye to Leadvilles [*sic*] old-time prosperity . . . John Chinaman means $1.50 a day and only a dime of that amount spent here. This can and must be prevented, and it can be done by the great body of miners going to work at once."[72] The appeal, under the guise of supposed racial solidarity—one to which the Irish did not subscribe and the nativist Anglo-Americans adhered to only selectively—was a direct threat against the Irish miners and saloon owners and an indirect threat against the Irish women who owned boardinghouses and laundries. In case the hint was too subtle, a later story in the *Leadville Chronicle* carried a personal account of an Irishman, whose wife was a laundrywoman, who helped "a crowd" kill two Chinese men who came to set up a laundry in Leadville. The man's tale is filled with regret over the murders: "I thought I was doing poor Norah a service."[73] Whether invented or true, the tale was perhaps intended in part as prophesy of the violence to come if the strike did not end. Moreover, it offered a warning to the Irish that even if the Chinese arrived and the Irish decided to drive them off with violence, they would be haunted by the experience.

In other mining towns with Irish populations, we find no concerted Irish effort to dominate the laundry business, which was usually Chinese-controlled.[74] Heavy Irish participation in the laundry industry in Leadville was the result of a virulent strain of nativist hostility toward the Chinese, which led to fewer of their number in the town; because native-born Americans refused to fill this gap in the labor market, it was eventually filled by Irish women.[75] Ensconced in this lower-class industry, Irish women's defensiveness reflects their low pay and tenuous occupational security—here again, economic competition prompted the Irish response and ethnic defensiveness as compared to the racially motivated hatred against the Chinese rooted in the formal structures and racial theory of groups such as the Workingmen's Party of California. Newspapers disingenuously feigned surprise at Irish women's hostility toward this competition.

The governor of Colorado declared martial law and labeled the striking miners "vagrants." Even when they were attacked by soldiers, Irish women stood side by side with their protesting men in the defense of their community. The *Leadville Democrat* hoped for "full punishment" for the men but pleaded for leniency for the women: "A married woman is apt to blindly follow where her husband leads."[76] The unsurprising portrayal of Irish women as passive or gullible figures robbed them of their agency and obscured their sense of awareness of the threats these actions had for the entire Irish community. Without another available avenue to protest, they willingly joined their husbands, relatives, and friends in aggressively challenging the invasion of troops and militia on their streets and homes. The miners demonstrated their distrust of local newspapers early in the strike. After parading through town the day after the strike was called, 2,000 miners supported Michael Mooney's suggestion that they boycott local papers and instead cheered the *Irish World*, Patrick Ford's New York–based Irish-American newspaper.[77] Obviously, they felt that the local press was not only failing to speak on their behalf but was in fact responsible for demonizing them and for increasing tensions within the region.

The Strike Fails

The use of the term *Molly Maguire* exemplified the aura of suspicion that clung to Irish activists and validated popular stereotypes of the violent Irish brute. Some newspapers noted this propensity, including the *Leadville Weekly Herald*, which defended Michael Mooney against such accusations in stark terms: "The [agents of the company] might call him a communist and a Molly Maguire . . . Persecution always follows an assertion of right."[78] Mooney offered an interesting response to these accusations: "I am not a Molly Maguire in the common implication of the term, although a sympathizer of that faction at heart."[79] Mooney had likely developed a nuanced understanding of the circumstances surrounding the Mollies in Pennsylvania from Irishmen who came to Leadville from the anthracite region (as had his soon-to-be wife's father), as well as from Patrick Ford's newspaper reports on the events. When mine owners received anonymous threatening letters, Mooney both disavowed them and promised that the miners would attempt to protect property and life.[80] The mine companies may have been using these letters as false flags in hopes of turning the miners and the wider public against the strike. At one point Mooney himself claimed that a mine superintendent was caught leaving threatening letters in a shaft filled with scabs.[81] In a speech in nearby Denver at a Greenback and Workingmen's meeting, Mooney ably defended the legitimacy of his identity and Irish loyalty to the United States: "I don't know that the whole American flag covers me, but I think I have a corner. It is cast in our teeth that this is an Irish movement. They say the Irish element is getting too strong. When our country's flag was assailed the Irish element was not too strong."[82] He tied the blood spilled by Irish soldiers in service of America to the ethnic group's loyalty to the nation and used the flag itself as a powerful symbol of his rights.

In hopes that conversation might lead to some sort of compromise, a reporter from the *Leadville Weekly Herald* sat in on a meeting of Michael Mooney and W. S. Keyes and George Daly, managers of the Chrysolite and the Little Chief, respectively. Daly lectured Mooney on how American capitalism worked:

> If the miners had taken the proper course and presented their wants to us in a proper way, more consideration would have been given the issue. You men have the right to make the demand, and we have the right to deny compliance with it; and, furthermore, to put men to work in our mines at what wages we see fit to pay. When we decline to make a concession to your demands, you make yourselves alien to your country and its individual and constitutional rights, by trying to enforce the issue, at the point of riot and public disorder.[83]

Mooney responded to the charge that the strikers were disorderly by asking, "Mr. Daly, don't you think we have conducted ourselves intelligently," which drew the grudging response from Daly, "I have never seen so respectful a body of strikers in all my experience with such movements." He then, however, reaffirmed his right to use armed force if necessary "to protect our rights." Notice that Daly does not say "to protect our property" but says "our rights." Daly's unmistakably paternalistic attitude reveals a popular tenet among American mine managers and owners at this time—a belief that they knew best what was right and good for the company and their workers. They viewed themselves as American individuals and entrepreneurs, and with this sense of superiority they perceived the workers as something akin to misbehaving children. Mine managers and owners expected deference from their workers ("proper course . . . proper way") to whom, in their perceived munificence, they offered employment. As such, in their view they deserved nothing less than total control over their company and workforce. Labor historians have referred to this paternalism as a "false love," a freedom based on the "freedom to control" workers, as ultimately, mine managers and owners sought to maximize profits.[84]

Surprised by the less-than-submissive Mooney, the mine manager's polite facade quickly disappeared: "You cannot deny that among the miners at this place, there is a preponderance of 'grasshopper sufferers' and 'potato and grave diggers,' who have formerly commanded wages of ninety cents a day, or perhaps a dollar, and who are thoroughly incompetent as miners, and undeserving of even three dollars a day."[85] Mooney ignored the goad and asked if they discriminated "between practical miners and those who are not." In his answer to this question, Daly went further than confirming a skill-based bias; his statements also demonstrated a discriminatory attitude against Irish workers: "For you take twenty potato and grave diggers—not one of whom can tell the difference between a drill hole and a pick handle—place them side by side with the same amount of practical miners, attempt to pay them three dollars a day while you pay the experienced miners four dollars a day for the same time, and they would everyone strike."[86] When Daly used the phrase *practical miner*, he meant "Cornish." This was reinforced in his next sentence when he stated his preference for placing the Cornish in "light positions of trust." Thus, the overwhelming preponderance of Cornish mine foremen throughout the American West was indicative of an ethnic bias, reinforced by both the use of the phrase *potato digger* and the stereotyping of the Irish miners as lacking practical mining skills.[87] Moreover, his inability to treat the workers respectfully, his insulting their culture, and his arrogant dismissal of their grievances typified his arrogant tone throughout most negotiations. This haughty tone doubtless only entrenched workers' antagonism toward management.

Daly's statement contained a further ironic layer because he himself was the son of Irish Protestant emigrants to Australia.[88] He arrived in San Francisco as a young man and worked as a printer, eventually purchasing a newspaper press. His unfaltering support of business interests and dogged hostility toward trade unions and immigrants drew the attention of wealthy mine stock investors, who hired him as a mine manager in Bodie, California, and later in Leadville. The course of his business career seems like a desperate overcompensation for his uncertain credentials as an Anglo-American. He was an effective speaker but ceaselessly antagonistic; as the sympathetic *Bodie Standard* wrote, "[He] was a born leader, but died with a myriad of enemies—all of his own making."[89] The *Homer Mining Index* had a more comprehensive and robust critique:

> We wonder what kind of people those Leadvillians must be to worship this dirty, little cub, whom every decent man in California and Nevada could not pass without an involuntary desire to kick him? He is a toady by nature, a scrub by instinct and a bully on general principles, as all men of his stamp are. He is always ready to lick the boot of a superior, and is just as ready to kick one of an inferior position . . . His success shows what a man can obtain by sycophancy, cheek and a willingness to do any sort of dirty work for his masters . . . He knows no more about a mine than a pig does about a blow-pipe—and if he had not accidentally fallen in with men who needed his debased services, he would have naturally become a pimp or barkeeper in a cellar dive.[90]

The publication rightly pointed out his questionable mining expertise, which many later historians accepted uncritically. Daly was awarded the position of mine manager due to his rhetorical flair and ceaseless self-promotion, which his associates doubtless found a useful quality when it came to pushing the sale of mine stocks. His combative nature was too much for his business partners, though, who sent him from Leadville to the distant south. There, he was supposed to begin promoting promising silver mine sites in New Mexico, but his antagonistic bluster finally clashed with reality after he led a posse of townspeople to hunt Chief Nana's Apache. The Apache ambushed the party, killing Daly.[91] His muted origins were hardly responsible for his death; instead, his belligerent attitude, as seen in his dealings with the Irish-led unions, propelled him to his eventual foolhardy demise.

Back in Leadville, the nativist anti-union forces continued to threaten miners and their leaders. The protest and sporadic violent incidents played into the hands of the mine owners who sought to capitalize by launching a new organization, the "Committee of Safety." This organization was largely a rebranding effort on behalf of the Gentlemen Vigs, whose title lacked even a veiled attempt to inspire security or inclusiveness. On June 11, 1880, a "committee of citizens" launched its manifesto, warning miners that if they made any attempt "to interfere with, intimidate or threaten any miner willing to work, the undersigned will hold the leaders of the union *responsible with their lives*."[92] However, even vehemently anti-union individuals such as the editor of the *Leadville Chronicle*, Carlyle C. Davis, believed there was widespread use of agent provocateurs, perhaps even responsible for instigating the strike in the first place: "It was afterwards generally believed, [they] sent their emissaries into the ranks of the Miners' Union to sow seeds of discord and discontent, inducing the leader, Mike Mooney to inaugurate a labor strike."[93] Davis himself was elected president of the new Committee of Safety, which by his own account in *Olden Days in Colorado* he admitted was "patterned after the San Francisco 'Vigilance Committee,'" and his first act was to hire Pinkerton detectives to infiltrate the miners' ranks.[94] He wrote, "The 'spirit' of his new nativist organization . . . was well understood by the lawless element and resort to violence did not become necessary."[95]

One historian suggests that the San Francisco vigilantes marked the emergence of a type of vigilantism that "found its victims among Catholics, Jews, immigrants, blacks, laboring men and labor leaders, radicals, free thinkers and defenders of civil liberties."[96] It would probably be more accurate to suggest that these new nativist and Anglo-American "vigilante" groups desired the label *vigilante* to legitimize their cause. Notably, Davis adopted the title of vigilante as an attempt to appeal to public support in Leadville and the outside world as well as to repeat the popular narrative that the frontier remained lawless, thereby justifying any violent measures taken to tame and civilize it or, more accurately, its inhabitants. Davis went further than justification, demonstrating pride in both brutality and the subversion of judicial process: "Little wonder that this Association should have been dubbed 'the Stranglers' by the turbulent element, since, among other objects declared was 'the punishment of crime,' a literal interpretation of which would leave little for the courts to do."[97] Violence, inherent in the mythology of winning the American West, was believed to be an effective civilizing force against the urban frontiers there and the hordes of foreign workers.

It is worth quoting in full the Committee of Safety's goals, stated in their pledge:

> Whereas, it is believed that there is *an organization existing in our midst whose objects are detrimental to the best interests of Leadville and the surrounding industries*; and whereas, this *lawless* organization has assumed such vast proportions that *the civil and military authorities cannot adequately control it or sufficiently punish the offenders*; and whereas, it is the duty of every citizen who has in view *the prosperity* of Leadville, Lake County and the entire State of Colorado to band together for the protection of these interests; therefore be it resolved that we the undersigned citizens of Leadville and Lake County organize ourselves into a Committee of Safety, whose objects shall be *maintenance of order, the punishment of crime, and to take cognizance of all lawless acts that may transpire within our midst and come within the objects of this organization.*[98]

In the committee's own words, it was a crime to challenge a certain class of Leadville's population and to interfere with the property or profit of the mining companies, regardless of what it cost others in terms of their health, their living standards, or their very lives. Those involved with the Committee of Safety sought to imitate lawful government functions as well as to usurp them, since the committee lacked democratic and political legitimacy. Such legitimacy was irrelevant anyway, since the committee could not deliver on its intended purpose. It was, by its own proud admission, a thinly veiled paramilitary organization intent on imposing its uncontested will on the people.

In an effort to demonstrate their dominance over Leadville and perhaps antagonize the Irish miners, hundreds of "business men," an "army of snobs" as the miners termed them, marched through the town.[99] The marchers had influential friends, such as Governor Frederick Pitkin, who sent a cavalry unit to escort them, although the force was delayed for several hours and missed the march. The march was a provocative act, and its goal of provoking others was made clear when participants attempted to parade through Sixth Street on the east side, the center of the Irish community. In the end, they were turned away by a huge, enraged crowd. Civic order throughout the region was collapsing, and so Major General David J. Cook, ordered to the city with the state militia by Governor Pitkin, decided to hire his own private detectives and infiltrate Leadville's groups to get a clearer understanding of the situation.[100]

The miners knew this chaos exposed their leadership to extra-judicial murder, so they sought to protect their leader. To prevent the committee from lynching Michael Mooney, they set 200 men to guard his house. Nativists used their political influence to issue a warrant for his arrest on June 14, along with a reward of $5,000, after which he went into hiding in nearby Denver.[101] That same day, the governor—always sympathetic to the nativists—declared martial law, and several hundred anti-union citizens were sworn into a state militia. Meanwhile, Major General Cook's detectives had uncovered a committee plot to kill Mooney and five other leaders by arresting them and handing them over to soldiers loyal to the Committee of Safety, whereupon a mob would pretend to overpower them and lynch the men.[102] Even with all the power of local and state authority on their side, the nativists pursued illicit methods of intimidation, up to and including murder. And while they sought to kill the leader of the workers who were opposing them, it was equally important that the nativists lynch him publicly to press their ownership of the public space and send a bloody message to any potential recusant.

Once the union leaders were in hiding, the nativists pushed their advantage against the strikers by issuing a vagrancy law allowing for the arrest of any man who did not have a "visible means of support."[103] The punishment was a $100 fine or road building, and 250 arrests were made in just three days.[104] With most miners thrown out of work, the results were predictable, and one journalist noted the preponderance of Irishmen among those in the chain gang, writing that the prisoners presented "an exceedingly heterogeneous aspect . . . [of] the hard-working Celt."[105] Coupled with blacklists, the intense legal and lawless pressure led to an exodus of Irish miners to the surrounding region and other mining towns, in particular Butte, Montana.[106] The core of a weakened Irish community remained in Leadville, largely composed of those who would not or could not leave their homes, businesses, or investments.

Two or More Unions

The story of Leadville was more complex than a duel between two opposing forces, with the Irish on one side and nativists on the other. Nativism was broader than xenophobia or anti-Catholicism, and its targets included labor unions, individuals, and groups that opposed Anglo-American efforts to dominate the social and political order in the American West. The groups' class dimension allowed them to be supported, propelled, and directed by businessmen in an effort to weaken all who opposed the profits of owners and shareholders. They were keenly conscious of their public image and attempted to shape popular accounts through contemporary and historical records.

The union difficulties during the 1880 conflict revealed important distinctions and tensions between different nationalities of miners. To more clearly see the personal aspect of these distinctions and track the efforts of the Miners' Union to control public discourse and outreach, it is useful to compare the earlier 1880 strike with a later strike in the same city in 1898. Key firsthand accounts of the 1898 strike from a perspective close to the unions are found in the records of Thiel detectives. Both strikes involved the Pinkerton Detective Agency (PDA); as mine companies continued to spend vast sums on spies and guards, competing spy agencies emerged, such as the Thiel Detective Agency (TDA). Mining companies hired them to infiltrate the Cloud City Miner's Union (CCMU). In Leadville, the Thiel agent reported to mine managers an 1898 conversation by members of the union:

> In the evening Frank Moore, C. L. Knuckey and Sullivan discussed the Cornish question. Moore thought the Cornish would come around alright. Sullivan said that if they did not join them in their next struggle they would fix them. Knuckley [*sic*] thought there would be no difficulty and thought that as there were such a large number of Cornish it would not do to attempt to drive them out of the camp, but Sullivan said: "We have really no use for them anyway and the quicker we get rid of the S+++s of B+++s the better it will be." Moore expressed the same opinion, saying that *while they had no use for the Cornish they would have to cater to them for awhile at least in order to get their aid in defeating the mine owners.*[107]

The English-born Charles L. Knuckey was once a laborer in Danville, eastern Pennsylvania, but had graduated to become a miner when he came west to Leadville.[108] Throughout the spy's records, Knuckey urges caution when dealing with both the Cornish miners and the more recent immigrants from northern Europe into the camp, particularly the Finns. Through the CCMU, an alliance quickly developed between the Irish and the Finns, with both ethnic groups supporting each other to strengthen their positions in the town. The halfhearted Cornish support of the Irish-dominated CCMU marked them as a group that needed to be driven out, if possible, after their *modus vivendi*. Various ethnic groups vied for prominent positions in the mines, and a few months before the above exchange, the Thiel agent reported that Frank Moore at the CCMU offices was celebrating the failing fortunes of their evanescent Cornish allies: "They have not got a Cornish man on at the Ibex, so that we need have no fear of the Cornish up there in the future."[109] Alliances, even under the tent of working-class solidarity, could not wash away long-standing cultural and religious divisions.

One reason for the Cornish miners' reluctant support of the union was their fear that company spies had infiltrated it. A Mr. Ahern noted at a meeting attended by the Thiel detective: "I think the Cornish are afraid while they think the mine owners are watching them."[110] Given that one of the best sources of information available to historians on the inner workings of unions are the detective reports from archived company records, the fears were evidently justified. Another reason for their lukewarm embrace of miners' unions was that because of their Anglo-Protestant background, mine managers preferred the Cornish for supervisory roles; by challenging the established order, the Cornish feared the loss of their privileged position. They were more accepted, sought after, and better paid than other miners in the American West; and by forming an alliance with other alien workers against mine owners, they stood to lose more than did other workers. Irish puissance within unions likely deterred them further. A cursory look at the rosters of union leaders in Colorado (and indeed across the US) shows numerous Irish surnames.[111]

Irish miners frequently utilized ethnic solidarity as a means of generating communal support and intimidating rivals, but in the case of the CCMU it adopted an inclusive attitude, counting German, Cornish, Scottish, and Swedish nationals as its members.[112] The language of the union leadership reflected this inclusivity—during the 1880 strike, Mooney avoided ethnic or cultural references to any one segment of the union, calling the strikers *freemen* or *workingmen*.[113] However, ordinary members were still cautious of trusting one another. During a major strike in 1896, one detective report stated: "There is a great deal of talk about the Swedes wanting to go back to work . . . all the leaders were men from Montana and Idaho, and were not Leadville men."[114] Here, the company sought to divide the workers from the union leadership. It also tried to sow ethnic divisions: "The Managers had tried to get the Swedes and Irish to fighting, but could not make it work and now they were trying to get the Catholics and the APA's [American Protective

Association] mixed up."[115] Although these efforts failed, the attempts to sow dissent between the various ethnic groups were carefully planned, heavily funded, and based on detailed information about the different factions. For example, a Thiel Detective Report detailed the nationalities of workers in Lake County—"700 are Carolians, 300 Italian Austrians and 200 Hungarian Slavonians"—and went as far as to state that the local Austrian Catholic priest sympathized with the union.[116] The change of tactic meant that rather than pit nationalities against each other, the managers were whipping up nativists' hatred of Catholics. The American Protective Association (APA) was a virulently anti-Catholic and anti-immigrant society, the successor of the Know-Nothings and the predecessor to the Ku Klux Klan. Interestingly, this association formed a few months after another mine manager in Idaho established a local branch of the APA for the specific purpose of undermining support for the Irish- and Italian-dominated mine union (see chapter 5, this volume). Mine managers seem to have been sharing with each other various strategies about how to deal with unions and the Irish, highlighting a parallel web that stretched across the American West.

While ethnic divisions are evident throughout the records, the union managed to hold a remarkably united front despite adversity and suspicion. A Thiel detective wrote about the difficulty union members had deciding who to trust, and they based their assessment on ethnic groups' historical allegiances in labor disputes: "Conley said they would be all right if there were no one but Irish men in the camp, then they would soon win the fight, but that there were so many other people here, they did not know who to trust. He said the Americans were no good unless they had a little Irish blood in them, but that the Cornish and Swedes were not worth hell room, and were better out of camp than in it."[117] Conley's affection for Irish-Americans (of whatever degree) was matched by his enmity toward the Cornish and Swedes. His views might have been personal opinions or may have echoed the voices of many Irish in Leadville, but such feelings did not prevent ethnic alliances from forming. The complex and shifting relationships between ethnic groups existed in many mining towns and changed over time, which challenges the concept of a singular "American-born" category. While this report highlights divisions, it also reveals how the Irish thought about their wider Irish-American identity. Remarkably, their loyalty was not questioned and their degree of Irishness was not quantified. Irish-America had a sophisticated self-awareness of the diverse shades of green that formed its community and of the necessity to include the descendants of the Irish-born in its ethnic group.

Still, grudges that remained from earlier strikes were difficult for miners to dispel, and at a CCMU meeting in 1899, when Joe Sowa reiterated the importance of bringing the Cornish back into the union and facing the mine owners with a united front, another member, Duffy, responded: "What do we want with the Cornish S+++s of a B+++s anyhow. The quicker we can get rid of them the better it will be for us."[118] Another CCMU member, John McKane, said, "We know that, but at the same time it will be better to have them on our side until we have brought the mine owners to recognize the union. After we have conquered the mine owners we can then turn our attention to the S+++s of a B+++s and attend to them."[119] Sowa reiterated the union's priorities, telling them "I hope you fellows will not get the question of running the Cornish out of the camp mixed up with the mine owner's question. We don't want to have to fight both outfits at the same time."[120]

The information in the detective reports ranged from banal to insidious details of workers' lives. Details of meetings constituted one source, and noting workers' membership in other organizations was another, such as the fact that many members of the Miners' Union were also members of the local Irish fraternity, the Knights of Robert Emmet [KRE].[121] This group appears in the general index at the start of the Leadville City Directory under "Secret and Benevolent Societies," a contradiction in terms that made sense for nineteenth-century public listings. As a benevolent society, the KRE's focus was to support members who were injured or their bereaved family, if they were killed, through a monthly subscription, the promotion of abstinence from alcohol, and—befitting its namesake—support for Irish nationalist causes.[122] In general, the information the company gathered was kept secret to safeguard the continued supply of gossip and the safety of the spies, but in its propaganda war against the union the company sometimes utilized this information to intimidate union members or those sympathetic to their cause. The wealthy Leadville mine owner John Campion decided to leak information about meetings to a local newspaper known to be sympathetic to the mine owners, the *Herald Democrat*. On August 3, 1898, the newspaper ran a critical editorial titled "The Miners' Union Picnic" that detailed the recent CCMU meeting, including intimate details about the union's forthcoming picnic and its plans to organize a union-affiliated Lake County Rifle

Club.[123] The realization that they were being spied on upset members, and one named Sullivan said, "I do not see why the S++s of B++++s cannot let us alone now that we are quiet and not saying a word or doing a thing."[124] But the Leadville mine owners viewed any union as a threat to their control of the region. As one historian noted, in the early days when law enforcement was weakest, the miners were at their most peaceable: "Only as the law grew stronger and the owners began to manipulate it as a tool of repression . . . did the miners in their frustration turn to violence."[125]

The spy reports contain a wealth of information on the daily workings of unions, as well as the priorities of its members, the spies, and the mine companies. Largely underutilized by historians preoccupied by either economic history or labor union formation, they remain a valuable source that requires careful use. Common sense is enough to guide the wary reader, once they keep in mind the fact that a private detective had a vested interest in continuing his fieldwork and thus could sometimes include exaggerated reports. For example, after sending several reports to mine owners about low attendance at union meetings, the Thiel agent noted one member saying "that will make the mine owners think that the union is to be abandoned, they will relax their vigilance and it will then be possible for us to accomplish the end we have in view."[126] The detective might have embellished this account, slyly promoting eternal vigilance on the mine owners' part out of his own financial self-interest. Similarly, some other phrases ring hollow, such as this statement by Ahern: "You will see that we will be in control of everything in Leadville a short time after operations are begun" and this comment by James McGowen at another meeting: "and we will show them sometime that they are right in being afraid of us."[127] Even if the detective was not incentivized to exaggerate and instead accurately reported statements, there were certainly many idle boasts from men who believed they were chatting in private. Reporting such bravado as factual statements by the detective was doubtless scaremongering.

These firsthand accounts also emphasize the importance of ethnic identity in nineteenth-century mining towns. The detectives possibly sought to expose exploitable internal divisions among workers, but such intent does not mean these details are less important in revealing the story of how people lived in mining towns throughout the US. Stratified by class and occupation, people in the American West were also sharply defined by ethnic group. Their identities determined where they lived and attended church, the fraternities they joined, and the bars they drank at; but few contemporaneous records provide their hidden conversations. The language in these reports, cursing and all, is probably accurate, since it lent authenticity to the reports when they reflected the language the miners themselves used. The reports also list union members, details of union sympathizers in the town, and membership fees or donations to the Miners' Union. A list of union members reveals that over 70 percent of those named were Irish-American.[128] While not a comprehensive or even completely trustworthy figure for total Irish-American CCMU membership, which was revealed as 1,200 in one meeting, it indicates the central importance of Irish membership to the union.[129]

As news from CCMU meetings continued to make its way into the newspaper, the miners slowly realized that this information was not idle gossip overheard from union men. They then grasped the reality that the union was infiltrated by spies. Charles Knuckey stated that he believed there was likely more than one spy, and the detective dutifully noted in the report that "it is someone who is well posted regarding the union and its members."[130] The leadership agreed, and notice went out for members to be on the lookout for any suspicious activity.[131] Over the next few days, the union focused its attention on the forthcoming workers' picnic, collecting prizes from local business and encouraging workers, their families, and the wider public to attend the festivities. The operative also reported which businesses supported the picnic with money and prizes: "Rocking chair for winner of girl's race, by F. Smith 128, E. 6th St."[132] To understand the union's and the mine company's preoccupation with the picnic, the event needs to be understood as more than an outing. It represented a symbol of solidarity among the workers as well as between workers and the public. It also symbolized defiance against company control of the town, the region, and its populace.

The uncertainty over public support for the union wore heavily on members the day before the event. Mine owners posted intimidating notices forbidding workers to go to the picnic under threat of being fired.[133] CCMU member Philip Finley said, "Even if that is the case it won't do the union any harm, as it will open the eyes of the public and show them that the mine owners are down on us and then we will have the full support of the community."[134] John Dooley, a union organizer who lived at 709 E 6th Street, also noted that persecution by the mine company would probably backfire:

"There certainly won't be any harm done to the union if they discharge some of the boys for going to the picnic, as that would bring a whole lot of the boys back to the union that are now on the outside."[135]

Although they demonstrated strategic awareness of the importance of public opinion and the ways unfair dismissal could work in their favor, union members also demanded the opportunity to avenge such practices and to defend themselves against mine company guards. They hoped to organize their rifle club soon, but as the CCMU was affiliated with the Western Federation of Miners (WFM), they had to request approval from the executive board. Dooley mentioned, "If they do not do something about that soon they will find that the boys will attend to the matter themselves."[136] While the WFM had a reputation for radicalism, it seems that the WFM executive board was trying to restrain its local members from engaging in violent actions. Adding more guns to an already tense situation could lead to disaster and might enable the mine owners to provoke an armed conflict that the miners would surely lose, given the tradition of state and federal authorities siding with the companies.

This militant edge was not limited to the Irish element in the union. Later in the day before the picnic, the Thiel operative met with Knuckey (an Englishman), Evan Owens (a Welshman), and Edward Moyle (a Cornishman), who stated their determination to go to the picnic and have a good time. Moyle added that he hoped to see shooting added to the list of sports, as "it would show the d+++d mine owners just what they might expect when they refuse to accede to our demands the next time we make them."[137] Moyle warned Owens, "That would not do at all because we don't want the G+d d+++d mine owners to know that we are trying to organize a rifle club in the future," but Owens replied, "D+++n the mine owners. The S+++s of B+++s will know all about the rifle club anyway, and I don't think it will do any harm to let them know just what they can expect from the boys."[138]

Union members estimated attendance at the picnic at somewhere between 3,000 and 4,000 people, of whom 1,500 were miners.[139] The operative spoke to one miner at the Redneck Saloon, James Amburn, who was reported as saying, "I think the crowd [that] is here today ought to show the mine owners that we have the sympathy of the public and if they are wise they will treat us a d++n sight better than they have been doing."[140] The saloonkeeper, Simon Rogers, added, "There is no doubt that those notice[s] did stop lots of the men who at work for them from attending the picnic, and therefore we can count on nearly every man who went out to it today as being members of the union, or in sympathy with it."[141] The event proved an important morale boost, dispelling some of the prevailing gloom and sharply demonstrating the family and community aspect of the workers' struggle.

The following day, the *Herald Democrat* infuriated the union by reporting that only 1,000 people attended. At a meeting of union members, reaction included this statement: "D+++n the H-d. It is down on the union and tries to give it a black eye every time."[142] At the same meeting, Mike Sexton added an interesting assessment about the newspaper's perception of the union: "You may say what you please, but you know that they think we are devils and S+++s of B+++s."[143] Their attention turned to the spy, and Sexton mentioned that he was watching the mine owner, John Campion, whenever he came to town to see if he spoke to anyone.[144] Joe Sowa piped up and said the mine owners "think they can do just as they please with the poor miners, and they will never learn differently unless the boys get together and give them another dose of Coronado." This referred to an attack on the Coronado Mine by union members nicknamed "the Regulators" on September 21, 1896. In response to this bravado, miner Ernest Nicholas snapped, "Joe shut your mouth and do not make a d++n fool of yourself."[145] Perhaps Nicholas felt violence was counterproductive, or perhaps he was more sharply aware than Sowa that there was a spy in their midst.

At this meeting, John T. Gallagher mentioned that some mines were opening downtown, and he "would not be surprised if they would skip in a lot of scabs from Missouri."[146] Sowa, undeterred by Nicholas's admonition, said, "They won't work their mines with scabs from Missouri, because we will attend to that matter. Union miners will work those mines or the mines won't work."[147] Amburn turned the conversation back to the picnic, noting "I did not see as many of the d++d Cornish at the picnic as I expected," but Ahern replied, sticking up for the other ethnic group, "There were quite a number of Cornish present and altogether I think we made a grand success of it."[148]

The following summer, several men discussed the unusual case of a mine superintendent working at the Johnny Mine who had once been run out of Meaderville, on the outskirts of Butte, Montana, by the powerful Butte Miners' Union. Apparently anti-union, the men recounted amusingly that the superintendent had discharged many Cornish, kept the Irish, and hired a lot of Swedish and Austrian union members: "I wonder what Campion and Mudd will say when they find out that he is hiring union men in place of the bloody scabs he is discharging."[149] Two miners, Mike Trainor and Paddy Connors, who were working in Leadville because of the strike in "the Coeur d'Alene country," also joined the local union meetings.[150] Much of the discussion centered on Trainor's and Connors's appeals and which occupational groups the union was going to include in its demands for three dollars a day. In response to whether the eight-hour day was as efficient as the ten-hour day, a man named Turnbull replied, "Oh hell yes. We hoist just as much in eight hours as we formerly hoisted in ten. I know that we can make it pay and the rest of the s++s of b++++s operating big mines in the camp can make it pay, and if they don't pay $3.00 to the miners for eight hours work they are looking for trouble."[151] "Trouble" was vague enough to imply a variety of violent and nonviolent protests.

Another class of worker in the mines were trammers, who transported the material along the tram lines out of the mines. They were classed as unskilled labor; as such, "the work of a trammer is not worth as much as that of a miner."[152] For the sake of worker unity, most of the men stayed quiet when dealing with questions of labor classification, even when they believed they would get a better deal on their own. As Pat O'Hara argued, "The miners ought to look out for their own interests."[153] He also claimed that the ethnic divisions had split their labor organization into "two unions, one composed of Irish, Welsh and Swedes and the other of Austrians," but in spite of his assessment of the fractures within the trade union, he still firmly believed "the men would be able to take care of themselves."[154] The shifting relationships between ethnic groups meant that alliances across such divisions proved difficult but not impossible.

After all the information leaks, members of the union were able to guess the Thiel agent's true identity, and in late July someone broke into his room. It is unclear from the reports whether it was the person who broke in or a policeman investigating the robbery who spotted incriminating evidence, but regardless, Captain Cunningham and deputy Sheriff Jack Bowman spoke to people on the street telling them the spy was "a s+n of b+++h and a would-be detective and you fellows had better look out for him, as you can be sure he is in the pay of John F. Campion and working against you."[155] At the next union meeting, the agent was surrounded and shoved around the room as the men accused him of being a detective. He was warned to leave the camp before morning or he would regret it.[156] The detective decided to stay, and over the next few days he protested his innocence, pointing to the *Herald Democrat* articles that labeled him a dangerous radical and telling the others that he was surprised that they would believe Bowman, since "the Sheriff's office had never had any use for any of them."[157] James Amburn began to voice his doubts, believing the men may have acted "a little hasty," and by the end of the evening, the operative had also managed to convince Mike Sullivan.[158] Although the miners threatened and blustered, it amounted to little more than macho posturing—it was rare that they resorted to serious violence even when they blamed the right man.

The Ethnic Hue of Hiring Cards

By the end of the nineteenth century, most mining companies had developed some form of employee hiring card. From 1900 through 1906, the Portland Gold Mine Company's card system kept a list of all its workers. Like other hiring card systems, it has its origins in blacklists that were used to keep unions out of company mines and intimidate workers from joining or establishing a union. These cards continued to be used as such, with suspected membership in the WFM resulting in firing. "H.g." on the reasons (for discharge) line meant high-grading or ore stealing, and "n.g." meant no good, which covered any other reason. The information on a selection of cards reveals interesting details about Irish miners and highlights their mobility. Irishman Eugene Murphy worked at the Copper Queen Mine in Bisbee, Arizona, for a year before going to Victor, Colorado, to work for the Portland Gold Mine Company.[159] Daniel Comerford's last place of employment was England.[160] Twenty-four-year-old Irish-born Michael Clair lived in Cripple Creek and worked as a mucker, giving his last workplace as the Homestake Mine, South Dakota.[161] He left there because the mine "shut down," and he stopped working for the Portland Gold Mine Company after half a year because he was injured on July 27, 1905. His emergency contact was a relative, Thomas Clair, whose address was 131 Walnut St., Hartford, Connecticut. James Sweeney's emergency contact was Mrs. Margaret Smart in Lewistown, Pennsylvania, highlighting the mobility of the workers and the breadth of locations linked to this one mining site.[162]

The cards sometimes contained irregular details. Thomas Conway worked as a machinist helper, with ten years' experience.[163] Hired on January 3, 1905, he was laid off or quit (both are listed on the form) on January 14, 1905. He returned to work as a mucker on February 1 and was discharged a week later because he was "no good."[164] The Irishman's age was given as twenty-six when hired in January, while on the second form his age was twenty-five. This rapid turnover of workers was fairly typical, depending on the company. Pat Hurley, another Irish mucker with thirteen years' experience, worked for the company less than a week before he failed to report for work.[165] J. J. Hand, an Irish shift boss, earned four dollars a day—a dollar more than the muckers—and was discharged after three months for being "drunk on shift."[166] James Battell worked half a month as a trammer and was fired because he was "no good" and "lazy."[167] James O'Connor quit to become a farmer in Greeley, Nebraska, while nineteen-year-old John J. O'Brien worked a year-and-a-half as a mucker, then "quit to go to school."[168] Irish machine worker Pat Sullivan simply decided not to show up to work one day, perhaps quitting the camp quickly and quietly in search of better work.[169]

Evidence of ethnic discrimination can be seen in the different attitudes toward Irish and English workers. Philip Pascoe's employee card is typical. The Cornishman left "to better my condition" after three months spent as a mucker, and the remark his boss made on the card states that he was "a good man."[170] No similar positive statements are seen on Irish hiring cards. Irishman O. O. Philips was hired and fired the same day because of an unknown dispute, and the boss who fired him managed the get in one last insult with a remark on the hiring card that read "left mother too soon."[171] Although the company used the term *nationality* in the employee cards, in practice it actually meant ethnic identity. Charles E. O'Conner worked for the company for two days in October 1905 before leaving because his wife was sick (his card also reads "no good," further hinting at possible anti-Irish bias); his nationality reads "Irish-Am."[172] Whether it was O'Connor who said this when he was interviewed for the job or the person interviewing him who noted it on the hiring card is largely irrelevant because regardless of who wrote it, the information was understood by both parties and the ethnic identity it revealed remained an important defining feature that affected the daily life of the miner and his relationship with his employer.

While the Irish miners in Leadville struggled against the nativist forces arrayed against them at a local level, these forces were supported by the dangerous and powerful combination of state government agents and vigilantes. In the case of Leadville, the scales were tipped particularly heavily against the Irish workers. The attacks on the Coronado Mine and the Emmet Mine during the 1896 strike reveal miners' increasing limitations on protesting against mine owners. As Walsh succinctly put it: "Striking miners never had the chance to prove themselves as peaceful activists because at every turn their solidarity was viewed as an insurrection against law and order . . . For [the mine owners] there was no such thing as a lawful and peaceful display of strength from poor immigrant Irish people."[173] As in other regions, Irish miners found a combination of nativist business and government forces willing to use legal and illegal means to intimidate workers. While organized labor in Leadville, and, by extension, the Irish, suffered defeat after defeat further to the north, an industrial city was beginning to take shape that would become a haven for the Irish, even more so than Virginia City. In Butte, Montana, a mining town was blossoming under the control of an Irishman, and it was there that they would finally be able to establish their strongest presence in the American West. Movement to the Promised Land of opportunity started in earnest, and this onward migration was most aptly represented by Michael Mooney—the impromptu leader of the 1880 strike—who was reported to be living in Butte with his family in the 1900 census.[174]

5

In Search of Respect

> In the mining region of the Rocky Mountains . . . I could get away from a crowded labor market, with the wages a mere pittance, just enough to maintain a miserable existence; it would open to me a new field.[1]

The wandering Irish chased economic frontiers of opportunity throughout the American West. Networks of communications and urban nodes where they established themselves and developed their communities were the key to prosperity in the US and a way of strengthening their position at the bargaining table with and against other groups, as they did with the union in Leadville. In Idaho, the transition between phases of mining development followed much the same course it had in other states. However, in the nineteenth century, the violent confrontation between labor and business was even more sharply defined in Idaho than in Colorado's strikes. The blueprint for the later bloody Colorado Labor Wars of 1903 and 1904 emerged in Idaho where the opportunities and hopes of earlier years, expressed in the letters of James Mullany, contrasted with years of violent corporate repression. Still,

https://doi.org/10.5876/9781646422517.c005

Mullany experienced the roots of these future corporate battles in the adversarialism faced in the trading town of McMinnville, Oregon, which Mullany abandoned, and explicitly in the records of a mining company in the Coeur d'Alene mining district in northern Idaho.

The shallow panning of placer miners quickly gave way to the hard-rock miners who drove shafts into the towering spires of the Rockies. Between 1861 and 1866, Idaho's rich gold mines produced a staggering 2.5 million ounces of gold worth $50.6 million, amounting to over 19 percent of gold production in the United States during that period.[2] The occupational demographics of the territory of Idaho in 1870 reflected the importance of mining there. One-third of native-born Americans worked in mining.[3] For the Irish population, the figure was two-thirds. The quote that opens this chapter was written by Thomas Higgins in a letter to his relatives wherein he passionately expounded that this "new field" offered less competition from other workers, and so he felt that Idaho represented his opportunity to earn a stable income. This was not a romantic vision of the West, it was purely an economic calculation. Even more starkly, he felt this was his last opportunity: "Besides it was the only thing left open to me, as I had tried to find something I was capable of doing that would be reasonably profitable in old settled communities, but without success."[4] More than in the breadth of the West, it was in the high wages that he found a dignity and fulfillment that had eluded him elsewhere, whether he worked as a prospector or a hard-rock miner.

Prospectors migrated further east after the gold deposits closest to the surface played out, leading to the next large gold rush in the Black Hills of Dakota Territory, a sacred site to Indians. Traces of prospectors' nationalities linger in place names such as Wild Irishman Gulch, named after a camp of boisterous Irishmen, and another gulch east of Sheridan, South Dakota, that earned the name Dead Irishman—unsurprisingly, because an Irish miner fell to his death there.[5] The end of the gold-rush movement caused some miners who remained in Idaho to change their occupation, sometimes leading them to become skilled hard-rock miners, while others left mining by using savings from their earlier days. They opened businesses such as saloons, hotels, stores, and farms that supported the growing mining towns. On rare occasions the reverse occurred, when Irishmen who had traveled west to earn money in a stable job found themselves tempted by the high wages of mining and the chance of striking it big.[6]

One such figure was James Mullany, an early prospector to the area, who was born to Irish parents in New York. He traveled west and eventually settled in McMinnville, on the Oregon Trail between Oregon City and Portland, where he decided to open a furniture store. He wrote to his sister on September 11, 1858, telling her that business was hectic: "I have got work enough to work six hands from now until winter if I could only get them but I have only got one hand now I pay him $4 dols per day."[7] He viewed the labor shortage in the exact opposite way from Higgins, noting that the lack of workers combined with high wages hindered his store's growth. Mullany had invested heavily in his business, leaving him with little cash reserve, and he asked his sister to let him know the situation at home regarding "money matters."[8] Obviously, his family hoped their son would send remittances to ease their financial difficulties on the East Coast; detecting their sore tone, he tried to alleviate their worries and explain his situation: "I should send you some [money] this fall but I will just tell you how it stands."[9] Mullany pointed out the complexities of the business and his future hopes for a better return from his investment: "I assure you it has troubled me many a time since I left home that I could not help you some for I also assure I had the will if I had the means but I hope soon to be placed in better circumstances."[10] The dual emotions of guilt and hope were some of the most common themes of Irish emigrant letters and point to the economic insecurity of most of the migrants.

Mullany found himself isolated in the newly formed trading town of McMinnville: "I have not seen a man woman or child that I have ever known or heard tell of since I came here nor have I not seen an Irishman or woman for the last five months and I will bet that is more than any of you can say."[11] The next time he wrote to his sister, in August 1860, he gave more details about why he felt so alone in the trading town, stating that the exact distance to the nearest Catholic Church was thirty-five miles: "I intended going last Christmas but I was sick at the time and I actually believe I have not seen two Catholics in two years. [There] is a strong prejudice against them here on account of the people here thinking it was the Priests that caused the Indian war three or four years ago."[12] He was referring to the Whitman Massacre in 1847, when a Protestant mission was attacked by Indians and wiped out. A Catholic missionary, Father Brouillet, came across the massacre and warned others to leave the vicinity—one of whom was Reverend Spalding, a Protestant preacher attached to the now-destroyed mission. Spalding, reduced to a paranoid wreck because of the loss of his friends, never forgave the Catholic priest for saving his life and wrote scathing polemics arguing that the Catholic clergy, fearful of the spread of Protestantism and hoping to reinstate British control of the region, had compelled the Indians to attack the mission. Brouillet argued vehemently that this was not true; however, it was 1901 before Spalding's lies were debunked comprehensively by a historian.[13]

The repercussions of Spalding's actions were directly felt in the following years by Catholics living in the region, such as Mullany. The venom was so intense that he hid his religious and national identity from everyone he lived with: "There is not one here that know[s] what I am and I intend that they shall not as long as I cannot go to mass."[14] Mullany was a second-generation Irishman and had grown up in New York, so it was easier for him to blend in with the Protestant American population, but he did not hope to do so for long: "I hear [there] is a church where I am going to this winter at least I hope there is for I never want to be as long again without [at]tending church and I will not if god [*sic*] spares me my life and health."[15] The Catholic Church mentioned in his letter was probably in Portland, which had a sizable Irish population.

Mullany's isolation from other Irish Catholics in McMinnville raises two issues. First, even in the West, the Irish were concentrated for mutual assistance, company, and protection; Mullany understood that he was isolated and exceptional. Second, it appears that some towns developed a distinctly hostile attitude toward Irish Catholics; in response, Irish miners seem to have avoided those towns, aware that they were persona non grata. If McMinnville was one such town, that could explain why Mullany failed to meet many Irish or other Catholic people. Although Mullany regretted not having attended a Catholic mass for so long, he seemed reluctant to move because of his business and had no immediate fears of being shunned or exiled. He responded to the hostility of other townspeople by hiding his identity, and he warned his sister to keep his Catholicism a secret: "When you write to Mrs. Yearywin I wish you would not say anything about church as she has asked me several times what church my folks belong to and I would not tell her and I wish you not say anything about it at least while I am here."[16] It was possible that Mullany feared she might inadvertently reveal his Catholicism to others, exposing him to a loss of business and possible danger from anti-Catholic sentiment in the town.

In the same letter Mullany told his sister "times here are very dull," reporting that morale in the town was at a low ebb: "I think I heard more complaining of hard times than [there] is here at present."[17] Indeed, the citizens of McMinnville had realized it was a dying town. McMinnville was originally founded as a supply center on the old Oregon Trail, but other towns had emerged that were closer to the new centers of activity—the blossoming gold mine camps in Idaho. When John Mullany wrote to his sister three months later, it was from Walla Walla, Washington Territory, and he informed her that he had sold his business and joined a group of friends on a mining expedition through the Cascade Mountains to Orofino and into Idaho. Instead of the fear that clouded Mullany's letters from McMinnville, his excitement and sense of adventure jump from the page as he describes the sights: "We would be down in little gorges of the mountain so that the lofty peaks would apparently touch the [sky] on either side of us and I think I never witnessed anything as sublime in all my life."[18] He recounted stories of Indians massacring settlers and the most recent gold rush news: "There is considerable excitement here about gold mines about 150 miles from here this place is on a line to all northern mines and it makes things verry [*sic*] lively here."[19] Yet these attacks did not trouble him, possibly because he was with a group of friends on his adventure or perhaps because he did not see the Indians as a threat the way Protestant Americans did.

Back East, it seems his family was experiencing difficulties and was unconvinced by his cheerful optimism. He wrote to tell them he was not ignoring their problems: "I assure you I have been more sincerely uneasy since you wrote before on account of the way things were when you wrote."[20] It is clear that his sister remained upset by his change in occupation and, perhaps more important, by the fact that he had failed to return home after selling his business.[21] He told her in a boastful tone, "We worked about 4½ days this week and took out (179) one hundred and seventy-nine dols [*sic*]."[22] He had done the same in an earlier letter, in September 1858, when he wrote home, "You tell Cousin Carey if he thinks I never amount anything to come out here this summer it would open his eyes."[23] This effort to prove himself to his family stemmed from both his guilt over remaining in the West and his attempt to justify to himself that his family's predicament would be worth it in the long term. His talk of wages lacked the context for his family to understand it: what were his costs, what amount was enough, and what was his timetable for returning? These unanswered questions haunted his family, and the venture may represent an interesting example of the "false impressions" that Father O'Hanlon, in *Irish Emigrant's Guide*, warned readers often appeared in emigrant letters.[24]

Mullany used his investments and the cost of travel to try to explain his delay in returning to Philadelphia: "If I had been situated so I could leave when I got your letter I should have started right home but I had a job of between five and six hundred near half completed so I could not go so by the time I got through."[25] He went on to suggest that his change to mining stemmed from the temptation of friends passing through McMinnville: "Some of my friends came up from Oregon coming to the mines [in Idaho] and indused [*sic*] me to postpone going home and come to the Mines and as you see I acted on their advice and have become a Miner."[26] His loneliness, boredom, and fear in McMinnville contributed to his decision; but it was also possible that Mullany was searching for an excuse not to return home. He certainly felt somewhat guilty for not returning to his relatives, which obligated him to write such excuses to his sister reaffirming that his loyalty remained with his family.

Near the end of the letter, he told his sister that he would not return home in the immediate future, writing that it would be "madness" with so little to show for his hard work after being away for so long: "I suppose I could sell out and rais [*sic*] $1000 but I do not think it would ever pay me for what I have gone through and if by remaining another season and make as much more I think I had better do it."[27] Mullany followed the mirage of prosperity embedded in the idea of prospecting; one more season and you might strike the mother lode, one more season and you could retire. While some failed in their perpetual search for the mother lode, others quit with a tidy sum to buy a farm or start a business. Some, lured by the possibility of anonymity, simply disappeared into the American West to escape their wives or families back East or in Ireland. That possibility existed, and Mullany could have simply walked away from his past, practicing the quintessentially American idea of reinventing yourself on the frontier. He did not, however, and the multi-generational loyalty the Irish in America had toward their ethnicity, religion, and family proved to be an overpowering bond. Evidence such as this suggests that these connections generally prevailed, and it was rare that the Irish concealed those ties or left them altogether.

Irishmen sometimes drifted out of their major enclaves like Mullany did, but it was rarer for Irish women, who more often remained centered in established urban pockets. Demographics show females' reluctance to settle in the West, and Irish women proved especially wary of the mining frontier compared to other female ethnic groups. The example of Boise County, Idaho, in 1880, where Irishmen outnumbered women 285 to 37, represented a pattern replicated throughout the American West.[28] Only the Chinese population had a more imbalanced gender ratio, and theirs was a unique example as women were brought over primarily as sex workers to serve the urban centers of the West Coast.[29]

While the Irish gravitated to mining, they were also found in every occupation in Idaho. They made up 25 percent of southern Idaho's shoemakers, grocers, saloon keepers, butchers, and livery operators in 1870.[30] The same percentage of southern Idaho's miners were Irish-born in 1870 and 1880, but only 10 percent of those who classed their occupation as laborers were Irish-born.[31] These figures suggest that it was the more skilled Irish who journeyed west and also that placer miners, generally less skilled than their hard-rock counterparts, filtered into other occupations as the easy gold played out and mining developed into an increasingly industrial activity. Despite the catchall hard-rock miner title, some certainly gained their mining experience elsewhere, such as in anthracite and bituminous coal mines.

Individual accounts illustrate the early role migratory Irish miners played in the settlement of Idaho. John A. O'Farrell was born in County Tyrone, Ireland, and mined in California. After marrying Irishwoman Mary Anne Chapman, he arrived in the newly established Boise City in 1863, then little more than some ramshackle wooden huts. The couple was credited with building "the first log cabin" in the town and John left mining to become a blacksmith, deciding to move into a more stable and safe occupation.[32] Others shifted occupation but remained involved in the mining process, for example, by becoming stamp mill owners. James O'Neal's ten-stamp mill near Boise City could process sixteen tons of ore a day, and he made a small fortune when he sold out in 1866 for $27,500.[33] The Irish-born Mike Leary chose a different path, again tied to mining, and he became a landowner of sorts. After he arrived in the Boise Basin in 1865, he spent years accumulating over 1,000 acres of claims, and by 1898 he had earned the nickname "the Placer King of Idaho." Leary managed his many claims by mining the miners.

He leased his land to them for 55 percent of their findings. Many who took up this offer were Chinese; while the cost was steep, this arrangement protected them from predatory anti-Chinese groups.[34]

In the Boise Basin, the strength of the Chinese community enabled it to claim fuller participation in society than was possible in many other parts of the American West or the country as a whole. In one interesting case, Anthony McBride, an Irish Civil War veteran, fatally shot a Chinese man. After he was convicted in court and publicly hanged, the local paper wrote, "If only he had killed a white man, he might have been acquitted," highlighting the influence of the local Chinese population.[35] Another commentator provided further details on his lack of social capital, writing that McBride "had few friends and no money, otherwise he would have escaped punishment," suggesting that both the individual's ethnic group and his position in society played vital roles in his treatment by local authorities.[36]

Attempts by the Idaho legislature to impose special taxes on the Chinese support this point; the taxes proved spectacularly unsuccessful at either pressuring the Chinese to leave the state or creating an additional stream of revenue for the territory.[37] Although overwhelmingly male, when the Chinese did settle in mining towns rather than repatriate, the increased contact between them and other ethnic groups greatly improved their standing and access to government.[38] In Idaho City, their influence was represented through interethnic demonstrations of solidarity and respect, such as an Irish brass band leading the funeral procession for a Chinese man through town to the grave site. As historian Liping Zhu points out, "Although this mixing of cultures was unusual, nobody thought it too improper."[39]

Mining development in Idaho stalled during the Panic of 1873, which was followed by a recession and then the Indian War (1877–1879)—all of which added to the disruption throughout the region. However, the arrival of widespread railroad construction in 1878 heralded the advance to large-scale mining in Idaho. More permanent industrial towns developed in the northern part of the territory, and by the 1880s, Irish mining populations had become firmly established in the Coeur d'Alene mining district—particularly in the towns of Wallace, Wardner, Kellogg, and Burke. A letter from a priest assigned from the Catholic mission in Cataldo to the town of Burke stated that "fully three fourths of the people in this city are from the Emerald Isle," and others in northern Idaho called people from the town "Burke Irish."[40]

The famous union leader Big Bill Haywood noted the solidarity illustrated by Irishwoman Margaret Fox's boardinghouse in the town, which was as "well known to most of the miners of the West. I have heard many stories of her warm-heartedness. A miner coming to Burke was always welcome to a meal at her place. She caused much amusement among her boarders. One time when new cabbage was just in, the old lady had cooked up a lot. It was just what the miners were longing for and they kept asking for more. Bringing up the last plate full, she said: 'Take that, ye sons of batches, and I'll bring yez a bale of hay in the marnin'.'"[41] Her sarcastic wit and hospitality endeared her to the many workers who boarded at her business.

The priest positively referenced the "strong river of faith" among the people there, writing that the Irish were "sincere friends" and "generous benefactors to the Catholic Church."[42] Indeed, one former Irish miner decided to join the religious orders; upon being accepted, "he celebrated his 'rich find' in true genuine miner's style by drinking poteen to excess."[43] Drinking to excess proved problematic in other aspects of life, including religious practices. Alcohol was widely available, usually in the form of whiskey or rye, and its consumption was doubtless a factor in many violent and work-related incidents. One priest in Coeur d'Alene decided to send his congregants away without receiving Holy Communion at mass if their breath was, in his words, "strong enough to drive a nail."[44] Although this small Irish community sustained his mission in northern Idaho, the priest felt he had to draw a line somewhere, for the sake of respectability. This mind-set reflected a deeper cultural collision that was occurring, one that was arising back in Ireland simultaneous to the one in the American West. At home and abroad, a new generation of young clergy, representing a revitalized Tridentine Church, sought to hone (primarily through education) the mentality of Irish Catholics by transforming "rough" working-class cultures into ones with more "respectable" manners.

Throughout the American West, the "upstanding" element felt the uncomfortably close proximity between the respectable and the scandalous as a direct taunt: "The priest and the saloon keeper jostle each other on the sidewalks, and the gentleman's wife must walk around the trail of the courtesan who lives next door, and does her shopping at the same counter."[45] From the perspective of the miners who had just arrived, their drunkenness was neither an effort to challenge the attitudes of middle-class citizens nor at odds with their faith and their right to attend the most sacred Catholic event: the mass. Their attendance was the key point. They understood their religion in terms that were different from those of either the lace-curtain middle-class Catholics from which most clergy were drawn or the puritanical spiritualism of the Protestants. The most important part of their religious experience was their presence, an affirmation of who they were in wholesale imperviousness to surrounding circumstance—be that the geographic location in the form of the Rocky Mountains, danger from nativists, or their physical inebriation. They were Catholic, so they were there; what did being drunk have to do with it?[46] Whereas in larger cities prostitutes, drunks, and the poor existed out of the sight of respectable citizens, mining towns revealed the full cosmopolitan spectrum of American society in a narrow enough vision to see it all on one or two streets. Instead of accommodating these realities of human existence and perhaps offering constructive assistance, this closeness propelled the growth of the Progressive and Prohibitionist movements throughout the American West. One effect of these movements was to try to legislate undesirable traits out of existence. This effort failed as social and public health policy, but it did the drive the saloons and the sex workers out of sight and underground, which the respectable element saw as a success.

The Coeur d'Alene Mining District

Not all confrontations between these two elements of western society took place along distinct class and cultural lines. Still, they proved useful for business leaders who utilized the distinctions between townspeople and workers or between one ethnic group and another for their own ends, as seen in the cases of Randsburg, California, and Leadville, Colorado. Whereas the former provides an example of the slipping Irish and union political influence and the latter of the ethnic fissures within the union movement, the Coeur d'Alene mining district represents another distinct showcase where the company had two intertwined goals driving its mission—the Progressive-tinged hope of creating a more respectable workforce controlled by the company and the expulsion of the Irish community from the area.

The Bunker Hill Company originated from several simultaneous claims on September 10, 1885. One claimant was Irishman Philip O'Rourke, but Noah S. Kellogg remains the best remembered because he told the story that he was out prospecting when his donkey wandered off in the night. When he found the animal in the morning, he noticed an outcrop of silver and lead nearby. The nearby town of Kellogg still proudly proclaims the story to visitors on a large billboard at the town's entrance: "The town discovered by an ass, and inhabited by its descendants." The disputed claim resulted in a rowdy trial and rapidly descended into a farcical entertainment spectacle for the miners, who one observer noted were mostly Irish.[47] Later, in the 1910s, the claim became the largest lead- and zinc-producing mine in the US, emerging as one of the few fully integrated mining and smelting operations in the American West.[48] A comparable mining location was the copper metropolis of Butte, Montana, and its smelter sister-town, Anaconda. Yet the difference between the two areas could not be starker: Butte had thirty-six years of uninterrupted labor peace between its largest mining company and its largest miners' union, while Coeur d'Alene gained notoriety for illegal internment in bullpens and the "dynamite express."

The Bunker Hill Mine faced initial difficulties due to its relatively isolated location, and the first shipment of ore from the mine in 1886 was forced to take a convoluted route through the Rocky Mountains, loading and unloading five times before reaching the newly constructed Northern Pacific Railroad ten miles away. That same year, Simeon Reed purchased the mine, and he created the Bunker Hill and Sullivan Mining and Concentrating Company (BHS) in 1887 (figure 5.1). He also hired a professional mine manager, John Hays Hammond. Hammond was one of an emerging class of mine managers who were university-educated, which enabled them to bypass a lifetime in the mines and created a sharp divide between workers and management within the company. This new elitist group arose as a solidification of the established Protestant Anglo-American upper class.[49] Expansion followed quickly on the heels of the capitalization, and in 1889 businessmen Daniel Chase Corbin and Samuel T. Hauser invested in two rail lines through the Coeur d'Alene district, connecting the area to the regional hub of Spokane, Washington. In 1890, the mine produced a remarkable $3.5 million in ore and concentrates.[50]

In 1891, the expanding workforce formed the Miners' Union of the Coeur d'Alenes, combining several smaller unions in the region.[51] Soon after, local mine owners formed their own organization, the Mine Owners' Protective Association (MOA). Other syndicates had been created between mine owners earlier in the nineteenth century, notably in the anthracite region of eastern

Figure 5.1. The Bunker Hill and Sullivan Mines, Kellogg, Idaho. Bunker Hill Company Records 1887–1984, Box 109, MG 367, Special Collections and Archives, University of Idaho Library, Moscow.

Pennsylvania, but they were hidden from the public and government. The MOA, in contrast, was public, which was a statement of intent in itself.[52] The board in control of the MOA had one notable member, the Irish-American Charles Sweeny.

Sweeny had been a miner in his youth and worked in Virginia City until a workplace injury persuaded him to confine his risks to the mercantile and investment sphere. Trailing a prospecting party to northern Idaho, he purchased land, property, and mine claims in the Wardner area, in particular the Last Chance claim located on the slopes above the town springing into existence.[53] Sweeny needed funds for his mine operations and looked to ingratiate himself to a wealthy New York businessman, Jeremiah O'Connor, by exploiting their shared Irish Catholic heritage. In one example he wrote to O'Connor describing the death of a business partner in subtle terms that were familiar to a Catholic: "He was the very best disposition . . . everything was fixed up in a very satisfactory manner during his last illness," implying that the spiritual as well as business issues were tended to before his death.[54] O'Connor's enthusiasm for the young investor proved so contagious that he convinced Father O'Dwyer, his own parish priest in Elmira, New York, to purchase Sweeny's mine stocks.[55] Sweeny's valuation of their friendship and heritage was revealed when he eventually squeezed O'Connor out of the mining business. This led to years of litigation, resolved only by O'Connor's death in 1912.[56] In contrast to other Irish-Americans, Sweeny's work life, personal life, and identity were all negotiable elements in the most direct sense in his true motive in life: the pursuit of wealth.

The MOA believed the unions were weakest in the winter and thus acted on New Year's Day 1892 by closing all the mines in the region, thereby throwing all the miners out of work. When they reopened in April, the companies seized the opportunity to lower worker pay, unexpectedly triggering a strike.[57] The union found strength from its ally in Montana, the Irish-dominated Butte Miners' Union, which offered a $5,000 loan and voted to assess its 6,000 members $5 a month per member to support the strike as long as needed.[58] Now, the union set out its own demands: first, reinstate the general wage of $3.50 a day for every underground worker (a major step for workers' solidarity, since it abolished the difference between a mucker and a miner), and second, end the compulsory payment of $1 a month for company insurance. Workers were attempting to wrest control of their healthcare from the company. Part of this was based on religious grounds. The large Catholic workforce wanted to patronize their own Catholic hospital; but, more important, the company had repeatedly failed to honor its own mandated worker insurance coverage (figure 5.2). It stubbornly limited or refused care and compensation to injured miners and their bereaved families. Instead, the company used the mandatory insurance payment as another revenue stream. When a shift boss named Patrick Curran died in a mine accident along with two other workers, the management ordered that his widow be paid as little as possible, ruthlessly exploiting her vulnerability in the days after the funeral. In thinly veiled management-speak they wrote, "She held out for more than we thought we should pay."[59] Management finally offered her a paltry sum after ensuring that following the payout the company would "not be assuming responsibility in the cases of the others that were injured at the same time."[60] The company responded to the public outrage through a sympathetic new Republican-leaning publication, the *Coeur d'Alene Press*, which stated dismissively that compared to other mine companies, only the BHS offered sufficient protection for its workers.[61]

Figure 5.2. Bunker Hill and Sullivan miners next to a compressed air drill, nicknamed "the widowmaker" due to the dust-filled air the miners had to breathe in. Undated, approximately 1890s. Box 109, BHM.

The strike dragged on, and the BHS started hiring scabs. Sweeny, on behalf of the MOA, imported armed guards from Washington State and miners from Michigan who were from Finland, Denmark, Poland, Austria, and Sweden (specifically not from Ireland or Italy), whom historian John Fahey described as "the least stable element" because "they might be expected to move on."[62] Union attempts to reach out to these miners or to block their arrival were hampered by legal injunctions issued by a Judge Beatty, who had been elected thanks to company support. On July 11, 1892, a firefight erupted at the Gem Mine and the Frisco Mill in which three union members, a Pinkerton detective, and a non-union miner died and part of the Frisco Mill was destroyed. Before federal forces or state militia arrived, the mine manager of the BHS, Victor Clement, closed the mine. Rather than signal defeat, this was an effort to prevent further damage to company property. A week later the mines reopened in what appeared to be an attempt to establish a temporary truce.[63]

The 1892 labor unrest in the Coeur d'Alene mining district has been the focus of many historians, in particular Katherine Aiken, whose effective use of BHS records was limited only by overlooking the ethnic shades that colored the labor struggle. The company records highlight the influence of Frederick W. Bradley, who replaced Victor Clement as BHS manager in March 1893 and managed to raise tensions in the region to new heights with his adversarial nature and disdain for the largely Irish Catholic workforce of the BHS. His comments on the battle in the neighboring state of Montana between the Scots Irish William Clark and the Irish Catholic Marcus Daly over the selection of the state capital reveal his prejudice. During the battle between the Copper Kings, money flowed freely as both sides bribed and cajoled officials and voters for the glory of establishing the state capital in their respective cities. The selection of Anaconda would represent a victory for the Irish, the unions, and the miners; whereas the selection of Helena would represent a vote for the Anglo-Protestant, Masonic, anti-Catholic forces in the state. Helena won by a narrow vote. Bradley reported triumphantly to the new company president Nathaniel H. Harris, "This is a great victory by the decent element of Montana over the despotic Butte labor unions backed by Daly's money."[64]

Back in Idaho, elections had gone poorly for the Republican Party. Democrats styling themselves as Populists (referred to as "the Union" in Bradley's letter) won six important positions, to which Bradley sarcastically commented "we will be represented in the next state legislature by six 'honest workingmen' . . . two of whom are ex-convicts."[65] The re-election of Republican William J. McConnell as governor was a "great consolation," though Bradley saw his victory as symbolic, "more a slap in the face to the Miners Union than is Helena's victory." He hoped the Republican representatives would pass strong "anti-dynamite laws," which can only be presumed to have meant harsh laws directed at strikers.[66]

In letters to Harris, Bradley wrote that his workforce in 1894 consisted of "84 Americans (many of these of Irish descent); 76 Irish; 34 Germans; 25 Italians," and an array of twelve other nationalities in smaller numbers.[67] The parenthetical statement is original to the source, emphasizing the priority the company placed on ethnic identity, similar to the hiring card systems used in Colorado. The overwhelming numbers of Irish-Americans working in the mines remained a sore point for the mine manager, and

Bradley wrote to a colleague a few months later regarding the possibility of importing Italian workers: "Our crew is not one tenth part Dago and is about 80% Irish. The Dago is not non union, but makes as good a union man and as good a striker as an Irishman does."[68] In the Coeur d'Alene district, an Irish and Italian ethnic alliance developed, likely out of the wider treatment of Catholics. Although the Irish dominated the payroll of the BHS, they included the Italians rather than excluding them from the union and formed a mutually beneficial front against the company. The workers' drive for a local Catholic hospital seems to have encouraged cross-community solidarity between the two Catholic ethnic groups.

When the Miners' Union declared a strike in November 1894, it caught Bradley by surprise: "After all their idle threats and after all the false alarms we have had, I had no idea they would go so far at this time right in the beginning of winter."[69] Winter was the most difficult season for the union to strike because workers could not hunt for food and had to rely on the union for support. In an unusual twist, union demands centered on other miners, as it ordered the company to fire the scabs who had worked for it during the strike two years before.[70] The union warned that unless the company acceded, "men would be driven out."[71] A week later the demands shifted to pay—$3.50 for all underground workers, regardless of status.[72] While Bradley balked at the thought of the added expense the wage increase would create for their operations—over $2,000 a month more than under the old scale system of $3.00 for muckers and $3.50 for miners—he reserved his greatest enmity for what he saw as an inevitable consequence of caving to union demands: "This simply means that the leaders of the Union would have no one in our mine but their friends and relatives—what a motley Irish crowd of old men and boys we would have on our hands as pensioners!"[73] In Bradley's mind, the strengthening of the union was inextricably linked to the strengthening of the Irish community in the Coeur d'Alene; while lowering workers' wages remained important, he believed the union's hostile stance would be its undoing, writing that the "independence of the Union has saved us as much more."[74] He believed the union was dangerously exposed by striking at that moment.

Other mine managers throughout the American West sent Bradley messages of support regarding his plight, and one reply highlights the broader effect of his crusade: "I hope you will black list a few of the worst of their agitators."[75] Believing Butte was the source of the contagion, one manager suggested that Butte "ought to be made into a reservation and those kind of people be required to live in there and never be allowed to step outside the line."[76] His use of coded phrases reveals his prejudice against Indians in addition to his cold, calculating hatred of a specific group. "Those kind of people" meant the Irish, and his use of "reservation" refers to Indian reservations and implies that the Irish were beyond the pale of Anglo-American civilization.[77] Charles G. Griffith, a mine manager in Marysville, Montana, where there was a strong anti-Catholic presence, sent his best wishes to Bradley in early December, writing, "I am very sorry to hear of your strike; it is an outrage upon you and your company and I sincerely hope you will find it in your power to *teach the Irish a lesson they will not soon forget*."[78] Clearly, many mine managers were not interested in compromise; they believed the Irish community and the labor unions were one and the same and wanted to break the power of both.

Over the following days, Bradley discussed with Harris whether the BHS should try to stay open during the strike with a much-reduced workforce or shut down completely and wait until the strike broke and the price of metals rose; eventually, they decided to shut down.[79] Focused on his political career, Governor William McConnell wrote to Bradley complaining about the labor disruptions and blaming him for the poor Republican vote in the region. Bradley laid the blame squarely at the feet of the Irish and Italian workers: "We have made every effort to get in a good class of employees here; and at the time of our shut down, we had among our underground crew a large number of Cornish and Americans and if we had been let alone [we] would have during the winter weeded out the bulk of our Irish and Dagoes."[80] Given Bradley's observations, it appears that in Coeur d'Alene a Catholic alliance existed between Irish and Italians versus Protestants, who were being used to undermine the union. Bradley did not explain how he planned to hire his "good class" without upsetting the Irish and "Dagoes" that he knew made up 90 percent of his workforce, and he added a further excuse that the "Canyon Creek Unions had long since determined to shut us down."[81] He went on to write that "90% of our men would have continued working for us if we could have afforded to have thrown out armed guards to protect them from a Canyon Creek mob" and that "now our former employees are willing to go to work and to cut the Wardner Union loose from the Canyon Creek Unions, but the agitators in Burke will not permit this."[82] Bradley added that the Canyon Creek Union had total control of that area of the Coeur d'Alene mining district: "No men but the Union men can work there and none but Irish Catholics are eligible to membership in the Union."[83] Again, Bradley traces the root of all his problems to the Irish rather than accept any failure on his part to mediate a settlement.

Bradley exaggerated the ability of the Irish-Italian alliance to exclude other ethnic groups, and in doing so he positioned the Irish as scapegoats. Bradley's actions forced the Wardner and Canyon Creek Miners' Unions to merge and then strike because he was "gradually Americanizing our crew to such an extent that the demand for our surrender, to be effective, had to be made at this time or never."[84] Although he touted his efforts to increase the number of Anglo-Americans and Protestant Cornish in the mines and promoted his active discrimination against the Irish, Bradley's successes were limited. His problems did not stem from any want in the fervor of his favoritism; instead, his inabilities resulted from the notoriously difficult conditions in the wet mines of the BHS and its low wages.

The Cornish of the BHS discovered that they were becoming tainted with the brush of radicalism due to the activities of the Miners' Union. One Cornish man wrote home to Cornwall that "the minister of the Wesleyan Methodist Church was around to the boarding house this morning making inquiries about me . . . Somehow or other he's heard of my being accused as being a member of the Western Federation of Miners."[85] The quote, in particular the use of the word *accused*, shows that the Cornish in the Coeur d'Alene stigmatized union affiliation within their ethnic group. Richard Thomas arrived in Wallace, Idaho, in 1904 and noticed the distinct lack of Cornish in the region: "Cousin Jacks are a scarcity."[86] He further complained about the difficult conditions in the mine: "I quit my job on the Bunker Hill property which was very much against the grain but it was the case of working in a wet place and I wouldn't stand for it."[87] He then traveled to Burke, where he again noticed an unusual absence of Cornish in the mining town: "I don't know of any Cornishman in the mines."[88] Yet Thomas did not describe any discrimination or intimidation from Burke's predominantly Irish and strongly unionized population in his letters.[89] Thomas eventually drifted to the Cornish bastion of Grass Valley, California, in 1906. He died in 1909 at the Grass Valley train station on the morning of his return journey home to Cornwall, collapsing from a heart attack.[90]

In early 1895, leaders from the Western Federation of Miners (WFM), among them the famous labor leader Edward Boyce, arrived to support the strike and boost the members' morale. The Donegal-born Boyce was present for the strike in 1892 and was jailed for a time because of a court injunction. Bradley described Boyce as "bigoted and vindictive" and wrote, "he is the only one that imprisonment did not tame."[91] While the president of the company, Nathaniel Harris, leaned toward a compromise at the rate of three dollars a day for all underground workers, Bradley wrote disparagingly of the possibility of arbitration: "Even if a compromise could be made, we would have no guarantee of its stability other than the promise of a class that we know to be revengeful, unreliable and unscrupulous."[92] In a note a few days later, he set out the alternative labor conditions in the nearby Poorman Mine, "the only Union mine in the District that apparently has no friction with the Union."[93] This might ostensibly have been the goal for companies, but Bradley revealed the unacceptable price of peaceable labor conditions: "This is because each member of the whole crew from Manager down is an Irish Catholic and the Union has anything and everything it asks for. Were we to accept Coeur d'Alene Union conditions, I am of the opinion that *we would have as much trouble as ever unless we were officered, manned and managed as the Poorman Mine is*."[94]

For Bradley, management of the BHS demanded absolute control of the company, and his bigotry toward Irish Catholics—denying them both representation and managerial positions—prohibited any possibility of reconciliation. His liberal use of a blacklist was bad enough, but for some, the BHS's refusal to even recognize the Miners' Union showed such a fundamental disrespect for workers that they refused to consider any compromise without it: "I understand that Flannigan, the Master Workman, K of L [Knights of Labor], has said that if the County starts up even at $5.00 a day, without recognizing the labor organizations, they would blow up everything from Kellogg to the Mill!"[95] Burbidge, one of Bradley's superintendents, wrote those lines; while he could have simply warned Bradley that a miner had threatened to blow up the BHS, he instead tried to alert Bradley to the dangers of pursuing such an intentionally antagonistic approach with their own workforce. He inevitably couched his warning within a statement about the irrationality of the miners' unions, the only language a fanatic like Bradley understood.

Bradley believed non-miners were "the better class" and took a further interesting step against the Irish: he oversaw the organization and fostering of local branches of the American Protective Association [APA].[96] As early as 1891, one Catholic priest commented that "the APA's [were] working hard in northern Idaho," and even mine owner Charles Sweeny wrote to his Irish benefactor in the East, Jeremiah O'Connor, that "the APA are making it hard for me and the Catholics."[97] Bradley believed the growth of the nativist group, which had 100 members by early February 1895 in Wardner and Kellogg, "will prove quite an important factor in settling the labor troubles here."[98] In a letter to Harris, Bradley claimed that thanks to the APA's actions, two leading members of the miners' union were "caught with stolen goods . . . and to avoid arrest have left the state."[99] Evidence of intimidation against one another by members of the union and the APA appear throughout letters that month, and these details highlight the slow fracturing of the community into adversarial groups. Rather than comprising some sort of organic nativist response to the existence of ethnic groups in the community, the intensified emergence of the APA was a very deliberate and insidious company tool devised to isolate the miners and their families from the community in case of a strike or any other disruption and to drive out specific ethnic groups, in particular Irish Catholics. Bradley showed no regret for sowing discord in the local community, and he ended his letter by distancing himself from the goals of the APA, explaining that he was simply using them as a means to an end: "I do not believe in the movement generally, but think that in our case it will accomplish much good."[100] The "general" parameters Bradley disagreed with are unclear, but given the context of the rest of his communications, he was in full agreement with their anti-Irish prejudice.

Writing to Harris again in March 1895 on the continuing growth of the APA in the region, he reported that "at last the best citizens of Wardner and Kellogg have organized into a secret society for the preservation of law and order."[101] In April, the BHS "loaned" forty-one revolvers to the leader of the organization, leading Burbidge to idly observe that if there was any possible violence, the "A.P.A. will have a hand in it."[102] The so-called secret society soon became more brazen, advertising a provocative public lecture "to expound to the public the principles of the A.P.A."—an intimidating expression of the group's newfound strength.[103]

The union responded to the growing threat by expelling all members of the APA from its ranks. One expelled miner, Sam Brown, spoke to Burbidge and showed him the letters he received from the union. The first told him to appear for trial, and the second informed him that he was "expelled for conduct unbecoming a member of the Western Federation of Miners."[104] Unsurprisingly for a member of the APA, Brown blamed his expulsion on Catholics: "He said the judge, prosecutor and the juryman were all Roman Catholics!"[105] He also reported that "they called him a traitor and said that the Union would not only expel traitors but would kill them."[106] The hostility between groups in the region was escalating toward a violent confrontation, just as Bradley had planned.

By June, the strike had begun to weaken and the BHS had partially reopened its mines with imported scab labor. Bradley wrote triumphantly to Harris that he had sent men to recruit unemployed miners in neighboring states: "In order to hurry matters up I have sent two missionaries over to Butte where there are about 2000 idle men."[107] In response to these efforts, the unions posted notices in mining towns urging miners to stay away from the Coeur d'Alene mining district (figure 5.3). Neither the union nor the Catholic community formed a rock-solid bloc, and even the BHS operatives noticed fractures. The local priest—perhaps fearing the influence of the WFM, blind to the company's efforts to remove Catholics from the region, or acting as an advocate for the "respectable" class of people in the area—urged men to return to work while Edward Boyce preached the opposite.[108] Meanwhile, further fractures appeared in the Irish-American community as another nearby mine owner, Martin Curran, reportedly "speaks highly of the APA, although a Catholic himself."[109] The somewhat unusual appearance, first of the priest and second of Curran, increasingly suggests the emergence of a widening class divide. This was seen as a betrayal of both the union and the ethnic group. Curran reported to Burbidge that the union threatened "to blow his head off," a threat Burbidge ignored and interpreted as an attempt by Curran to garner sympathy and financial support from the BHS because Curran's mine company "was in a bad way financially, owing $17,000 in Mullan alone."[110] Even if this was possibly a stunt, it still indicates the diverging priorities of the working-class Irish community vis-à-vis other privileged Irish figures.

Bradley's efforts to extend company control in the region also included opening a company store. Workers were forced to purchase goods at the more expensive company store either with the threat that they would be fired or the issuing of company scrip. This was one more step in controlling

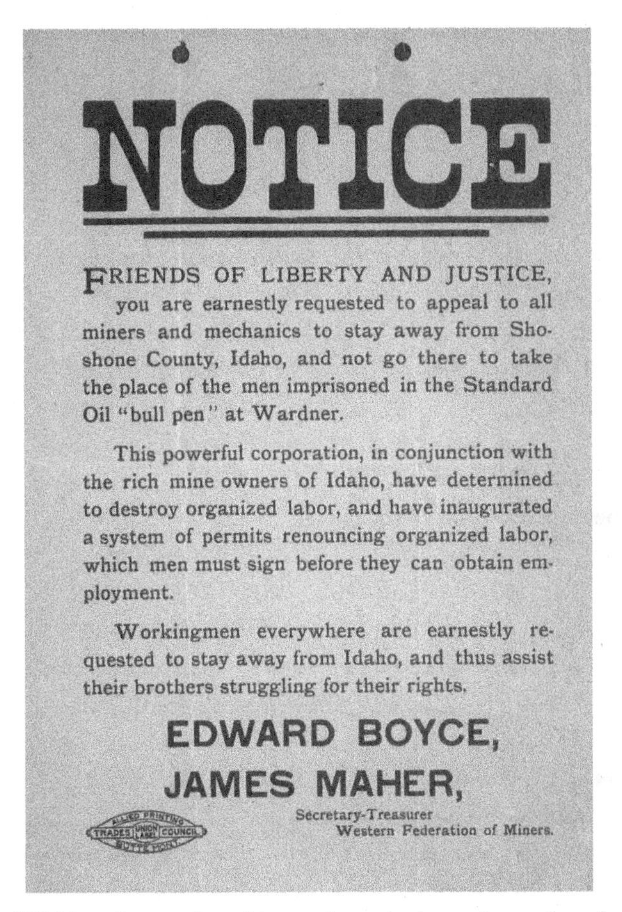
NOTICE

FRIENDS OF LIBERTY AND JUSTICE, you are earnestly requested to appeal to all miners and mechanics to stay away from Shoshone County, Idaho, and not go there to take the place of the men imprisoned in the Standard Oil "bull pen" at Wardner.

This powerful corporation, in conjunction with the rich mine owners of Idaho, have determined to destroy organized labor, and have inaugurated a system of permits renouncing organized labor, which men must sign before they can obtain employment.

Workingmen everywhere are earnestly requested to stay away from Idaho, and thus assist their brothers struggling for their rights.

EDWARD BOYCE,
JAMES MAHER,
Secretary-Treasurer
Western Federation of Miners.

Figure 5.3. WFM notices such as this one from the later 1899 strike asked men to "stay away" from the Coeur d'Alene district and illustrate union efforts against the mine company bringing in scabs from other mining camps. Thiel Detective Service Reports 1899–07, Folder 9, Box 19, Campion Papers.

their lives, with the added benefit of increased profits. As a famous mining song recounted: "You load sixteen tons and what do ya get? / You get another day older and deeper in debt. / St. Peter don't you call me cause I can't go. / I owe my soul to the company store."[111] Aware that such a move would alienate local merchants, he tried to establish the store through a third party and gave the business its own name rather than simply calling it the company store. He then created a credit system BHS workers could use to get items at cheaper prices than those at other local stores; this move perhaps indicated that Bradley was sympathetic to the union, and it had the added benefit of increasing workers' dependence on the company.[112] Bradley also requested help from the Pinkertons. Harris was convinced that he had the right man in mind, a Scotsman with a proven record in the Pinkertons.[113] In response, Bradley urged Harris not to send that man and instead to find a skilled Irish miner so he could successfully infiltrate the union; in a letter dated May 15, 1896, he specified: "In no event do we want a Cornishman."[114] Burbidge, the mine supervisor, wrote to Bradley telling him that workers often found the detectives out because they were "above the average in intelligence," which differentiated them from most miners.[115] At face value, the comment demonstrates management's bigotry and anti-Irish stereotyping, but perhaps it refers to formal education—after all, if they really were smarter, they might not have been so easily discovered.

Having already armed the APA, the BHS encouraged expansion of a state militia, which local authorities sympathetic to the BHS could deploy when needed. In May 1895, Burbidge received a private letter from Governor McConnell recounting a complaint from an unidentified person who called the lack of funding for the state militia a "disgrace." McConnell claimed that "if he were a private citizen he would Know what to do—as he 'did in Boise Basin years ago,'" but that in his role as governor he "should say nothing."[116] Burbidge was uncertain what McConnell meant but clarified later in his correspondence that the Boise Basin experience was "something in the vigilante line"; thus, he was assured that McConnell was "prepared to support the law and order element here in its conflict with the dynamiters to almost any extent within his power."[117] The "law and order element" referred to a specific kind of vigilante activity that used capital punishment against undesirables, that is, other ethnic groups.[118] Many involved in early vigilante movements in the American West had become prominent political figures by the late nineteenth century, and those same men helped establish the myth of vigilantes that linked them to "law and order." They were also often involved in the establishment of pioneer fraternal organizations and historical societies throughout the West.[119]

The targeting and isolation of foreign and Catholic miners, particularly Irish-Americans, was part of a process Bradley started in an effort to strengthen the BHS and weaken the union. When Bradley made his move in 1899, firing seventeen workers for being members of the union, workers knew full well that negotiation was pointless against such entrenched bigotry and responded by dynamiting the mines. Federal forces were requested and sent and they, in turn, illegally detained over the winter approximately a thousand men—mostly miners—in large wooden sheds called "bullpens." During their incarceration, three of the men died. Only a few were eventually charged with crimes.[120] The official government report after the event displayed little regret over the egregious breaches of its citizens' most basic rights: "Pending the discovery of the actual perpetrators of the crime, it was necessary to hold some innocent men in order to prevent the escape of the guilty, and to hold some as witnesses."[121]

Accounts from those who lived through the events at the turn of the twentieth century detail how union membership became secretive because of company persecution. As one woman recalled, "My mom hid the union card under the carpet of the house."[122] BHS policy resulted in a migration of skilled miners out of Kellogg and Wardner to other mines, including the new Hercules Mine near Burke: "The gentlemen who owned these mines were not opposed to the strike so much and they hired the rest of these miners, and they were real lucky to get them because they were really professional."[123] These Irish hard-rock miners were highly skilled, and the mine became one of the most profitable in the region. By 1912, there was a glut of workers, with plenty of men looking for work throughout the area—"ten or twenty men [competing] for every job"—as the BHS led local mine owners in keeping down wages.[124]

The onset of war brought about higher wages in other mining towns, such as Butte, but the Coeur d'Alene district lagged behind primarily due to the mine companies keeping a united front through the MOA. Soon, other industries such as manufacturing and shipbuilding were paying higher wages; and British and Canadian recruiters signed dozens of men for military service, shrinking the potential labor force. After the US joined World War I in 1917, Eugene Day, the owner of the Hecla Mine, wrote to the state director of the US Employment Service, complaining: "If more men are taken from this district, the mines will be unable to operate. I appeal to you to secure more men to keep the mines producing . . . no further recruiting in this district."[125] This contraction of the labor pool forced the Coeur d'Alene mine owners to institute a sliding scale of pay and set the starting level at $3.50 a day. Finally, in late 1917, the $3.50 mark had been reached for workers, but numbers continued to decline and wages soon rose to $5.25 for underground workers at the Bunker Hill and Hecla mines.[126] An indication of the mine owners' weakness at this point was the closure of the Mine Owners' Employment Bureau on January 31, 1917, which had blacklisted all union workers from MOA mines throughout the region and was described by the *Engineering and Mining Journal* as "the instrument that effectually broke the power of the Western Federation of Miners."[127] With the labor shortages, the elimination of the blacklist, and rising wages, unions reestablished themselves in the region. Further changes wrought in Idaho at this time included passage of a Miners' Compensation Act that established an industrial accident board to inspect safety conditions at mines and also provided for medical insurance plans funded by employers.[128] However, the increasing strength of labor brought on by wartime conditions combined with the companies' desperation, not any corporate benevolence, is what brought about these changes.

World War I ultimately undid Frederick Bradley's efforts to destroy unionism. But through his earlier actions, his promotion of nativist organizations, his influence and infiltration of political office, his use of private detective agencies, and the manipulation of state and federal forces to illegally detain large numbers of the local workforce, the company became the source of significant discord in Idaho in the late nineteenth century. Whereas other mine companies (even in the Coeur d'Alene district) negotiated with their workers and recognized their labor organizations, Bradley and the management of the BHS singularly opposed settlement in pursuit of profit because a negotiated settlement would have legitimized the union cause and solidified the Irish presence in those towns. Mine managers were not the only ones to link Irish workers with labor agitation, and newspapers chimed in against unions, with a few claiming a connection between recent troubles and the Molly Maguires: "The mining troubles that so long disturbed the Coeur d'Alene district in Idaho . . . are recognized at once as Molly Maguireism in an aggravated form."[129] As in Colorado, the term *Molly Maguire* became a sensational catch-all for any labor resistance, tying it to violence and the Irish in one fell swoop.

J. Anthony Lukas neatly summarized the American perspective of the entire mess when he wrote: "If the westering experience was expected to produce the Jeffersonian ideal of self-sufficient yeomanry, each sturdy citizen rooted in his own land, church and family, the silver mines of Coeur d'Alene—like those of Silver City and Leadville—produced the polar opposite: a wage earning proletariat at the mercy of absentee mine owners and their managers, helpless in gut-wrenching cycles of boom and bust, never sinking roots in permanent community, destined to drift from one ramshackle mining camp to another in a futile quest for their long-lost dream of western autonomy."[130] They were not powerless, and they did not lack roots. But western autonomy was a mirage for most Irish-Americans, frequently denied by a bellicose managerial class. Identified as the strongest source of organized opposition to corporate ascendancy, the Irish earned the particular enmity of mine managers who believed they alone should and could control their businesses. Irish autonomy found its most obvious expression in the ability to form their own communities, as demonstrated by the earlier concentrations of Irish in the 1860s and 1870s in the Idaho gold mining towns of Bullion, Wood River, and, later, Burke.

The ethnic alliances that formed between the Irish and the Chinese early in Idaho's mining history and later between the Irish and the Italians in the Coeur d'Alene district suggest that nuanced and varied relationships between ethnic groups existed throughout the American West. This system of unspoken but innately understood alliances likely led to difficulties, as in the case of James Mullany hiding his Catholicism; but it had proven sufficient to govern the relationships between the workers themselves, between miners and laborers, and between employers and workers until the arrival of a new paradigm in the form of Frederick Bradley's anti-union, anti-Irish management style. This confrontational approach had been tested in the East and found wanting, in particular by Franklin Gowen, the man who managed to bankrupt the mighty Philadelphia and Reading Railroad twice during the course of his war on his heavily Irish workforce. In addition to its corrosive effects on society, adversarialism could, counterintuitively, nurture the broader-based unionist sympathies already present in the Coeur d'Alene district. The Coeur d'Alene labor dispute between the Irish-Italian union on one side and the BHS and APA on the other marked a shift within miners' unions from advocating for a tiered pay system, "an aristocracy of labor," to a general wage for any man working underground and corresponded to mechanization and a process of de-skilling. While the issue remained divisive in many mining camps, where the skilled miners proved unwilling to cede their privileged position, the emergence of a more egalitarian unionism reflected the widening social basis of organized labor driven by a heavily Irish union leadership that realized the precarious position of labor and the Irish communities supported by these workers.

Although ethnic alliances would remain important throughout the early twentieth century, they faded as trade union organization superseded the importance of workplace group identity and the earlier familial and community ties that bound ethnic groups such as the Irish-Americans together further melded with newer local and regional ties. Ed Boyce, who was involved in the Coeur d'Alene miners' strike in 1892, wrote in *Miners' Magazine* in April 1900 that all workers should aim to "banish forever that miserable contemptible, religious bigotry that has proved the ruination of the laboring people for centuries."[131] While it would take time, this effectively marked the beginning of the end of the exclusive ethno-centric labor groups that had dominated the nineteenth century and the rise of a more inclusive labor unionism in the US in the twentieth century. This departure did not mean that Irish-America would forego its entrenched leadership of unions any more than it would its leadership of the American Catholic Church. Given a head start due in part to a culture rooted in a tradition of political agitation and ethnic solidarity, Irish-Americans continued to dominate the leadership of mining unions in the early twentieth century.

The hostility of the peoples of McMinnville toward Mullany and Bradley's hostility toward the union and the Irish reveal the different aspects of Anglo-American clannishness. Irish worker organization, class difference, cultural difference, and Catholicism were antagonistic challenges to their narrow definition of "American" as well as to their dominant position in the US. They responded to these perceived challenges with intimidation and violence, usually through vigilantes or groups such as the APA. The emergence of figures such as Charles Sweeny and, earlier, George Daly reveals a new aspect of Irish-America. Focused on his profitable mines, Sweeny hired hundreds of scabs from other states and in the middle of a strike wrote, "I personally have guaranteed them employment and the [mine owners'] association will see that they get it." Simultaneously, utilizing his cultural background as a shield, he met union leaders in person to feign negotiating a settlement.[132] He manipulated others using his Irishness, be they business partners such as Jeremiah O'Connor or union leaders, and his example foretold the direction of most of the Irish-American business class in the twentieth century and traced the limited purview of its ethnic loyalty. Those exact limits were seen in the later examples of Marysville and Butte in Montana, where Thomas Cruse was unable to assist in the development of the Irish community at his mines at Drumlummon and where Marcus Daly was unable to extend the ethnic loyalty of businesses from one generation of Irish to the next. In Butte, the class and generational fractures of Irish-America exploded figuratively as well literally, in the form of dynamite blasts destroying the headquarters of the Butte Miners' Union—the organization that had done much to help the struggling Irish strikers in the Coeur d'Alene district.

6

Oro y Cobre, Gold and Copper

Come all ye Corkonians, come all ye Kerryonians,
Come all ye good people from the land of the green,
Come all ye big wheelers, who ain't got your papers,
You'll have to go mucking behind a machine.[1]

Ethnic alliances and mine ownership were important factors in determining the course of fortunes for Irish individuals and communities in the mining towns of California, Nevada, Colorado, and Idaho. Ownership of the mines might appear to have been the most important factor, since in Virginia City it led to the formation of the most secure Irish mining community in the nineteenth century. As the gold mines of Marysville, Montana, showed, however, it was not sufficient that the person who discovered a mine or retained ownership of it was Irish; the main prerequisite for the formation of a community was stable employment. Another location in Montana offered such employment to a vast number of Irish: the copper metropolis of Butte. The song above directs the Irish to come to Butte and orders them to work, a sustaining direct path of migration from Ireland to America.

https://doi.org/10.5876/9781646422517.c006

Of all the resources mined by the Irish in the American West, it was Montana copper that generated the foundation for the largest and longest surviving Irish mining community in the United States—Butte. Copper's tarnished green hue serves as an apt metaphor for the emergence of this beating heart of the Irish-American West. The sea-foam copper greening of the Statue of Liberty, an embodiment of this process from her construction in 1886 to 1920, greeted the many Irish on their path from Ireland to Butte. She stood at the entrance to the New World as a manifestation of the change possible in the US and of the emerging Irish-American city, sustained by the same ethnic alliances and family networks that ran through Smartsville, California, and Leadville, Colorado, in previous decades. For most of the Irish who came from Ireland, Britain, or elsewhere in the United States, their eventual arrival in smoggy Butte heralded the emergence of a shining Irish city in the Rocky Mountains.

A Golden Mirage

Geographic movement accompanied occupational movement, especially in the earlier placer days, and for Irishmen such as Michael Lane, the low price of land in Montana facilitated his conversion from placer miner to farmer.[2] He had first arrived in the territory in 1865 and in less than two years had earned enough to buy a farm in Gallatin County and marry his compatriot, Sarah Kelley. Thanks to the remittances they sent, his father and four brothers soon joined him from Ireland, whereupon they also bought farms. The closing of the placer mining era in Montana caused an exodus similar to the end of other rushes. The *Montana Post* lamented, "Daily we see teams loaded with men that have made their pile, leaving the country, some to return in the spring, some never to see this fair Territory anymore."[3]

Some prospectors remained defiant—perhaps they had grown comfortable with what they knew—and continued to hunt for their big strike. Support for such men came from saloons and wealthy individuals who "grubstaked" the miners. Unsurprisingly, the richest man in Deer Lodge County in 1870 was Irish bachelor David Welch, described as a "Grocer and Banker," who held property valued at $26,500.[4] The Irish dimensions of these hidden support networks made possible the ability of one Irish miner, Thomas Cruse, to stubbornly persist until 1876, when he struck a rich vein of gold and silver near Helena in what he named the Drumlummon Mine. This became the largest gold discovery in Montana.[5] He was exceptional in another vital regard; unlike many others, he decided to stay with the property rather than sell out. He spent years developing it with six miners, finally driving a tunnel to a depth of 500 feet by 1880. The year 1882 brought almost $77,000 in profits, which drew the attention of investors from London.[6] At this stage, Cruse had bought out the owners of the nearby five-stamp mill and had a great deal of control over his operations during the negotiations.[7] Cruse's canny negotiating, backed by a glowing report on the state of the mines by English mine inspector John Darlington to London investors, earned him a vast fortune—$1.6 million, wisely coupled with a one-sixth stake in the new venture, the Montana Mining Company.[8]

Stories of his life regularly recall that before Cruse discovered his mine in 1872, he was a poverty-stricken, insignificant prospector—"his frail constitution broken down from exposure and hardship"—who was denied credit from merchants. These accounts might be exaggerated a little to heighten his rise and the fulfillment of the American dream.[9] Another account notes that his discovery of gold meant his nickname changed from "Old Tom Cruse" to "Colonel Tom," and it was certainly true that with wealth came respect and prestige.[10]

Cruse shared many similarities with the other Irish-born Montana mining millionaire, Marcus Daly. Both men were born in County Cavan and as young men immigrated to New York, moving west after a few years. Instead of settling in a major city as Daly did, Cruse placer mined in California, Nevada, and Idaho for three years before arriving in Montana Territory in 1866.[11] Both Cruse and Daly became fabulously wealthy, not by selling up but by holding a significant interest in the mines during their profitable lifetimes. The rapid expansion of Cruse's gold mines created the town of Marysville but, unlike the patriarchal Daly, he remained a distant figure from the mining operation and its workforce.

By 1890, Marysville had a population of 1,480; within a few years it also boasted electric lighted streets, a Catholic church, an Episcopalian church, a Methodist church, six hotels, an opera house, a grade school (with nine teachers), and two newspapers: the *Mountaineer* and the *Berry Picker*.[12] According to one newspaper observer, there was also a sizable Chinese population and stable employment; the account further promoted the town by stating that people "had everything they desired. There was peace everywhere, no class strife, no lodge or religious dissention."[13] However, a closer examination of company records reveals a very different story.

The Montana Mining Company (MMC) records contain Pinkerton Detective Agency (PDA) reports from 1892, when the company hired the PDA to investigate an accident in one of the mines. The company suspected foul play and consequently decided to infiltrate its workforce to discover three things: whether the fire was malicious, the nature of worker sentiment in the town, and the presence of union sympathizers. The PDA began by sending "Operative J. F. F." to shadow a company contact in Butte, a Mr. Whitmore, who pointed out Tom Ryan, a suspected union organizer for the Butte Miners' Union.[14] Whitmore then entered a series of saloons with Irish-sounding names—Harrington's and Driscolls, P. S. Harringtons, McNultys, Hayes, and Miners Resort—as well as "all other saloons where miners are in the habit of going."[15] Failing to see him in Butte, Whitmore and the operative traveled to Walkerville to see whether Ryan was at Driscoll's grocery store. He was not, and they believed Ryan had returned to Marysville, so Whitmore ordered the detective to travel there "under the pretense of looking for work" and spy on the workers.[16]

Arriving in Marysville, the operative visited Farrell and Murray's saloon and overheard a number of miners discussing the troubles in the Coeur d'Alene district.[17] The Pinkerton agent noticed that all the men agreed that the miners would lose; moreover, they did not believe in the Miners' Union because it advocated for equal pay for unskilled and skilled laborers. The operative quoted one miner named Casey:

> What would you think if a few men came into this camp and from some personal or money motive demanded that all the men working in the Drum Lummond mine should quit work unless the company agreed to pay every green-horn that came along as much as a man that spent a great part of his life in learning the mining business. I believe he said and always think that mining is a trade and I have no use for a Union or any man who is in favor of giving a common laborer as much pay as a skilled miner, and I think when a company such as we have here that acts squarely with its men, ought not to be put to any trouble.[18]

The miners were unaware that the strike in the Coeur d'Alene district started because the company tried to class skilled miners as unskilled labor due to increased mechanization, revealing the misinformation circulating about the cause and course of strikes. Their view of mining as a "trade" and their resistance to allying with "green-horns" demonstrates the persistence of the elitist miner mentality. By focusing on the idea that only "a few men" might make such demands out of personal or monetary gain, Casey attempted to undermine any appeal by union proponents for an egalitarian approach to other mine workers. The appeal to a "skills gap" simultaneously differentiated miners and discouraged wider unionization. The Pinkerton agent later visited the McCammis saloon and asked Rory, the owner, and several miners about the fire. None believed it was malicious, and they suggested that it had been caused by a worker carelessly leaving the stump of a candle burning near the timbers.[19]

When the operative returned to Rory's saloon the following day, he overheard one man say that "Tom Ryan had skipped town" and that if he succeeded in getting a job in Butte, "he would take his family with him and then his creditors here could go and whistle for their money."[20] The bartender railed at the suggestion: "The son of a bitch he said owes me $27.00 and there is not a saloon, grocery store or butcher shop in town that will not lose perhaps more than that amount on him."[21] The movement of the Irish across the American West was certainly opportunistic, although in this case it means something a little different. Some seized the chance to leave their debts behind as they moved on to the next location. The bartender need not have feared such a scenario this time, however, as later in the day a man named Sullivan came into the saloon and told Rory he had just seen Ryan walking down the street. The bartender blustered, "I am damn glad of it I will make it hot for him pretty soon if he does not come and pay me."[22]

On May 17 the Pinkerton detective reported receiving a letter from Whitmore telling him to stay alert for any of the "prominent men of the Miners Union from Butte around town" as well as anything else of interest to the company.[23] The operative returned to Rory's saloon and heard a conversation involving Jim Campbell, "a fellow named Sinott," and a few others who stated that to get work in Marysville, "a man must belong to the order of the Patriotic Sons of America otherwise he has no show" and that in a year or two the organization would completely control the town.[24] Campbell challenged this assertion, saying, "I tell you that if I wanted work in the mine tomorrow morning I would go there and get it and I would never be asked what order I belonged to."[25] In response to Campbell's bravado, Sinott gave the example of Minkel, a man recently killed in a mine accident in the Drumlummon, who applied for work but was refused and "was obliged to change his religion and join the Patriotic Sons of America in order to get a job that would pay him $3.00 a day."[26] "If you doubt what I am saying," Sinott said to Campbell and the others around him, "you have only to go to Hanley or some of the other bosses there and you will soon find out that if there is a man of the order looking for work you will be told you are not wanted."[27] Campbell may well have being talking from a position of privilege, as the sources do not reveal his religious affiliation or nationality.

The Patriotic Order of the Sons of America, to use the full name, was a virulently anti-Catholic and nativist secret society that emerged in the northern Midwest and in California. In Montana it was visible as a serious presence only in Marysville.[28] Many entries in *The Progressive Men of Montana* include boasts of prominent membership in secret societies, but unusually, this organization only appears once—in the entry for Milton T. Small, a rancher listed as a member of the "Patriotic Order of America" as well as many other fraternities.[29] Unlike other fraternities, whose meeting times and leading members' details were often advertised in town directories under the heading "Secret Societies," this group took the secrecy part seriously, perhaps fearing retribution from the strong Irish presence in other parts of Montana.

While many nativist American groups organized themselves around anti-immigration positions, the Patriotic Order of the Sons of America stated overtly that its goal was to "crush the foreign elements."[30] It shared sentiments and probably membership with other Masonic organizations in shunning Catholics, Jews, and others from membership. Its aggressive stance positions it as a close precursor of the Klu Klux Klan rather than a general fraternal organization. One notable incident of fraternal overlap involved William A. Clark, a copper king and a Presbyterian, Scots Irish, and thirty-third degree Mason who invited the Patriotic Order of the Sons of America to a public beef barbecue on a Friday, thus excluding Catholics—a move David Emmons described as "insensitive," an overly generous assessment considering the Patriotic Sons' fairly specific platform.[31]

Likewise, its presence was an indication of dampened Catholic presence and of stimulated Protestant immigration to Marysville. MMC records include letters of recommendation on behalf of Cornish miners written directly to the manager, indicating his ethnic sympathies.[32] It appears likely, then, that the emergence of the Patriotic Order of the Sons of America was rooted in several years of anti-Catholic hiring practices enacted by mine manager Richard T. Bayliss, in effect formalizing an atmosphere hostile to Irish and union influence in Marysville. This episode further demonstrates the limitations of Irish union influence in Montana. Despite the close proximity of Butte, the "Gibraltar of unionism," its sway did not radiate throughout the surrounding mining towns, although it may have made the Patriotic Order more secretive. With the spread of railroads, these communities became less isolated in one obvious way, but that did not mean they were inclusive. The same opportunities that had enabled the Irish in Butte to create an ethnic enclave were used by other groups, such as the Cornish and nativist Americans in Marysville, to form their own enclaves.

The anti-Catholic fraternities may have been bolstered by the strong historical Confederate presence in Montana. Indeed, it was said that the "left wing of the army of Confederate General Price in Missouri never surrendered; it retreated to Montana."[33] When Alder Gulch was renamed Virginia City, locals tried to name it Varina City after Jefferson Davis's wife.[34] Anti-Catholic sentiments could also be found among clergy in Protestant denominations. Methodist reverend Daniel Tuttle witnessed with disappointment the increased Catholic presence in Virginia City in 1868 and wrote: "Two Romish priests are here. One has gone down to Oregon to bring up three or four 'sisters' who are going to open up a school at once. The people of the town have subscribed $2,700 for starting such a school. This sum or nearly all of it comes from Protestants."[35] Tuttle's disapproval drips from the letter, yet his account illustrates the ecumenical outreach between Catholics and Protestants.

Irishman Patrick Kearney noted the strength of certain non-Catholic fraternities in the US. As he wrote to his family about limited economic opportunities, Kearney initially blamed the many other immigrants who were willing to work for "mere nothing," then he blamed the networks of Protestant fraternities. He claimed they were gatekeepers to secure employment: "If a man today in America . . . doesn't belong to a secret order or society, such as the Odd Fellows, the Masons and numerous others, he can get nothing. I am sorry to say there are several Irishmen turning over every day."[36] When Irish communities were too weak to offer stable employment and insurance

Figure 6.1. Queen's court. Fourth of July Carnival. "Queen" Mamie Cruse, towering over her royal attendants, surveys the crowds atop her horse-drawn float on Main Street, Helena, 1907. Mulvaney Real Photo Postcard Collection, Montana Historical Society Research Center Photograph Archives, Helena.

benefits, some Irish apparently turned to organizations that could make, and of necessity made, the difficult choice to abandon their faith and heritage. The subtle historical impact of figures such as Tuttle and organizations such as the Patriotic Order and other fraternities should not be underestimated in the development of Montana's history.

Cruse's role in the company was more discreet, as were his expressions of Irishness. In 1886, Thomas Cruse married Margaret Carter, the sister of Montana Democratic senator Thomas H. Carter. She died shortly after they were wed, giving birth to their only child in December of the same year. Cruse chose to live in Helena, where he established the Thomas Cruse Savings Bank on September 15, 1887. Backed by his income from the gold mines, this was the first savings bank in Montana and it thrived, surviving the turbulent economic conditions of the 1890s.[37] Whether Cruse could have developed an Irish-dominated community in Marysville is debatable, but he turned his attention away from the town and focused on his bank and his daughter. A subtle stab directed at monarchism and the Anglo-American well-to-do of Helena can be seen in a photo of the Fourth of July parade where Mamie Cruse sat on an extravagant float as the so-called queen of the festival, an Irish-American woman elevated above all others and leading the parade (figure 6.1). Another representation of his stubborn Irishness was the Catholic Church in Helena. As a symbol of his faith and identity, he generously donated $250,000 for the construction of the remarkable Helena Cathedral and the bishopric residence attached to it, a lasting reminder of the Irish-American presence in the town.[38] The imposing Catholic cathedral is an impressive display of Cruse's strong faith throughout his life. It stands towering over the city, a very direct message from Cruse to nativists in Marysville and their sympathetic allies in Helena.

A short walk from the cathedral, another Irish symbol stands in front of the state capital, serving as another defiant message of the Irish presence in Montana. When Acting Governor Thomas Francis Meagher drowned under suspicious circumstances at Fort Benton in 1875, rumors circulated among the Irish that responsibility lay with a not-so-secret organization of nativist vigilantes. Whatever the truth, after the overwhelmingly Irish town of Anaconda lost out to Helena in the vote to choose the state capital, the Thomas Francis Meagher Memorial Association was established with the publicly stated goal to build a statue to the lost hero who had led the Irish Brigade during the Civil War.[39] Its surreptitious objective was plain to all—the placement of an armed Meagher sitting proudly atop his horse in front of the state capital so the "APA's [American Protective Association members] . . . must look in his face and salute his glorious memory for all ages."[40] Despite the difficulties the Irish faced during their lives, the overwhelming majority remained loyal to their ethnicity and their religion, identities that intertwined within Irish-America. Even in towns where they were few in number, physical proof of their constant loyalty to their identity exists through buildings such as the Helena Cathedral and monuments such as Meagher's statue. Both are representations of obstinate loyalty standing in diametric opposition to the American Protestant idea of reinvention on the frontier.[41] Cruse's distance from the mining town he helped found represented a distinct difference between himself and copper mining magnate Marcus Daly, who in contrast created an Irish enclave through direct patronage and employing Irish people. It is possible that Cruse remained oblivious of the managerial bias in Marysville, but, more likely, he felt that his work was done and decided not to emulate Daly's hands-on and confrontational approach—in business at least.

The Copper Hues of Butte

Nach óg sa tsaol nár ceapadh mo leas dom
Nuair a d'fhágas Éire ghrámhar,
Chun a bheith ag siúl na dúthaí mBúit Montana
Is gan pingin i dtóin mo sparáinín;
Mar obair ní bhfaighir ná aon tslí bheatha
Ná bean tí fhairsing fháilteach,
Ach bladar go leor is fós drochmheas ort,
Mara bhfuil dollars agat le háireamh.

Wasn't I too young in this world to consider my good fortune,
When I left my loving Ireland.
To walk the townlands in Butte, Montana,
Without a penny in the bottom of my little purse,
I won't find work or any job,
Or a generous, welcoming housewife,
Only ample flattery and feel only disrespect for you,
If you haven't got dollars for the counting.[42]

The first major copper mines in the US were in Houghton County, Michigan, on the Keweenaw Peninsula, beginning in 1845. The county's location on Lake Superior facilitated connection to eastern trade routes and encouraged the westward spread of the railroads. Keweenaw copper was found in its metallic (or native) form in the rock, which meant it required far less reduction and smelting than was typical.[43] For decades, this area provided the majority of US copper; but by the 1870s, Arizona had begun to extract large amounts of copper. Soon after, the mighty copper mines of Montana superseded those in both Arizona and Michigan, increasing production from 33,899 tons in 1885 to 157,375 tons in 1905.[44] Centered in Butte, the "richest hill on earth," a series of honeycombed mines surrendered an outstanding quantity and quality of copper ore used to feed the rapid electrification of American cities. Butte was the largest copper-producing location in the US for two decades until technological advances in ore refining allowed Arizona to return to the top again by 1910, a position the state held for the rest of the twentieth century.[45] Not all Americans warmly welcomed Butte's rise: "Butte is the black heart of Montana, feared and distrusted . . . in appearance and population Butte was the antithesis of rural and Protestant America."[46] Unsurprisingly, the large urban Irish Catholic community lay at the heart of the negative perception of the city, contrasting sharply with the pastoral aspect of the frontier mythos, By 1900, Butte was "one of the most overwhelmingly Irish cities in the United States," and a full 25 percent of the residents of Silver Bow County were Irish-born or second-generation Irish.[47] That year, 8,026 Irish-born lived in the city with a total population of 30,470, a larger proportion of Irish than in any other city in the US. Their presence stood as a powerful symbol of their physical endorsement of the American West.

Some Irish immigrated to Butte directly from Ireland, particularly from the former copper mining region on the Beara Peninsula in Cork. Others came from mining regions throughout the US. The records of the Ancient Order of Hibernians (AOH) from 1884 to 1892 show that AOH members transferred to Butte most frequently from Michigan, Pennsylvania, Colorado, and California.[48] Branches as far away as Ireland wrote letters of recommendation for their members. Owen Morgan, Division 298, Rostrevor, County Down, wrote in March 1907 that "John Connelly is a member of our Div and we alway fond him to be a true and faithfull member and we highily reccomend him to the frendship of you brethern. The members all join in wishing him success in his jurney."[49] This letter is one of the very few remaining hints of a transnational bureaucracy of the AOH stretching from the US to Ireland.

"[The Irish] got there first and theirs was the dominant culture," but neither early arrival nor numerical superiority inherently equated with stability, as seen in the cases of Leadville and the Coeur d'Alene district, respectively.[50] The only town to rival Butte's numbers of Irish in Montana was its sister town, Anaconda. Built later by Marcus Daly to smelt the vast quantities of rich copper mined by the city, it represented an extension of Butte's Irish influence within the Anaconda Company (AC) as much as a growth in the number of Irish in the region. Butte's development as a center of permanent Irishness occurred largely thanks to the influence of Daly and his preference to hire his compatriots as workers in Montana.

Daly was born in County Cavan and emigrated to New York at age fifteen.[51] He worked odd jobs for several years before following his sister to

California, where he worked for a time as a ranch hand, logger, and railroad worker before falling in with another Irishman, Thomas Murphy. Daly followed Murphy to the Comstock Lode in Virginia City, Nevada. By 1871, Daly had become a foreman for the Walker Brothers and married Margaret Evans in Salt Lake City.[52] He spent the following years honing his mining expertise.

In 1876, the company sent him to Montana to locate worthwhile silver mine prospects, and he purchased the Alice Mine, cannily retaining a 20 percent stake in the claim. Michael Hickey, an Irish Civil War veteran turned miner, had originally discovered the rich lode of silver. After the lode appeared to play out, he agreed to sell it to Marcus Daly in 1881.[53] Daly discovered a massive vein of copper. Capitalizing on the growing demand for the metal, he sold his other interests and quietly bought up nearby mines. His operation grew until he owned the largest mining company in Montana, which, in turn, gave him immense influence in the region. Daly used this power to further expand his business interests and to solidify the strength of the Irish community in Montana.[54]

One priest in Butte, Father Brosnan, stated: "Marcus Daly was the man that made Butte an Irish town . . . He did not care for any man but an Irishman and [he] did not give a job to anyone else."[55] While this was an exaggeration—Daly did hire skilled workers from other backgrounds—the priest echoed the popular perception that any Irishman who came and was willing to work could find work with his company. Secure employment and good pay were all the Irish needed to hear to flock to Butte; the fact that Marcus Daly was an Irishman only sweetened the deal. Daly responded to their intense loyalty by treating the Irish well. He sponsored Irish causes and supported Irish fraternities, and the Democratic Party in the area became so synonymous with him that supporters earned the name Dalycrats.[56] By 1890 he had become one of the richest men in Montana, granting him a degree of security against industrial and political competitors in the state.[57] When the so-called War of the Copper Kings began among Daly, the Scots Irish William Clark, and F. Augustus Heinz (who, incidentally, had an Irish mother), Daly emerged as the eventual winner with the formation of the Amalgamated Copper Mining Company, which became the Anaconda Copper Mining Company in 1910.

The centrality of mining to the success of the Irish community in Butte cannot be overstated—in 1900, more than 70 percent of Irishmen in the city were miners.[58] Douglas Hyde toured America at the turn of the century to raise funds for the Gaelic League and stopped in Butte, another sign of the city's importance in the network of Irish-America in this period. He noted: "This is almost an Irish city. Most of the people in it are Irish. The mayor is an Irish man and the governing of almost everything is in the hands of the Irish."[59] And yet he had reservations about the position of the Irish in the city, writing, "I wouldn't say the status of the Irish is satisfactory here, they depend entirely on the mines."[60] Hyde advocated a move from mining to agriculture and implied that Daly had tried to settle the Irish in the Bitterroot Valley, but a "bundle of yanks came from Missouri and settled" and anyway "you could not buy a valley now for $40 million."[61] Hyde knew the time had passed for a major Irish settlement project, but his reference to the influx of Missourians was interesting, as he might have been implying that the presence of former Confederate soldiers in the region made it unsavory.

Labor Dangers

Silver Bow County, Montana, bucked the broad trend of declining Irish numbers in other US mining regions, which showed the success of Daly's policies in drawing the Irish to Butte. Butte's unique position, illustrated by growing numbers of Irish-born between 1890 and 1900, is important; however, the continued increase in the number of Irish into the twentieth century is perhaps a more important trend, since it demonstrates the stability of Butte's Irish community and its unique longevity as an industrial mining town in the US. Riobard O'Dwyer's genealogical work provides hundreds of examples of chain migration by women and men from Beara to Butte: 1,700 people emigrated from Kilcatherine parish between 1870 and 1915, and of those, 1,138 went to Butte.[62] Economic difficulties in Butte temporarily stymied these paths of migration from Ireland and the rest of the US. Such patterns were notable to most observers throughout the nineteenth century. As one British report noted, "The spirit of emigration fluctuates according to reports received from America."[63] An economic crisis could have a range of effects. One crisis that occurred in 1873, when Butte was in its infancy, merely slowed the growth of the city whereas an 1878 effort by some mine companies to cut wages at the Alice Mine and Lexington Mine resulted in an economic shock and a walkout as workers joined the Butte Miners' Union (BMU) en masse. This led indirectly to a more stable labor and economic situation, and thanks to Irish control of the union, it encouraged further Irish migration on the foot of the victory.[64] One event that caused brief large-scale unemployment for half a year occurred in mid-1886, when the AC closed its mines to improve smelting and refining techniques, but by January 1886 they had opened again.[65] Even the Panic of 1893, the most serious economic downturn in American history to that point, was largely shrugged off by the AC as demand for copper propelled the local economy through the recession.[66] Statistical evidence offered by Michael O'Connell shows that 69.5 percent of Irish-born miners in Butte in 1900 had arrived at the mining camp between 1880 and 1895, based largely on chain migration that took place in a relatively short space of time.[67] These Irish provided Butte with the core of its skilled mining strength.

However, Butte's rapid development came with problems, including high worker turnover and great demand for lodging and food. Irish workers could avail themselves of the opportunity to get accommodations in an Irish environment, either by boarding in Irish-run hotels or, more commonly, lodging with an Irish family. When Chinese entrepreneurs attempted to enter this market, they had little success, illustrating the institutional strength of lodging with compatriots. This was an important thread in the web of the Irish diaspora. As Mary Murphy states, "Women could retain cherished parts of Irish culture while taking advantage of the opportunities America offered," but so too could men.[68] The boardinghouses were places where information could be shared and newcomers could establish a foothold. Hugh O'Daly viewed Butte as akin to an Irish Mecca; he judged other mining towns as insufficiently Irish and complained that they lacked Catholic churches.[69] This comment might seem strange, as even Irish communities in smaller towns such as in Lander County, Nevada, were able to request and retain Irish-born clergy to serve their small congregations, but O'Daly longed for the excitement of the larger crowds and unshakeable Irish influence saturating Butte.

The increased size of the working population created high demand for female labor, evident in the many job postings in local press, especially in accommodations. By 1880, while almost a third of the Irish owned their homes or stayed with relatives, a majority lived in large Irish boardinghouses that catered to those needing temporary lodging (table 6.1). By 1889, there were sixteen Irish-owned boardinghouses in Butte, mostly in Dublin Gulch, and by 1900, half of Butte's Irishmen lodged and ate in "one of the many miners' homes."[70] Ellen Mulkerin, born in Galway, followed her sister to Anaconda where she helped her secure accommodation as well as a job in the Gavin house, a large Anaconda Mining Company boardinghouse; two years later she married Tim Tracy, an Irishman born in County Waterford.[71] Between 1880 and 1930, the *Anaconda Standard* advertised heavily for housekeepers, waitresses, chambermaids, seamstresses, laundry girls, and dishwashers.[72] Butte offered women as well as men the opportunity to earn a decent wage.

Women workers organized into various unions. One of the most notable was the Women's Protective Union, representing boardinghouse and restaurant workers. Established in 1890 to promote the "dignity of women's labor" and led by Irish women, the union expanded to include any women workers not already covered by other extant unions, such as the Teachers' Union or the Laundry Workers' Union.[73] The unions enabled women to strike in 1920 and to use their bargaining powers to demand eighteen dollars a week, which was lower than a man's income but much higher than the fourteen dollars estimated as the minimum weekly income a single woman needed.[74] The attractive opportunities for both genders meant that by 1895, "Irish migration had become self-sustaining, [with] Irish following Irish," both men and women.[75] The network-based nature of Irish emigration fused with the critical mass of Irish present in Butte, forming an established Irish-centric community whereby the arrival of children likely helped promote lifestyles that had relatively greater stability.[76]

Even with this core of stability, there was a transient edge to the population, and the massive turnover of workers in the mining camps accommodated the large numbers of Irish who kept flooding into the city to replace workers who were leaving or dying. Of the 3,197 Irish miners in 1900, only 478 (15 percent), remained in Butte ten years later.[77] Emmons assesses this figure in a negative light, portraying the 15 percent as a small center of Irishness holding its own in the city, but it also reveals another interesting aspect of this smoggy Promised Land the Irish occupied—that even if Butte was not where most lived permanently, it offered them a stable step for their next move onward.[78] Many spoke of returning to Ireland—indeed, the *Butte Bystander* believed the Irish stories, songs, and boasts and claimed that 40 percent had returned home. These exaggerations belied the emigrants' uncertainty about returning to Ireland, and in reality the overwhelming majority stayed in the US, even if they moved on from Butte.[79]

Table 6.1. Accommodations of Irish-born in Butte, 1880[a]

Relation	*No.*	*%*
Lodger, roommate, boarder	1,066	61.7
Head of household	341	21.5
Nuclear family member	109	6.5
Extended family member	53	3.3
Prisoner	13	0.8
Unknown	6	0.4

[a] O'Connell, "Emigration from the Berehaven Copper Mining District," 73.

As stated, most Irish who came to Butte participated in the Irish community—there was no other way for it to maintain its stability. They, in turn, benefited from the most powerful Irish enclave in the American West through employment, friendships, and marriage partners.[80] Still, most Irish ultimately moved on. This movement meant Butte was a hub of ideas; and Irish-American socialists, trade unionists, and other radicals were forged in the maelstrom of the city. One such figure was Tom Hickey who became a significant publisher of socialist texts—including a newspaper *The Rebel*—and a leftist organizer in Texas.[81] This culture of transience even appears in the heartbreaking last words of Jim Moore, who scribbled a message to his wife when he was trapped underground during the Speculator Mine disaster: "I tried to get all the men out, but the smoke was too strong . . . If anything happens to me you had better sell the house and go to California and live. You will know that your Jim died like a man and his last thought was for his wife that I love better than anyone on earth. We will meet again." He ended his note the following morning as his strength failed, "9:00am In the dark."[82] Such horrors were the basis of the people of Butte's tough, unflinching reputation, but it was not something to be senselessly proud of; rather, it was one aspect of a conscious exchange for the Irish, a means to an end and not an end in itself.

Evidence of the high price industrial America extracted from the Irish is seen in the example of Moore, but there was much more scattered in other records. In his dying words, Moore urged his wife to travel to live with their relatives in California, and examples of flight from tragedy appear among the relatively few who repatriated to Ireland. Mary Flynn's census occupation reads "Copper Miner's Widow," and she had returned to her relatives on the Beara Peninsula shortly after her husband died. The census further lists the children's ages, five, three, and two, and their birthplace as "Butte City, USA America."[83] The lingering pain from her journey to America, her marriage, her bereavement, and her sorrowful return to her home and family in Ireland with her American-born children brings into sharp focus the importance of the lives lived along the transatlantic ties.

The Butte Irish coped with loss as they dealt with life, by leaning on their traditions and religion as manifested in an Irish wake.[84] This tradition kept the body of the deceased in the house over a series of days, during which the bereaved were visited by relatives and those offering condolences. Food and alcohol would be available for mourners who would talk and reminisce. As tradition dictated, at least one person accompanied the remains of the deceased day and night until the day of the burial. One cynic observed, "For the men, it was a lot of fun. For the women, it was a sorrowful, tedious, dragged-out affair."[85] This framing of the wake purely in the context of a patriarchic division of household labor obscures the centrality of community and family in the tradition as well as the reason for its length. The bereaved are supported by others through donations of food and other necessities, and the length of the tradition—several days—offers those who have lost someone close the time to come to terms with the grief of the loved one's passing and the opportunity to listen and share happy memories of the deceased. The exhaustion of hosting the event is in some senses a form of therapy to cope with the awful moments after death, ensuring that the bereaved are numb and kept busy; though tired from hosting and socializing with others, the grieving is made easier because they are given time to come to terms with their loss. The Butte Irish understood the importance of such a tradition, not in spite of the frequency and proximity of the Irish community to death but because of it.

The exceptional longevity of Butte's mines stood defiant in the face of a different fear in the minds of the Irish community: that the mines would not last. This was a consequence of their internalized sense of exile from their homeland and the psychic scars from repeated exposure to economic cycles of insecurity, resulting in a people "more inclined to invest in goods they could take with them when they had to leave."[86] This community was both cohesive and transient, and it enabled the Irish to earn enough for themselves or their children to move up and out of mining. This mobility was effectively a quality-of-life issue for Irish miners who had to constantly weigh the costs against the benefits. Butte offered the availability of stable work for any Irish miner who wanted it, the existence of an increasingly powerful Irish community, and simple respect from the company in return for the serious risks the men took while laboring in the bowels of the earth. This stood in stark contrast to the atmosphere of suspicion and hostility in other mining companies such as Marysville and across the state line in the Coeur d'Alene mining district. Irish miners were willing to accept the workplace dangers in Butte, since the advantages there outweighed the many disadvantages that prevailed in other locations. Thus, many came to the city from all over the Irish diaspora:

> *There were miners from Bisbee, and some came from Cork;*
> *Some from New Jersey, and some from New York;*
> *And a big-bellied Dutchman from over the Rhine,*
> *Got a job slinging muck in the Mountain Con Mine.*[87]

As the song states, the Irish came not just from Ireland or other mining regions of the American West but also from other parts of the US. The children of Butte's Irish-born were most commonly born in Michigan, highlighting the mass migration from Keweenaw, the copper mining area of the Upper Peninsula. The second most prolific birth state for children was Pennsylvania.[88] These same paths applied to the children of Butte's second-generation Irish.[89] The transfer request records of the Robert Emmet Literary Association (RELA), the radical Clan na Gael front, point to this movement path from Avoca in Luzerne County, Pennsylvania, to Butte.[90] The AOH also listed Pennsylvania as the second most common place from which members transferred between the years 1884–1892 and 1905–1916.[91] The anthracite coal companies' discriminatory hiring practices and vigilante-promoted violence against the Irish caused them to leave those states for Montana in the 1880s and 1890s, driven westward.[92] In both Pennsylvania and Michigan opportunity had disappeared, replaced by company hostility toward the Irish—making the choice to follow the promising letters calling them to a new land much easier.

In Butte the Irish, portrayed as unruly and disruptive by mining companies elsewhere, were remarkably docile once they garnered respect from their employers and a commensurately stable wage. Over the thirty-eight-year period between 1878 and 1916, the BMU was one of the most peaceable unions in America, with a membership as high as 6,000. Butte miners and their income provided vital financial support for other unions across the American West. The BMU's use of this power led to it being called "the greatest single social force of the working class in the western part of America."[93] Remarkably, it never led its members out on strike or challenged the industrial coalition that served as the economic bedrock for the community.[94] This flew in the face of a commonly accepted belief among labor historians that the further west one went, the worse the labor conditions and the more militant the unionism became.[95] A comparison of the peaceable labor situation in Butte with that in other mining districts with sizable Irish communities, such as Leadville and the Coeur d'Alene district, casts a dark shadow over the wisdom and motive of companies that spent vast sums on private security and armed militias instead of compromising with labor. The many examples of managements' private and public correspondence in preceding chapters demonstrate that it was not a mathematical equation measuring costs versus benefits. Their ideological and prejudiced attitudes predisposed them to an inability to accept any compromise. Their ambitions were focused less on profit and more on control, which in most cases they refused to share with their workers or unions—especially those dominated by Irish Catholics.

A multitude of dangers stalked the Irish in Butte, and the draw of high wages came at a terrible long-term cost. The arsenic-laden dust weakened and scarred miners' lungs and left them vulnerable to tuberculosis. One woman noted that most miners seemed to die at a young age. Watching her father crippled by "the miner's con," she saw how "he was like an old man."[96] He died at age fifty-two. The "rush for rock" heightened the dangers in the mines as the AC enacted new safety programs through the years, which management and workers dutifully ignored.[97] The conditions meant that Butte held the sad distinction of having the worst mining fatality rates in the world.[98] The language of one Irish lament was more direct: "I came to Butte city, to work for Jim Brennan, and now dig my grave."[99] The Irish were well aware of the risks. After weighing the dangers in his mind, an Irishman recalled, "The mine was very unhealthy on account of sulfur or brimstone that was in the copper but the pay was good and we worked away as well as we were able."[100] Metal miners earned $932 a year compared to an average worker's yearly income of $639.[101]

These dangers extended to families as well. Between May 1913 and May 1914, a startling 152 people died from tuberculosis, 5 of whom were women. Women were left "unhealthy from washing clothes saturated . . . with . . . bad air and giant powder smoke, and [being] compelled to inhale the deadly fumes."[102] During an American fundraising tour, Douglas Hyde observed Butte as his train pulled into the station. The Irish linguist noted the effect of the acrid air: "I am certain it is the most horrible place I ever saw. There isn't a tree a bramble, a plant, there isn't even a stalk of grass in or around it. There's miles and miles surrounding it without any growth. Everything has been burnt or eaten by the smoke that comes from the great chimneys in which Senator Clark smelts his copper."[103] The entire population was affected, including children who suffered a higher than average fatality rate, succumbing to sickness from the smoke and the dust that claimed their fathers. Across the industrial landscape, children played and predictably suffered accidents, sometimes falling down the mine shafts that pocketed the ground. The heavy toll of human life was infamous as far away as western Ireland, with one song calling Butte a town "where the streets were paved" not with gold but "with Irish bones."[104] The Irish faced down these dangers and continued to flock to Butte because they were respected there and they earned some of the highest wages offered to immigrant workers in the US.[105] Marcus Daly's notorious bias ensured that during his time as owner, the AC's bias was directed against known APA members and Orangemen. This was not simple Irish Catholic prejudice or collective memory but rather ethnic solidarity against those who, outside of Butte, usually had the advantage over them. Ethnic solidarity naturally meant a degree of ethnic exclusion.

Daly managed his mines closely, and by 1894 he employed 1,251 Irish-born (and an unknown number of Irish-Americans) in his mining business compared to only 365 English-born out of a total of 5,534 men.[106] Certain mines were known as "Irish mines," with handball courts built next to the miners' changing rooms and job notices posted in Irish.[107] Daly, proud of his mining background, often walked through the mines prowling the different levels of his tunnels while doing spot checks of his workers—an unwelcome physical manifestation of the patriarchal owner class infringing on working-class space: "It was understood if Marcus caught you with your back straightened up you was fired."[108] Such knowledge sharpened the class divide within the Irish community, as evident in this Irish poem:

Is mór mór go mb'fhearra bheith in Éirinn an ghrinn,
Ag éisteacht le ceolta na n-éanlaithe ró-bhinn,
Ná ag lorg lá oibre ar spriúndlóir beag cam,
Gur dhóigh leis gur asal tú a bhuailfí le feam.

It's far far better to be in Ireland where there's cheer,
Listening to the melodious bird songs,
Than looking for work from a crooked little miser
Who thinks you're only an ass to be beaten with a stick.[109]

Matty Kiely, a ballader from County Waterford, remarked that "no man need wear a copper collar but for the eight hours he's down in the hole."[110] Considering that these Irishmen willingly came to work in Daly's mines, their hostility might seem somewhat surprising, but they viewed Daly as both friend and oppressor, victim and perpetrator of the same system that exploited everyone. Although Daly shared an exile heritage with his workers, there were limits to ethnic loyalty, especially when it came to ridicule.

Harry McClintock recounted the life of wandering miner Paddy Burke to folklorist Sam Eskins. Burke was a gifted storyteller who lived off that trade instead of laboring underground and so, like an ancient Irish bard, he traveled from mining town to mining town entertaining people in barrooms. In the gold and silver mining camps "men were pretty generous . . . but he wandered into Butte, and fell upon hard circumstances," meaning he had to find work underground. It being Butte and Burke being an Irishman, he found a job mucking in Daly's Neversweat Mine.[111] During one night shift, Burke was telling a story atop his muck pile when Daly came across the workers. Noting that a chill had come over the group, Burke pushed on with another story, immediately telling his clandestine crowd to gather around once more. Since the men knew they were going to be fired regardless, they listened. Burke told them in wonderful, winding detail that he dreamed he had gone to heaven. The story built slowly to the moment he arrived at a great banquet where the hosts of heaven were assembled. He was about to take his seat when it was suddenly announced that Marcus Daly had died. Daly promptly arrived in the hall, and the climactic end of the story was reached: "And from the other end of the hall, where the great golden throne was, came another voice, 'Marcus Daly, welcome to heaven. Advance to the foot of the throne.' So the hosts of heaven parted like the Red Sea . . . And

Marcus Daly, all alone, walked down the aisle to the foot of the great golden throne. Almighty God arose, and He says, 'Marcus, I bid you welcome.' And turnin' to his right hand, He says 'Come, Jesus, get up and give Marcus your seat.' "[112] Burke spotlights Daly's vaunted status by having him on the cusp of arrogantly taking Jesus's seat. The story serves as a parting shot by the bard to the copper king's pretentious preeminence in Butte, implying that even if his position was endorsed by God almighty himself, it does not make his actions right.

The complex relationships within the Irish community did not mean the Irish lived in isolation from the wider diversity of Butte's population, and there were Irish who worked for other mine owners including Daly's rival, William Clark. One Butte Irish song commented on the mining town's remarkable diversity:

We do our work for an Englishman
And room from a French canuck
We board at a Swedish restaurant
Where a Finlander cooks our chuck
We buy our clothes off a German Jew,
And our shoes from a Russian Pole.
And we trust our hope in a Roman Pope,
To save our Irish Soul.[113]

These popular stories, songs, and poems reflected Butte's Irishness, its place within American society, and important local events. The people utilized a vast range of cultural muses for their purposes, as the modified lyrics for the famous William Jerome and Jean Schwartz song "My Irish Molly O" show. They were changed the following way by locals:

Oh Maggie dear, and did you hear
the news that's goin' round;
They're firin' all the Corkies
That are workin' underground.
I've rustled at the High Ore,
And I've rustled at the Con,
And the dirty blackguard Bohunks
Is all they're putting on.[114]

Judging from the specific reference to Corkonians and the strikebreaking Hungarians as "dirty blackguard Bohunks," this version appears to have been intended for an Irish audience. Other ethnic groups seem to have participated in a similar rewriting of songs, and one version of the above tune contains the third line "They're canning all the savages" and the last lines "But the dagoes are the only ones / That they are putting on." The use of "savages" indicates that the intended audience for this version was the Cornish, who would have approved of the use of "savages" as a reference to the Irish.[115] Though the origins of these parodies are unknown, they probably date from the 1910s, and the last Cornish version neatly echoed the beginning of the end of Irish dominance in Butte. Their faded privilege had disappeared, as indicated in this verse of a song in which a worker recalls being able to take a short break:

Oh there was a time in Butte Maggie,
When you could take five and hold your job,
But now its put the rock in the box Maggie,
And then put the waste in the gob.[116]

A Terrible Beauty

Daly died in 1900, and his memory lived on in the form of a statue built by the famous sculptor Augustus Saint-Gaudens, who not coincidently was born in Dublin. More important, the statue was publicly sponsored, echoing that of Thomas Francis Meagher in Helena. Matty Kiely, the man who had described his mine work as comparable to wearing a copper collar, visited the bronze figure of his former employer weekly, recounting the local news aloud to it. The only thing that upset Kiely about the statue was that it was placed facing the town, with Daly's back to the mines. Kiely commented: "'Tis no luck will ever come of it. In life Marcus Daly never turned his arse on the mines of Butte or the miners who dug them."[117] If there were limits to loyalty, then death was not one for Kiely. The sole surviving line of a song lost to time mentions that "ten thousand Micks laid down their picks when Marcus Daly died," reinforcing the point that the Irish could simultaneously have a healthy disrespect for their chiefs and an awareness of the importance of their contributions to the community.[118]

Figure 6.2. Ancient Order of Hibernians gather in front of St. Patrick's Church in Butte, circa 1910. Archives and Special Collections, Maureen and Mike Mansfield Library, University of Montana, Missoula.

Daly's eventual successor in business was John D. Ryan. An Irish-American cut from the same cloth as Charles Sweeny, Ryan owed most of his good fortune in Butte to his reputed charming nature and his personal connection to Daly's widow. Like Sweeny, he was a businessman first and foremost. Under his control, the AC would no longer limit its revenue by propping up Irishtown or yielding employment decisions to Irish unions and fraternities. Ryan was focused on profit and cared little that loosening social cohesion and removing control of Butte from these groups left the younger, disgruntled Irish with fewer reasons to join traditional organizations such as the BMU and the AOH (see figure 6.2) and thus fewer reasons for them not to join more radical organizations, such as the Industrial Workers of the World (IWW). The seismic rift within the Irish community took place over a tumultuous few years but can be traced to their increasingly insecure and desperate position and their anger at a class who had well-paid, established positions.

Figure 6.3. The Miners' Union Hall after it was dynamited on June 23, 1917. Archives and Special Collections, Maureen and Mike Mansfield Library, University of Montana, Missoula.

The changes began in 1912 with a major effort by the BMU and the AOH to solidify their grip on the workforce. They wanted to stop radical groups from establishing a foothold in Butte and so set out to preemptively destroy the possible base of their support with the mass firing of 500 Finnish miners from the company. Many of the men were members of the BMU, and together they appealed to the union's committee.[119] The BMU leadership appeared ready to stand with the Finns, but one member managed to get a motion to strike put to a vote of the Irish-dominated membership, which refused to jeopardize its employment to support the Finns or their radical element. The bishop of Montana, John Carroll, the national chaplain of the AOH and a mouthpiece for conservatism, thanked the miners for "voting them down," highlighting the broader social pressure the workers faced.[120] The AC then established a rustling card system through the BMU, purportedly to limit infiltration by subversives; however, by 1914, the AC was using the system against all workers to lower wages and create a company blacklist.[121] The ties that bound company to union to community sundered spectacularly when the main union office of the BMU was demolished by a progressive branch of the union that soon thereafter established its short-lived alternative, the Butte Mine Workers' Union (figure 6.3).[122] Fractures were becoming fissures.

As World War I began, the old guard continued to wander toward irrelevance, unable to find its footing in the new, unstable wartime landscape. The US may not yet have joined the war, but many Irish-American nationalists openly supported Germany as a means to help Irish independence, as did nationalists in Ireland.[123] In Butte, this support translated into an alliance between the German and Austrian communities, exemplified in the 1915 St. Patrick's Day parade when thousands of Germans and Austrians joined the procession. The *Anaconda Standard* reported that it might have marked the first time in the history of the world that the German, Irish, and American flags were carried in the Irish parade.[124]

For the Irish, this was not solely an expression of nationalist hope that a German victory over Britain might lead to Irish independence; the symbolism was directed at the local crowds watching the parade and emphasized the importance of recent changes in Butte's ethnic makeup. Irish loyalty and friendship toward these other groups presupposed the continuation of Irish dominance in Butte and forestalled German and Austrian attempts to challenge the established order. As Father Brosnan wrote to his father, the Irish were "beginning to get beat on account of . . . Austrians . . . Germans and Dagos" and thus made a strategic alliance.[125] The ethnic alliance might have succeeded had the Easter Rising in Ireland not preceded the US entry into the war. These two events forced a decisive declaration of patriotism from Irish-America and undermined commitment to its dual identity.[126]

Between the two events, the famous Irish labor leader James Larkin visited Butte, and the episodes that followed exposed deeper internal divisions in the Irish community. The Dublin Lockout of 1913 garnered the Irish labor leader international recognition, and in late 1914, "Big Jim" traveled to the US to collect funds for the rapidly expanding workers' militia, the Irish Citizen Army.[127] The dominant Irish-American establishment feared Larkin's incendiary presence during his first visit to Butte in September 1915, when at the last minute acting mayor Michael Daniel O'Connell withdrew permission for Larkin to speak at the city auditorium. The Finnish Workers' Club offered to host him. An angry Larkin told the assembled crowd, "I love my native land and I love my race, but when I see some of the Irish politicians and place hunters you have in Butte, my face crimsons with shame, and I am glad they did not remain in Ireland." This was perhaps the ultimate insult for leaders of an emigrant community.[128]

Larkin visited Butte again in June 1916, and his focus was on the recent Easter Rising in Dublin. In fierce language he reminded the largely Irish crowd that the rebellion was "a working-class rising to keep Irish boys out of the British army." Drawing closer comparisons he said, "You forget that the struggle you have here is the same you knew in Ireland—The struggle against economic and political tyranny."[129] Larkin's authority on these matters was unquestionable after the lockout. Men knew that if Larkin had been in Ireland during the Easter Rising, he would have been executed alongside Irish labor leader James Connolly.[130]

Larkin excoriated local nationalists as nothing more than "mercenary phrase mongers" using the cause of Irish freedom to help themselves. James Mulcahy, editor of the *Butte Independent* who was in the crowd listening, finally stood up to protest this abuse and challenge Larkin. Larkin turned his fury on Mulcahy: "Real Irish patriots would scorn to recognize the likes of you," silencing him instantly.[131] Larkin warned the men not to listen to leaders urging caution and directed them to "be true to the spirit which inspires the rebellion in Ireland, you must do your own thinking and not delegate it to any judge, lawyer, editor or priest."[132] He was appealing to the men's deep sense of cultural loyalty while simultaneously stripping local leaders of their own, thereby adeptly undermining their legitimacy.

Ireland was not some distant location; rather, it was a nearby reality, akin to the spirit world, one that spoke to them and to which they spoke back. This was expressed not just in letters or funds but also in manpower, as people traveled back and forth. Some in Butte answered John Redmond's call to join the British Army at the outset of the war in the hope of securing home rule for Ireland. Those who did suffered ridicule for their efforts:

But every day brings more news of the allies and all,
How the big wheel was wrecked by a Dutch cannonball,
And the long man from Cork bought a ticket from Butte,
To make slaves of the Irish scooping rock in the shute.[133]

As part of a much longer song, this verse referenced one person, nicknamed "the long man," who did exactly that. Interestingly, his effort to serve in the British Army was portrayed as simultaneously serving the enemies of both Ireland and the working-class Irish, a sort of double betrayal—economic and political—that Larkin wholeheartedly endorsed.

Larkin visited Butte once more in early 1917, and again he railed against the "parish pump form of patriotism" and leaders who were little more than "malignant beasts in human form . . . gombeens in their relations with their fellow Irish, shoneens in their slavish servility."[134] The use of the Irish words *gombeen* and *shoneen* added deep cultural weight to his insults. A gombeen was an Irish person paid to do the dirty work of absentee landlords or the British government; a shoneen was an Irish person who adopted the customs, language, and traditions of the oppressor. Irish traitors, both economic and cultural, were two sides of the same coin and were being used in the copper mining city 5,000 miles from Ireland. For younger Irish immigrants, Butte no longer offered secure employment, and they saw the AOH and RELA as relics of the recent past. To fully contextualize this shift, it is worth pointing out that the RELA was the local Butte branch of Clan na Gael, whose radicalism was so strong it had once refused to meet with Father Sheehy—the Land League priest who was imprisoned by the British for agitation—because of its opposition to clerics and the fact that they spread Anglicization and "respectability" among the working class in Ireland. Indeed, one member went as far as to say that the "deplorable state of Ireland" was due to "Catholicity, especially English catholicity."[135] By the 1910s, most RELA members were from an older generation that could not understand and was therefore unable to connect with the younger Irish whose radicalism reflected economic insecurity.[136]

A significant manifestation of this disconnect was the emergence of the Pearse-Connolly Irish Independence Club in Butte before Larkin's last arrival. A possible indication that Larkin's speeches were having an effect lay in the organization's dual purpose of tying the cause of labor to that of Irish independence and attempting to reclaim Connolly as a socialist hero rather than a purely nationalist one—a development that infuriated Larkin. The AOH protested lamely that there were too many Irish organizations in Butte.[137] By St. Patrick's Day, the AOH—the traditional organizer of the parade—abdicated leadership of the event to the Pearse-Connollys, ostensibly out of a fear that the heightened tensions would lead to a riot. The AOH replied to inquiries with the statement "we are going to mass," giving up on the effort to coordinate that year's parade and effectively surrendering its leadership of the Irish-American community for the narrower allegiance to the Catholic Church.[138] Acting for the establishment, Mayor Charles Lane refused to grant the Pearse-Connolly club the permit for the parade, but in spite of him calling it "an IWW affair," the parade proceeded.[139]

The strength of Butte's nationalist and socialist sympathies is illuminated by the unusual presence of a priest named Father Hannan as one of the founders of the Pearse-Connolly club. He even allowed the group to hold meetings in his church.[140] Not all of the Irish-American clergy in Montana were as sympathetic to labor organizations as Hannan. In Helena, Bishop John Carroll, the AOH's national chaplain, had made his position clear during the earlier BMU struggles in 1912; he further condemned the IWW as "a purely Socialistic organization."[141] The fact that the Pearse-Connolly club, the IWW, and the Finnish Workers' Club all shared the same address at 318 North Wyoming, Finlander Hall, speaks volumes about the affinity among the groups as well as the widening rift within the Irish community between the established, more conservative leadership and the newer Irish workers lower down the social ladder.[142] However, the presence of a priest in the Pearse-Connolly club helped weaken criticism of the groups from the Catholic hierarchy.

By April 1917, the BMU no longer existed; instead, Irish and Finnish workers found themselves fighting a battle on two fronts—protesting both company treatment and US entry into the war. On June 5, several Pearse-Connolly members helped found the Mine Metal Workers' Union (MMWU) whose headquarters was the same building as that of the other socialist organizations: 318 North Wyoming. These men felt they had more in common with their fellow Finnish workers than with their compatriots. On June 8, the Speculator Mine disaster resulted in the deaths of 165 men including Jim Moore, whose letter to his wife was mentioned earlier. The MMWU led the workers out on strike from June 11 until December 28 of that year. The United States had officially joined the war, but because of the strike, production of copper virtually ceased. As one company official told the Montana Chamber of Commerce, "The Kaiser had found one of his most effective allies' [*sic*] in the IWW and the Pearse-Connolly Club."[143] Certainly, a hatred for Britain was shared in both Butte and Ireland, but the more complex building of unrest and the curious shape of alliances make it a more complex story than that alone.

This complexity was further seen with the lynching of IWW organizer Frank Little in the early hours of August 1. Armed gunmen, posing as police, beat Little and dragged him from his boardinghouse room into a car. They then hanged him from a railroad trestle with a note pinned to his underwear that read "Others Take Notice. First and Last Warning. 3-7-77," followed by the initials of labor leaders in the town. In the nineteenth century, vigilantes in Montana used these numbers to identify their murders.[144] Little's funeral was one of the largest Butte had ever seen, and 1,000 members of the Pearse-Connolly club with their bright green sashes marched second in the procession only to the IWW.[145]

Rumors circulated as to who was responsible for the lynching. The perpetrators were never caught, but two theories are worth mentioning. The note points to a reformed vigilante organization, popular in the 1870s and 1880s, revitalized as a force against labor agitators in the twentieth century. Little had certainly made inflammatory comments attacking the government, capitalists, and soldiers, whom he called "uniformed federal thugs."[146] Yet another theory believed by locals suggests that the vigilante note was a cover for the Anaconda Company goons who carried out the lynching in hopes that Little's death would incite the workers to violence, whereupon the federal troops stationed nearby would be called into action to suppress the strike.[147] Copper, a wartime commodity, had greatly increased in value and the company was losing money every day the mines stay closed, so out of desperation, the company might have tried such a tactic.

Wherever responsibility lay, it was obvious that the unity of the Irish-American world in Butte had been shattered by the Easter Rising and the US entry into the war, and these events drove a deep wedge between the two sides of the hyphenated identity. Yet the events only provoked a confrontation that was already bubbling beneath the surface between workers and management, between lace-curtain Irish and shanty Irish, and between the company on which the town depended for its prosperity and the workers who were dying by the hundreds for their pay. John D. Ryan, the inheritor of Daly's mantle, and Cornelius "Con" Kelley, his vice president, seemed like the ideal candidates to perpetuate the Daly machine. Both were Irish-American, both had intimate ties to the local community, and both were experienced businessmen. However, the Anaconda Copper Mining Company under Ryan's rule was not the outfit it had been under Daly. Ryan and Kelley focused entirely on the pursuit of profits and willfully neglected the structures that had made Butte the most Irish city in America. In hindsight, the foundation for the violent eruptions during World War I had been laid during the preceding decade, if not earlier. In the same year the statue of Thomas Francis Meagher was being erected in front of the state capitol in Helena, the lot next to the courthouse in Butte was purchased by William Clark. It became the four-story headquarters of Butte's Silver Bow Club, where the rich enjoyed fine cigars and expensive drinks in opulent settings sufficiently distant from the plebs.[148] Its location made a statement identical to the one made by the presence of the Irishman's statue next to the capitol, and its elevated position above the court was a perfect encapsulation of the vaunted power of the wealthy. As martial law was enacted, the realm of the rich stood as the sole legal drinking establishment exempt from closure, thanks to the direct intervention of Governor Sam V. Stewart. Historian Mary Murphy notes that the New Year's ball was a particularly extravagant affair, and in that year of labor strife it was Mary Kelley, wife of the AC's vice president, who "outshone the other women in a dress of silver, with a girdle of gold cloth, marten trim, and a silver headdress"—an apt metaphor for the class chasm slowly cleaving Butte's Irish-America.[149]

Employment of the Irish became less certain. One disbelieving report stated, "Hibernians got discharged to make a place for members of the compass and square [Masons] and other organizations."[150] The AOH reacted as it traditionally did, by sending influential local figures to speak with the boss, Ryan, but it had little effect.[151] The writing had been on the wall for some time, and in 1910 the organization sent a group to Ryan regarding the "employment of our members"—Irishmen. The AOH records detail that they received a "very chilly reception, thanking this committee for calling upon him."[152] The confused AOH should have seen this as a sign of things to come. One AOH member noted that perhaps it was "not wise any more to be too Irish."[153] The AOH was unable to direct its anger at a management that was Irish-American but culturally disloyal. Whether management had thrown its lot in with the Masons or saw them as a useful tool to challenge Irish control of the workforce, it was obvious that it felt no loyalty to and had little respect for the Irish ethnic enclave and its representatives.

With the passing of the Sedition Act in 1918, that sentiment became ever more obvious. The *Anaconda Standard* published its opinion—with no sarcasm—that there were two types of people, those who were afforded rights and those who had no rights: "There is no longer freedom of speech for the disloyal or the pro-German. A man can talk all he pleases if he talks right. The loyal people of this country have and will have all the freedom of speech they want."[154] It was a remarkable shift for a paper established by Daly in 1889 to give voice to the Irish Catholic bloc of the Democratic Party in Montana.[155] The older generation heaped scorn on the younger generation, claiming that "their ideas drifted to pugefights [boxing], dogfights and brutality and not much above animal nature . . . houses of ill-fame had more attention for some of them than a Div. of the AOH."[156] Father Moran stood up for the younger generation and suggested that the AOH act constructively and provide dances, music, and other forms of entertainment to keep the young from "bad influences"; he admonished members to "not peddle so much hot air."[157] Their effort to maintain respectability trapped them in an unwillingness to adopt any policy more radical than speaking to the owner of the company; if the inability to guarantee employment for other Irish was a step too far, then it was truly beyond their ability to imagine changing their quality of life in Butte. The older Irish preoccupied themselves with attendance and roll numbers to maintain the appearance of importance—showing how languid their earlier success had made them, making them impervious to the reality of their ineffectiveness, and undermining the sustainability of the Irish-American community in the city.

Strikes and unrest continued in Butte after the war and into 1919. One military officer observed that the leaders were "Finnlanders, Sinn Feiners, and members of the Pearse-Connolly Club and IWW" and suggested that "the situation could only be resolved by the prompt deportation of undesirable aliens, mostly Finns and Irish."[158] This marked another departure that was representative of the weakening of the Irish community, not just an older nativist tendency reasserting itself. The fracturing of Irish Catholic ethnic solidarity in Butte between 1914 and 1923 was exacerbated by political complications in America and Ireland, strikes and civil war, so that Irish-American nationalist and labor sympathies now appeared incapable of marrying the diverging strands of identity. The growth of radical labor factions such as the IWW on one hand and an increasingly conservative, middle-class (and heavily

Irish) American Catholic Church on the other demonstrated this inability.[159] Éamon de Valera visited Butte twice in 1919. Speaking to crowds of workers, he carefully criticized Larkin, calling him "a labor writer" who wrongly believed that "the rebellion in Ireland was a social rebellion."[160] Pearse and Connolly could be joined by a hyphen in the afterlife as co-martyrs, but the living Irish leaders found it more difficult to ally.

Perhaps efforts to reconcile trade unionism with Catholicism were always doomed to failure, but the censure of the American Catholic Church's liberal movement sent a clear message to Irish-American Catholics to focus their attentions on the hereafter and not on improving their present station in life.[161] If the message was ambiguous, it was made clear on April 21, 1920. On the third day of a strike by the IWW and the MMWU, deputized company guards opened fire on the picketers, injuring sixteen and killing one, Tom Manning.[162] The event was dubbed the Anaconda Road massacre, perhaps an appropriate label not because of the number of casualties but rather because of the intense psychological blow it represented to the Butte Irish. Twenty-one years earlier, Marcus Daly had ordered a private investigation into the suspicious death of John Daly; now, the Anaconda Company would conceal the murderer of an Irishman through a rigged coroner's court.[163]

Irish-American miners in Butte sensed the overt and subtle warnings of the decline of their mining town and appeared to interpret the hereafter to mean an improved life for their children, as they encouraged them to move into other occupations. Less than half of the second-generation Irish were employed underground.[164] This occupational movement also changed Irish communities, as wealthier Irish residents left Butte or moved out of solidly Irish parishes such as St. Mary's in Dublin Gulch to the west and south sides of Butte.[165] The movement away from mining was not a result of Irish upward social mobility or the change in company favortism. While many saw opportunities elsewhere for themselves or their children, many Irish also hated the job or detested the urban setting:

Nil buaint ná rómhar ann, treabhadh ná branar,
Ná fuaim na speal ar bhántaibh,
Ach clismirt ag' motors gach ló ag tarrac,
Do scaipeadh do mheabhair chun fáin uait.

There isn't a reaping or digging there, ploughing or a fallow,
Or the sound of the scythes on the grasslands,
But the din of motors every day hauling,
It would drive your mind to stray from you.[166]

These sentiments were also found in the English language, as this song expresses:

Then who's all for Old Ireland, the land of good miners,
That dear little island, that I see in my dreams.
Ah go back to old Ireland, to the girls which are for me,
To hell with your mine and your mining machines.[167]

Powerful images of pastoral life that contrasted with the viciousness of America's industrial setting would have resonated with the Irish listening to these words and may have contributed to their mobility onward.

Michael MacGowan did not state his exact reasons for leaving Butte other than to follow new opportunities, although his ability to do so was funded by his employment in the copper mines. He followed the gold rush to the Klondike and, as he continued wandering through Montana, he arrived in Missoula late one night on his way to the coast:

> We gave a good loud knock on the door and it wasn't long till the owner stuck his head out of the upper window. He heard us down below talking in Irish—and the first greeting he gave us was to ask, in the purest Irish, what the devil were we up to at that hour of the night. He questioned us as to where we came from and, when he was told, he came down on the spot and let us in. He was as surprised and delighted to meet Irish speakers like this as we were ourselves to meet one, in such a place. But that's the way it was in America then. You'd never know the time or place when you'd come across an Irishman . . . I don't think anyone spoke a word of English on that long journey from Butte in Montana.[168]

For MacGowan and others like him, Butte was not just a bastion of Irishness but a staging ground for the Irish. MacGowan's retreat from Butte provides insight into how quickly some could move on to find other forms of employment in different US regions.[169] Some remained tied to the community in Butte even after they left. One man continued to send his dues to the Irish fraternity in Butte from California with "the wish to be remembered by all his old friends."[170] Such friendships played an important part in sustaining these networks of those who left and those who stayed, be it from Ireland to America or Butte to California.

The networks—be they family, friendship, localism, tribalism, cultural, religious, or any combination of the aforementioned possibilities through which the Irish defined themselves in the nineteenth and early twentieth centuries—enabled the creation of small and large communities. This explains how Irish villages such as Smartsville in the California foothills were connected to cities such as Butte nestled on the Continental Divide. This divide was in ways like the hyphen between Irish and American. It was not a division that separated things, in the way nativists saw it; rather, it was a feature that bound the parts of the thing itself together, allowing a new sense of self to emerge. The broad possibilities and definite limits of Irish-America are seen in what emerged from its contested space. Success in America is often crudely framed as the accumulation of wealth; but the Irish who had fortunes, such as Thomas Cruse and Marcus Daly, used the money to channel their Irishness, through the church or through business, respectively. Later, less-than-ripe Irish-American fruit, heralded by Charles Sweeny in Idaho and epitomized by John D. Ryan of the Anaconda Company, revealed the limits of cultural loyalty. As the twentieth century continued, it was clear that both Irish-America and the mining industry would require leadership, imagination, adaptability, and support if they were to survive in some familiar but reduced form. That did not happen.

Conclusion

One of the most vivid depictions of an early miner in the American West is the opening scene of the film *There Will Be Blood*, based on Upton Sinclair's book *Oil!* In the opening shots, we see the solitary protagonist drive a vertical mine shaft in the dusty scrubland and then chip samples away from the rock face of what might finally be a promising seam of ore. As he climbs out of the literal and metaphorical hole he dug for himself, he falls and breaks his leg. Alone and in agonizing pain, he crawls to the nearest town, miles away. From the floor of the assayer's shop, he intently watches his samples being tested, the scales of capitalism weighing his worth. His grimaced face reveals his inexorable focus on his prize even at the cost of life or limb. The scene has no dialogue, heightening the overarching sense of isolation and danger. The Irish-American Edward L. Doheny provided the real-life basis for this character.

It is a powerful scene that lacks historical accuracy in one key respect—the character's utter isolation. Placing him alone in the spectacular remoteness of the mountains, valleys, and deserts of the American West implies that he

https://doi.org/10.5876/9781646422517.c007

might be swallowed whole, never to be seen again, literally entombed in the stygian darkness of his own making—a powerful metaphor for the loss of humanity within capitalism. Historically, few Irish lost their way, thanks to the tight bonds between friends and families. Community offered definition, protection, and meaning to the Irish during this transient period.

Identity proved persistent, and intergenerational links endured. The second and third generations retained different aspects of the newly hewn Irish-American identity. Doheny's Irish-born father and Irish-Canadian mother met in St. John's, Newfoundland, and moved to the mill town of Fond du Lac, Wisconsin, where Edward was born.[1] At age sixteen he left his family home and became a mine prospector in the Rocky Mountains in the 1870s. He lived in Kingston, New Mexico, from 1880 to 1891, where he prospected in the surrounding area with varying degrees of success and married the daughter of the owner of the Occidental Hotel, a well-known Irish-American hangout.[2] His wife's poor health and a string of bad investments led him to move his family to Los Angeles, where his mining expertise served him well when, after visiting a nearby tar pit, he decided to drive a shaft and struck black gold.[3] Within a few years, he was a famous wealthy national figure and a symbol of success for Irish-America. Doheny demonstrated the importance of his hyphenated heritage through his generous support of the Catholic Church and the Irish struggle for independence.[4] The destruction of Doheny's political and historical reputation during the Teapot Dome Scandal hid his importance in American life and, more important, obscured the bonds that tell us what his life meant to him.[5]

Doheny was one of many thousands of Irish or people of Irish descent who shaped the mining frontiers of America—from the seismic Gold Rush, to the development of dozens of Irish communities including Virginia City, to the Colorado and Idaho labor wars and Butte's riots during World War I. They believed in the possibility of high wages worth risking life and limb for. Many others, not miners, also formed a vital part of these emerging communities. They were supported through the vast migration networks of information—often finances—that encouraged the movement from Ireland and Britain to the US and enabled the creation of Irish neighborhoods and communities in many mining towns. These Irish traveled to the US believing it was a more prosperous and egalitarian society, with better opportunities for them and their children. Reality clashed hard with these expectations, and Irish miners frequently found themselves embroiled in the painful birth pangs of America's industrialization.

Communications between miners and their families and friends in other parts of the US or in Ireland contained details about every aspect of their lives, and they often expressed loneliness and regret. In one such letter, after detailing the recent death of her four-year-old daughter from scarlet fever, Ellen Wogan continued to her brother in the far West: "Dear Brother you would like to hear from all your old school-fellows. There is not one here that I know of they are all dead and scatered [*sic*] away."[6] Such news understandably caused migrants to despair at their lot in life. "I have mined so long I am hardly good for anything else" wrote another Irishman, William Kennedy. He continued: "Oh dear oh dear how I can look back and see all the old familiar places and faces and imagine myself back there a boy hen [*sic*, probably meant boyeen] going to school with my old School mates—bare foot running races on the green grass and wading in the Lagan in the afternoons coming from School. Oh dear oh dear those times will never come back to me."[7] Despite his melancholy turn, he ended the letter determined that "I will see all the old places, once more before I quit this world."[8]

A few had the opportunity to fulfill this hope and return to Ireland. Thomas Walsh's mother died in Ireland when he was an infant. After he made his fortune in Deadwood, South Dakota, he funded a headstone for his mother's grave, which he visited: "a stone cross that he had paid for with Black Hills gold."[9] His daughter described their journey to Ireland: "The whole of our journey was simply part of a pilgrimage to Tipperary . . . he picked up the threads of many old friendships, and his hand was in and out of his pockets throughout the days; he made no show of the giving, not ever."[10] Others who had the opportunity to return home after a lifetime abroad often found their quest a disturbing experience: "In the township where I was born, every person or family living in it at that time, was related to me in one way or another; the day I visited it I could not find in it one soul of kindred blood; all were gone. The older generation of course had passed away, and doubtless most of the next had emigrated to foreign lands; anyways I met no person under fifty who had ever heard of me or any of my family."[11] Time and place had shifted beneath his feet.

Thomas Higgins, one of the few to revisit Ireland, pronounced himself "disappointed in the country and the people" and bemoaned the fact that it rained twenty-seven of the thirty-one days of his visit: "Even in general appearance I was disappointed, the hills and mountains did not appear as high as I thought they were, the river did not seem so wide as when I was a boy fishing on its banks."[12] He blamed emigration for the degradation of the Irish people, "the weaklings had not [emigrated]," implying that it was the strong who left and the weak who stayed and that those who left were better Irish—an interesting reversal of the accusation that those who emigrated "fled" Ireland and were no longer Irish.[13] Higgins detailed the pitiful forwardness of people who accepted gifts or begged from strangers. In his mind, Higgins had rationalized the dreadful death toll during the famine as a sort of noble sacrifice because people starved rather than receive relief food.[14] The persistence of the insult "souper" (implying that someone or their family took the soup) well into the twentieth century in Ireland shows the power behind this idea. In part, Higgins blamed the generosity of those who left: "As almost everyone living there now gets remittance from their friends in America, the habit has made a beggar of them."[15] He was correct that remittances were a widespread practice, but as they were used most frequently to pay for passage to America, it seems unlikely that families squandered them.[16] Ironically, his attitude is reminiscent of that of English administrators during the Great Famine, but his insults are also an expression of his feeling of neglect by his homeland. This was not a sentimental connection with the homeland, some vague diaspora; it was much more intimate and familial than that, and its pained complexities made it more real.

In spite of his complex feelings, Higgins revealed in another part of the letter that he had reconnected, perhaps reconciled, with his brother John after a twenty-year gap. John died subsequent to this exchange and left behind a widow and ten children: "I need hardly add they were left poor and unprovided for, it could not be otherwise, with a family so large, and a poor country. I came to their assistance at a most opportune time; I had them educated and made ample provision for each, so they can now enjoy an abundance of good things in the world."[17] This connection was not enough to reestablish his sense of place in the homeland when he eventually decided to return to Ireland. He felt out of place and out of time, and his words rang as a sad short reflection of this: "I was an utter stranger."[18] America was now his home and Irish-Americans were his people. His visit to Ireland reconciled him to this reality.

Emigrants attempted to make sense of their insecurity through cultural prisms. The poem "*Mo chiach mar a thána*" by Seamus Ó Muircheartaigh promotes worker solidarity through specific folk images of the Irish diaspora centered on English theft. After detailing his hardship working in the mines of Butte, his poem ends with this verse:

Nuair a thiocfaidh an lá is nuair a lasfaidh an spéir,
Go ngluaiseoidh na sluaite faoi bhrat uaithne na nGael,
Go mbeidh sibh an lá úd faoi ghradam sa bhruighin,
Ag ruagadh an tSasanaigh dhamanta thar toinn.

When the day comes and the sky is alight
And the hosts march under the green banner of the Gael,
That day you'll be glorious in battle
Which will scatter the cursed English across the sea.[19]

These anti-English sentiments appear across the Irish mining community, not just in Gaelic poems. The poem "What Ireland Was She May Be Again" began with the lines "Barbarian England groped in savage night, / While Erin beamed with Civilization's light," whereas the song "God Save the Bastard King of England" was taught to one miner by an "Irish miner in California."[20] The latter's origins are murky, though it is unsurprising that Irish-America would learn and propagate a ribald melody excoriating Europe's monarchs as venereal disease-riddled degenerates. John F. Kearney wrote these lines in his poem "Enigmatical Acrostic" for *Miners' Magazine*:

Beggared exiles we of Pirate England's make
Outcasts whose spirits will not bend or break–
Live! Live! To see a European earthquake![21]

Poems and songs were important methods for the Irish to share their culture with other ethnic groups and to impart memories and political views from one generation to the next. Finding such sentiments in a general miners' publication proves the resonance of these songs and the miners' sophisticated awareness of Irish grievances. Notably, these culturally specific images targeted Irish imagination as a tool for promoting worker solidarity.

The invisible distinction between culture and politics is further highlighted by the fear that prosperity would weaken the emigrants' cultural convictions. One returned "Yank" wrote, "I was only home a few months from America and during my absence, I may add, I did not learn to love Irish landlordism or English rule."[22] Emigrants understood that their time away from the homeland was seen as weathering their identity and their sense of fairness embedded in that identity. Irish culture had developed a sharpened sense of righteous adversarialism against illegitimate laws, exploitation, and disenfranchisement. This belief—gestated within a history of British misrule in Ireland—was the source that fueled Irish organizational skills and enthusiasm for protest, and it appears in many Irish emigrant accounts.[23] This relentless hunger for respect was intertwined with their identity and shaped by their image of themselves as exiles. In turn, it could be used to challenge injustice in their new home or even in their old homeland, a feat they accomplished in many different ways: through membership in unions, subscriptions to fraternities, and support for causes such as the Land League and Clan na Gael or the nascent Irish republic and its struggle for independence in the twentieth century.

Another Irish poem, "*An Spailpín Fánach*" by Seán Ruiséal, further echoed these same sentiments:

Tá mo shúilse 'rís le Rí na bhFlaitheas,
Is leis an Mhaighdin mhánla,
Go dtógfaidh Miléiseans arís a gceanna,
D'fhonn bualadh thabhairt do námhdaibh;
Beidh bodaigh an Bhéarla dá ruagairt feasta
As Oileán naofa Phádraig,
Is dlí na nGael arís dá spreagadh–
Sin cabhair ag an Spailpín Fánach.

My eyes are again with the king of Paradise,
And with the gentle Virgin,
That Milenaus again will lift their heads,
With the wish to grant defeat to their enemies,
The English language louts will be chased henceforth
From Patrick's holy isle,
And the laws of the Gaels propagated again,
That is a help to the Wandering Laborer.[24]

Irish miners understood their position as spiritual exiles like Adam and Eve, as well as their physical suffering in the sense of Catholic mortification, as a penance. Ejected from Ireland/Eden and forced to suffer, *gamentes et flentes in hac lacrimarum valle* (mourning and weeping in this valley of tears), they placed their hopes for the future on their children and hopes for themselves on the afterlife. But the *"dlí na nGael"* (laws of the Gaels) mentioned in the poem tell us of their awareness that there were alternatives to the current unjust system of laws, a paradise lost. This strong sense of injustice was passed down to the second generation, as demonstrated in the activities of Irish-Americans such as IWW activist Elizabeth Gurley Flynn in trade unions and other radical organizations.[25]

Mining in the nineteenth century was hard, miserable, dangerous work. Nonetheless, it was an important preference for many of the Irish who came to the United States. Like other sections of the Irish diaspora in the US, these Irish-American mining communities demonstrated organizational skills par excellence through their membership in and leadership of fraternities, unions, and the Democratic Party. Irish emigrants who grew up in mining regions in Ireland gained valuable experience there, and so would the many who worked in British mines before traveling to the United States and the American West. These mining skills proved valuable and would be a factor in establishing further frontiers of opportunity for relatives and friends, both at home and in neighboring parishes. The trails they traveled were journeys taken not only in search of jobs, wives, and comrades but also in search of respect and community.[26] Irish Catholics carved out a community in the grassy hills of Yuba County and the barren steppes of Virginia City, Leadville, and Butte in much the same way they tried to in many other mining towns in the American West, with varied success.

Butte was similar in many ways to its predecessor, Virginia City. Irish mine owners' sympathy toward their own people meant that the Irish community had a solid footing in both towns. In Virginia City, James Fair, James Flood, John Mackay, and William O'Brien were Irish-born or second-generation Irish; in Butte, Marcus Daly hailed from County Cavan. None of these men forgot their roots or their compatriots. They and their Irish workers refused to travel the violent and bloody path of industrialization many other Irish encountered in Colorado and Idaho, where professional mine managers relentlessly and unashamedly pursued control and profits over compromise and peace. Daly—through his dominance of Butte, his loyalty to the Irish community, and his powerful rebuking of eastern business interests—made the city an Irish (and by firm extension unionized) oasis of peaceable labor relations until the events of World War I sundered Irish dominance and the ties among the union, the town, and the company, ending the thirty-six years of labor peace in the city.

Simultaneously, these historical examples reveal the source of the problems in other troubled mining camps and towns throughout the American West, demonstrating that responsibility for repeated escalation of labor disputes frequently lay at the door of mine managers and owners. They orchestrated disputes as part of an effort to strengthen company control of mining towns while concurrently weakening workers' negotiating positions and wages. In these efforts, they specifically targeted Irish workers and communities.

The managers and owners, usually Anglo-Americans, viewed the Irish as disruptive, obstinate, and a barrier to increased profits and were well aware of their centrality in union organization and membership. Predictably, these tactics, in which state and federal authorities were often complicit, did not end in 1920 or with the disappearance of the Irish as a major mining population or with the decline of the ethnic enclave. Later events in West Virginia in the 1920s and in Harlan County, Kentucky, in the 1970s demonstrate that the brutal tactics used in the American West were not a historical aberration.

Irish miners struggled not only to improve working conditions but also to climb the social ladder—if not for themselves, then for their children. They further struggled to carve out their place among the Chinese, Welsh, Cornish, Italian, Hungarian, Polish, and American miners. Against the mélange of other workers and company exertions, they deployed the effective mechanism of ethnic solidarity and, conversely, exclusion. When the Irish saw that their goals aligned with those of other ethnic groups and when those ethnic groups agreed, alliances formed; but Welsh and Cornish hostility toward the Irish often precluded this opportunity. Even when such alliances formed, Irish-American communities remained remarkably endogamous throughout the nineteenth century. Many ethnic groups clustered to enhance their position, but few did so as successfully as the Irish. Because of the way society was organized, ethnic allegiances preempted unions or sometimes acted in lieu of unions. These were the messy compromises that caused the compartmentalized societies to flourish or flounder.[27] Miners relied on numbers and the internal cohesion of varied ethnic groups to stake out their claims on communities. The Irish did this in many towns such as Marysville and Smartsville in California and in the Coeur d'Alene mining district in Idaho, but there was nothing exceptional about this since the Cornish did the same in Grass Valley, California, and the Chinese managed it in the Boise Basin. Each clustering enabled the dominant ethnic group to leave its mark on the region. Over time, ethnic identities became less important, but it was not the shifting movement from opening one mining frontier to the next that diminished that importance; rather, it was the declining usefulness of kinship in an increasingly corporatized mining industry.

Mining and the railroads that followed enabled the American expansion west. Our popular image of the westward-bound pastoral settler ignores the more numerous and important industrial workers, particularly the miners. As Patricia Nelson Limerick wrote, "Mining set the pace of development": all other industries, both agricultural and commercial, were ancillary props to it—from the Gold Rush until World War I.[28] The frontier for mining was not geographically defined as the American West; rather, it was a series of developments, an industrial frontier that swept across the United States in different stages. The entrepreneurial system envisioned by Free Labor advocates existed for a time in the Rocky Mountains, in small mines established by groups of miners setting out and developing a claim. Large-scale industrial mining required vast investments of capital, technology, and labor to exploit deep veins.

As eastern business interests moved further west, this frontier gradually closed—not geographically but economically. Some mine companies and Irish-miners-turned-mine-owners remained independent and influential, sheltering the Irish community in towns such as Virginia City and Butte. However, by the early twentieth century, these mine companies had changed too, becoming the impersonal corporations we are familiar with today. They existed under the control of a professional managerial class, beholden not to some sort of social contract with workers and the local community but to distant shareholders far removed from the mining towns and the people who lived there. As such, the American West offered only temporary respite to the Irish as a frontier of opportunity.

The "analysis of networks . . . rather than all-encompassing communities" is central to understanding mining towns and how they existed in the

American West.[29] The integration of Butte into a broader narrative contextualizes the city's interdependent importance and helps tell an important aspect of the Irish experience in the US. Focusing only on Butte or on the larger Irish-American cities on the East and West Coasts somewhat erases the wider paths of mobility, both geographic and occupational, that defined the lives of many Irish-Americans in the nineteenth century—especially miners.

The Ireland of these exiles existed within both their traditions and their actions in America. It was a memory refreshed through the surrounding communities of Irish, letters from home, and their own sense of cultural loyalty. Their poems, songs, stories, and Catholicism expressed it. In this respect, the Irish differed from both other immigrants and the American Anglo-Protestant population. The English, Welsh, and Scottish merged seamlessly into the dominant Anglo-American culture: "British-Americans had no 'second-generation' . . . their children were simply Americans."[30] In this regard, it becomes more obvious that the Irish Catholics failed to achieve their perceived obligation in the minds of some Americans to simply become Anglo-Americans, and they succeeded in enriching the country by becoming a distinct group: Irish-Americans (figure 7.1).[31]

These stories of emigrant lives highlight both the persistence and the importance of Irish identity in American mining regions and the opposition they faced. Irish-America paid a tremendous price in blood to prove its loyalty to the Union during the Civil War, which greatly undermined nativist theories deployed against the Irish; but bigotry did not end when the guns fell silent after the Confederate surrender.[32] Anti-Catholic hysteria continued, and it was a thread of hostility the Irish repeatedly faced in the American West, along with continuing organized opposition.[33]

Over time, what it meant to be Irish-American would change, as indeed it differed for each person depending on circumstances, both personal and societal. The shift in the popular portrayal of the Irish in America in newspapers and other popular accounts did not necessarily reflect a change in their overall distinctiveness as a group in industrial America. The cases of Smartsville, Virginia City, Leadville, and the Coeur d'Alene district, to name a few, all have a different ethnic and cultural milieu that makes each Irish experience different.[34] However, the distinctive differences between individual encounters also highlight the similarities of these Irish communities through their religiosity, nationalism, and clannishness.

Figure 7.1. Uncle Sam sits facing away from a ballot box as a queue of figures vote. These figures are smaller than Uncle Sam and look childlike by contrast. They are each drawn half as European immigrant caricatures, half as typical Americans. The first person is clearly an Irish stereotype, with his Irish half dressed in green peasant clothes and simian-faced and his other half in checkered suit and no simian features. Uncle Sam looks backward over his shoulder, displeased and with his arms crossed. "The Hyphenated American: Uncle Sam: Why should I let these freaks cast whole votes when they are only half Americans." *Puck*, August 9, 1899, Print and Photographs Division, Library of Congress, Washington, DC.

Woodrow Wilson, describing the 1884 market crash in his *History of the American People*, observed, "It was evident, [that] the friction between laborers and employers grew, not less, but greater, as if some unwholesome influence were at work to clog the productive processes of the time."[35] He explained that workers adopted Irish strategies such as the boycott to enact "their radical programmes of social and political reform."[36] The Irish are the only group mentioned in this passage, and Wilson suggested that the workers terrorized anyone who "would not yield dictation of the labor organizations in any matter."[37] Thus, the courts were "forced to execute, sometimes very harshly, the law against conspiracy."[38] The use of "execute" might be a poor pun by Wilson, but his writing reveals the contemporary Anglo-American intolerance for any perceived "radical" action against business—whether concerning employment conditions, wages, or hours of work. It was this viewpoint that so frequently crashed against Irish expectations of America and their traditions of protest. Radicalism in this sense was any novel approach by labor to address the problems emerging with the increasing power and wealth of big business as the US carved a painful path to industrialization.

Wilson's comments represent the efforts of Anglo-America to come to terms with the Irish and the inevitable changes brought about by this process, one whose seeds had been sown in the previous decades. Historian Barbara Miller Soloman shows that by the 1880s, Anglo-Americans were attempting to make sense of this changed world by integrating the Irish into a new national narrative—in essence, to convince themselves that their narrow vision of the US was triumphant and to assuage their fear of pollution, which stated that most Irish were "English or Scotch in blood; Teutonic (Nordic) in type rather than 'Celtic.'"[39] This effort to accommodate Irish success in the US was simultaneously an effort to court support from the formerly unassimilable "Celtic": Irish Catholics. The shift in the nativist narrative also marked a new effort to exclude other so-called whites; German Jews were now classed as Slavic while Serb-Croatians had "savage manners," and though Poles "verge towards the northern races of Europe," they earned a mention with a caveat that "they are high-strung."[40] A government report authoritatively summed up its pseudo-scientific assessment of these newcomers: "All are different in temperament and civilization from ourselves."[41] And, of course, that was the point of its arbitrary assessments.

This Immigration Commission report detailed its new vision of immigration in 1911 by assessing the various nationalities in labor, class, and ethnic terms. It stated a preference for the "older immigrant labor supply" because

> [that group] was composed principally of persons who had training and experience abroad in the industries which they entered after their arrival in the United States. English, German, Scotch, and Irish immigrants in textile factories, iron and steel establishments, or in the coal mines, usually had been skilled workmen in these industries in their native lands and came to the United States in the expectation of higher wages and better working condi-

> tions. In the case of the more recent immigrants from southern and eastern Europe this condition of affairs has been reversed. Before coming to the United States the greater proportion were engaged in farming or unskilled labor and had no experience or training in manufacturing or mining.[42]

Earlier chapters of this volume have shown many examples of skilled Irish miners, but there were many who were not. To claim that all Irish immigrants who worked in American "textile factories, iron and steel establishments, or in the coal mines" were skilled workers is false, but it does expose the absurd lengths the commission was taking to try to bring the Irish into its new category. The Joint Immigration Commission's perception of the Irish, partly incorrect, reveals the subtle shift in the Anglo-American views of the Irish; and it represented the commission's efforts to come to terms with an increasingly influential, politically powerful, and proudly distinct Irish-America.

The mixture of varying degrees of skills and mobility complicates the story of Irish miners in America. Following the more in-depth and broader overview presented in this volume, it becomes more evident that Timothy M. O'Neil was correct in his assessment that "many scholars of American mining, erroneously [construct] a rivalry between skilled Cornish miners and unskilled Irish laborers, while statistically it is clear that the rivalry was between two ethnic groups competing for the skilled position of miner or mining captain."[43] Unfortunately for Irish miners, preferential hiring practices by company owners or through the use of shadowy fraternities such as the Patriotic Order meant that skills on their own were insufficient criteria for employment.

The Irish understood this preferential behavior and rapidly repurposed it to the benefit of their ethnic group. Skilled Irish miners who were hired as mine bosses or superintendents played an important role in acting as a gateway to other Irish gaining employment. The Irish were exceptional organizers—an aptitude that made them the target of company policy, of the prejudices of particular mine managers, and of forced contests between business and labor to decide who would control the workforce, as seen in the examples of Leadville, Colorado, the Coeur d'Alene district in Idaho, and Marysville, Montana.[44] As ethnic alliances weakened through class and generation differences in the twentieth century, the corporate managerial class arose. Irish-Americans such as Charles Sweeny and John Ryan, because of their vast wealth and celebrated positions as businessmen, felt little in common with the Irish families living in Dublin Gulch or the miners singing in Harrington's bar or the fears, struggles, and hopes of the Irishmen who were drilling and blasting in the darkened depths of the earth.

On the topic of the Cornish in America more generally, Ronald M. James states that their unity developed as a response to antagonism shown to them by the Irish: "Expressions of conflict [between Irish and Cornish] run the gamut from shooting, knifing, and other forms of fatal violence to bar fights and street brawls" in the mining towns of the American West.[45] James did not distinguish between murderous intent and faction fighting, which, however appalling to modern sensibilities, sometimes represented a form of competitive sport between ethnic groups and was typical of much of the history of the American West.[46] Rowland Berthoff is similarly mistaken in stating that "anti-Irish sentiment among British-Americans was quite spontaneous."[47] In fact, anti-Irish sentiment among Welsh and Cornish miners originated in their homelands in Britain—as seen most notably in anti-Irish riots in Camborne and Tredegar—and was carried over to the US.[48] One contemporary wrote, "The Irish have to encounter considerable prejudices . . . in almost every section of the Union, though in different degrees."[49] The present book contains many instances of this prejudice. As one historian of the Chinese in the West notes, however, we must be careful to avoid the group becoming "more celebrated for what happened to them than for what they have accomplished."[50] As important, to contextualize the hostility toward the Irish and from them toward others does not excuse their actions, in particular against the Chinese in Leadville and their participation in the expressly racist Workingmen's Party of California.

The Irish clung tenaciously to a cultural distinctiveness and a defensive sense of superiority over the English, "*Beidh bodaigh an Bhéarla dá ruagairt feasta* (the English language louts will be chased henceforth)," sentiments born from centuries of cultural persecution and political resistance.[51] As an oppressed and weakened people, they were a defeated underclass in their own land and a colony within a global and domineering British Empire. Rivalries were a cause of many faction fights, as a poem about a Cornish miner tells: "I hitchhiked to old Butte, Montana, and I met an Irish bloke, said 'Ireland should have her freedom, Queen Elizabeth was a joke,' I started to weigh into him, his partner he got sore. When I woke up next morning I was sprawled out on the floor."[52] Both ethnic solidarity and adversarialism were integral parts of the Irish nationalist narrative. These themes followed through in English-language songs, such as the closing verse of this song commemorating the death of an Irish rebel in County Cork during the abortive Fenian Rising in 1867 and sung by Matthew Hanafin in Butte: "God rest you Peter Crowly, you sleep beneath the clay / But someday you'll return again to lead us to the fray, / With a thousand men at your command, be they all both loyal and true, / That will conquer English, Dutch and Dane as Irishmen can do," harking back to centuries-old Irish conflicts involving William of Orange and the Vikings.[53] The Irish developed and mastered ethnic exclusiveness and were able to deploy it effectively against any and all, from low-level faction fighting to high-level strikes against challenges ranging from local demeaning comments to the actions of big business, thereby aggressively opposing every group that threatened them anywhere. This tendency, frequently referred to as clannishness, fails to fully describe the degree of insulation this cultural attribute lent to the Irish whenever they faced a cold welcome.

Similarly, the experience of the famine cannot be quantified solely by mortality statistics: the Irish who lived through those harrowing years experienced a cultural shock that unquestionably left a psychic scream echoing down through the generations. Whatever the effect of conditions in Ireland and Britain on the Irish Catholic emigrants, they knew they were an oppressed people, and the Great Famine encapsulated the evidence of this historical mistreatment. Evidence of the strength of their convictions can be found in the huge list of mining towns that donated to Irish nationalist causes and perhaps in the creation of schools, churches, and hospitals to take care of their own.[54]

The story of firebrand labor activist "Mother Jones" serves as an example of this sense of conviction and the growing sensitivity of Irish-Americans toward other ethnic groups. The Cork-born Mary Harris had an undeniable influence as a national organizer and agitator for workers' rights.[55] Harris utilized her Irish background in her speeches and writings to preach unity to workers of varied ethnic backgrounds: "Miners were considered a bad class. They came from different countries, and were of the kind that believed in settling all differences by force . . . [as] in Ireland years ago. The inhabitants of one county would fight those of the other until there was continual trouble."[56] She traveled throughout the American West acting as a labor unifier and a firebrand speaker during major strikes, earning the ire of businesses and local governments and resulting in her being forcibly escorted from Colorado in the months before the Ludlow Massacre.

Her work necessitated a keen sensitivity to ethnic identity. While the following lines are taken from one of her speeches in 1912 in Montgomery, West Virginia, it is easy to imagine her utilizing the same language in the West. She quickly realized that the crowd was mostly composed of Italian miners, and she pled for unity by recounting this conversation: "A fellow said to an operator, 'Why don't you prop the mines?' 'Oh,' he said, 'Dagoes are cheaper than props.' Every miner is a Dago in their estimation—every miner they can rob."[57] She repeatedly preached militant labor action, and she preceded her comments about "Dagoes" with "the day of human slavery has got to end. Talk about a few guards who got a bullet in their skulls! The whole lot of them ought to have got bullets in their skulls. How many miners do you murder within the walls of your wealth producing institutions? How many miners get their deaths in the mines?"[58] This woman, who was not a miner and who dressed as if she were the miners' grandmother, was able to deftly undermine their passivity and goad them into fury at both the economic injustices and structures of repression around them. One historian wrote, "The concept of Irish exile, Catholicism's renewed cult of Mary, antislavery doctrine, producer ideology, republicanism, separate spheres for men and women, the idealization of motherhood led her to in effect become Mother Mary, interceding on [the workers'] behalf."[59] This made her a remarkably effective labor activist as she masterfully subverted the expectations of society for her own ends, deploying her age, sex, Irish heritage, and Welsh alias as tools in her struggles for labor.

Mother Jones also represented a sea change in trade union organization in America. Mining unions in the nineteenth century often became exclusive organizations for both a specific class and ethnic groups, to protect their narrow sections of workers' interests. This occurred particularly when mine companies imported unskilled workers to try to break the power of the unions in a town. Later, unions operated on a basis of adversarial inclusiveness in their efforts to establish a decent standard of living through a living wage and to further the goal of worker solidarity. John Mitchell, the Irish leader of the United Mine Workers, clarified these sentiments and described a living wage as "a wage that would afford a worker a sufficient amount of money to enable him to live as we believe [a] workingman should live, educate his children, clothe them properly, and to save sufficient."[60] Mitchell told workers that "our constitution provides for their admittance as members; and in fact our obligation provides that we must not discriminate against any man on account of creed, color, or nationality."[61] He also believed workers in the union should raise immigrants' standards and oppose the entry of those who would not aspire to this standard, a qualified inclusivity.[62]

The hyphenated identity that Irish immigrants had done so much to make an acceptable fact in American society was also used by other immigrants to wrap their identity into a statement of loyalty to their new nation. But as labor unions adapted to ethnic rivalries, others continued to see these differences as inherently compromising their definition of "American." In 1919, President Woodrow Wilson toured the US to convince a skeptical public of the importance of US membership in his new League of Nations. Heckled by Irish-American and German-American miners during a speech, he dropped his mask of geniality, saying "I cannot say too often—any man who carries a hyphen about with him carries a dagger that he is ready to plunge into the vitals of this Republic."[63] An added irony was the fact that Wilson's Scots Irish grandfather was born in County Tyrone. At the heart of the Anglo-American nativist attitude toward the Irish was the underlying belief that not only Catholicism, which was inherently anchored to the Old World, but Irishness itself was a subversive form of otherness. Oaths of loyalty to the US, such as those given by the AOH in Nevada, could never convince those who believed Irish-Americanism itself was disloyal; such ideas, if less popular after World War I than in the late nineteenth century, continued to be widely held by men such as Wilson and guided their influential decisions well into the twentieth century. As in the case of most principles rooted in fear rather than reason, this one was self-defeating, and the irony of the hatred was that it solidified Irish allegiance to the structures of community, faith, and identity within the hyphen of Irish-America that anchored them in the new land and buttressed their descendants from the worst effects of this nativism.

The Irish-American Association in Salt Lake City, announcing a ball in honor of the "Father of Utah Mining," Patrick Edward Connor, may have gone too far in the opposite direction when it claimed that no Irish were there because of "love of greed but . . . [that they] sought only to promulgate the tenets of allegiance to the flag."[64] Irish loyalty was not quite as pure as that, since there was the expectation of opportunity. Some were disappointed or disillusioned by the American experience, as Michael MacGowan was after his return to Donegal. He vowed: "*B'fhuath liom Meiriceá agus ba mhion minic a dúirt mé i m'intinn féin, dá mba rud é agus go dtógfainn clann choíche, nach ligfinn aon duine acu anonn go brách. B'fhearr liom a gcur a chruinniú bratóg!*"[65] (I didn't care for America and I often said to myself that if I ever had a family, I'd never let one of them go there. I'd rather have seen them gathering rags!)[66]

Despite the existence of these heavily Irish towns, this American opportunity remained an elusive objective for many. The story of Irish miners in the US remains a complex one, with a diverse series of outcomes for the many immigrants who went there as miners or entered mining towns seeking employment. A few made their pile and struck it rich, such as Thomas Cruse and Marcus Daly. As Terry Coleman notes, "The emigration movement . . . is heroic to look back on . . . but for the individual emigrant it was often a personal tragedy," and Cruse's life certainly fits that description.[67] Their migration to and through the US shows the Irish traveling in search of opportunity and desperate for security. Despite the mid-century horrors of the Great Famine, most remembered their homeland fondly, with their memories rooted in the nationalist narrative that Ireland was subjugated to cruel British governance by an arrogant ruling class. The Irish miner John McCue wrote a ditty, "Although I'm in America, the land of liberty, I still live in expectation, old Ireland to see."[68] Their lives rarely resembled heroic tales; instead, they chiseled their way through life and pushed their children to attain better lives than the ones they themselves endured. The object of their blame for difficulties was circumstantial. For example, Patrick Kearney blamed hordes of immigrants in the US who were willing to work for "mere nothing," as he wrote in 1890: "I have done a good deal of running around in America seeking the best place, but all my sorrow I have lost by it. The American country is gone."[69] Highlighted in Kearney's aching disappointment was his initial belief that America had once offered hope of a better life, but it did no longer.

Irish miners came to America in the nineteenth century with this same belief but found reality at odds with it. Their hopes often clashed against a relentless drive for profit and the unprecedented expansion of private and corporate power in society. For Irish miners in the nineteenth and early twentieth centuries, the root of their insecurity lay not in the dangers of their work or the instability of the natural resources but rather in a model of business that viewed their conditions as unimportant, their organizational skills as treasonous, and their identity as an opportunity to isolate them from an American society they sought to become a distinctive contributing part of despite the great difficulties.

In 1916, after another series of vicious disputes between mine workers and businesses, the US government launched an inquiry that resulted in the publication of the *Final Report of the Commission on Industrial Relations*. This report stepped away from past political tendencies, which had blamed immigrants for not seizing the supposedly self-evident opportunities politicians and business leaders presumed were inherently available in America. Instead, the report announced a new vision for the country. In a startling departure it stated, "With the inexhaustible natural resources of the United States, her tremendous mechanical achievements, and the genius of her people for organization and industry, there can be no natural reason to prevent every able-bodied man of our present population from being well fed, well housed, comfortably clothed, and from rearing a family of moderate size in comfort, health and security."[70] This echoed the Irish-American dream of Nellie Cashman, the disillusioned hopes of Seán Ruiséal and Seamus Ó Muircheartaigh, the exhausted hostility of Patrick Kearney, and the stubborn resistance of Michael Mooney. The conflicts in mining regions forced the nation to look in the mirror and see its multi-ethnic features clearly and eventually to see them as a possible strength rather than a weakness, as Wilson did. This sentiment would find flowering in the New Deal. By then, the few remaining Irish miners had grown old. Most were buried in the graveyards of fading towns or were relinquished to collapsed mine shafts—their voices, their songs, and their prayers that once had echoed throughout the Irish-American West now silent.

APPENDIX I

Irish Poems, Songs, and Notes about Mining

Personal accounts of the experiences of Irish miners and their families are rare. These letters, songs, and notes are a few of the surviving records that still exist. The first three poems appear in a collection of folksongs and ballads from the West Kerry Gaeltacht (the Irish-speaking area of Ireland). These poems were created in a specific old Gaelic style (*sean nós*), as a sort of lament. The following two poems were written by Seán Ruiséal from west of Daingean in County Kerry, who spent twelve years in America: in Butte, Montana; Cetcucan in south Alaska; and Springfield, Massachusetts. He returned to Ireland and made his living as a fisherman.

The third poem, *Mo chiach mar a thána*, was written by Seamus Ó Muircheartaigh—who was from Smerwick, Ballyferriter, County Kerry—while he was living in the US. He left Ireland around 1900 and worked in the mines in Butte, Montana, as well as wandering through the American West. He lived in California and wrote Gaelic articles in newspapers for his local Irish community. He is the author of several ballads lamenting his economic exile from his native land. He was buried in San Francisco.

https://doi.org/10.5876/9781646422517.c008

Two additional songs are included: the folksong "Muirsheen Durkin," an upbeat tune that stands in contrast to most emigrant songs, and the original lyrics for "Spancil Hill," as written in a letter from emigrant Michael Considine in California to his family in County Clare. The final primary source includes the last words of Jim Moore who died in the Speculator Mine disaster in Butte, Montana, on June 8, 1917.

"An Spailpín Fánach"[1]

I

Nach óg sa tsaol nár ceapadh mo leas dom
Nuair a d'fhágas Éire ghrámhar,
Chun a bheith ag siúl na dúthaí mBúit Montana
Is gan pingin i dtóin mo sparáinín;
Mar obair ní bhfaighir ná aon tslí bheatha
Ná bean tí fhairsing fháilteach,
Ach bladar go leor is fós drochmheas ort,
Mara bhfuil dollars agat le háireamh.

Wasn't I too young in this world to consider my good fortune,
When I left my loving Ireland.
To walk the townlands in Butte, Montana,
Without a penny in the bottom of my little purse,
I won't find work or any job,
Or a generous, welcoming housewife,
Only ample flattery and feel only disrespect for you,
If you haven't got dollars for the counting.

II

Im spailpín fánach atáim le sealad
Fóraoil deacair cráite!
'S gur mhó 'gam pingin im póca thiar sa Daingean
Ná punt sna dúthaí fáin seo;
Mar ba shéimh an lá dhom féin 's dom charaid,
Bheith socair thiar I dTrá Lí,
Seachas bheith ag siul na dúthaí im scrúille bacaigh
's gan bean a thabharfadh grá dom

I'm a wandering spalpeen for a time now
(An excess) of grievous hardships!
And a penny in my pocket back in Dingle was worth more
Than a pound in these townlands where I stray,
When the day was mild for me and my friend
To be secure back in Tralee,
Instead of walking the townlands and I a miserly beggar
And without a woman to give me love

III

Do shiulaíos ó thuaidh agus arís ó dheas é
Soir agus siar gach áird ann,
Is ní fhaca móinéar ná fiú goirt ghlasa
Do thógfadh mo chroí le háthas;
Nil buaint ná rómhar ann, treabhadh ná branar,
Ná fuaim na speal ar bhántaibh,
Ach clismirt ag' motors gach ló ag tarrac,
Do scaipeadh do mheabhair chun fáin uait.

I walked it from the north and again from the south,
Every height to the west and to the east,
And I didn't see a meadow or even a green field,
To lift my heart with joy.
There isn't a reaping or digging there, ploughing or a fallow,
Or the sound of the scythes on the grasslands,
But the din of motors every day hauling,
It would drive your mind to stray from you.

IV

Mo shlán beo siar chun ciumhais an Daingin,
Is chun cuan Fionn Trá geal álainn,
Mar ar leagadh an garda bhí ag Dáire sealad,
Ag comhracle Fianna Fáile;
Is as súd siar chun cuan Bhaile 'n Chalaidh,
Chun Dún an Fheirtéaraigh ghalánta,
Mar ar chomhraic Piaras go dian Ó Donnell
D'fhonn bean Uí Loingsigh d'fháil uaidh.

That my health might see me back to the edge of Dingle,
And the lovely white harbour of Ventry,
Where the guard Dáire for a time was felled
By the encounter with the Fianna of Fáil,
And went from them to the harbour of Bhaile 'n Chalaidh,
To the gallant Fort of Ferriter,
Where Piaras, fought Ó Donnell diligently,
To win back the desire of Uí Loingsigh his wife from him.

V

Is as san mórthimpeall chun Béidreach an Fhearainn,
Is chun na mBinnteacha 'tá in airde,
Mar a gcodlaíodh Diarmuid, an laoch nár mheata,
D'éalaigh seal le Gráinne;
Is as san suas chun Gort na gCeanna,
Mar ar buaileadh cogadh ar Spáinnigh
Ag Aodh Ó Neill is Sir Walter Raleigh
Nó Lord Grey do chroch na táinte.

And from there right around to Béidreach an Fhearainn,
And the peaks high above Binnteacha,
Where Diarmuid, the hero who never sickened,
Escaped for a time with Gráinne,
Where the battle was won against the Spaniards,
By Aodh Ó Neill and Sir Walter Raleigh,
Or Lord Grey who hung the multitudes.

VI

Da mairfeadh Ó Conaill arís i gceannas,
Ba shocair suairc an lá dhuinn,
Mar do dhéanfadh sé íde ar gach Tory galair,
Is do bheadh Home Rule againn gan spleáchas;
Is mo thrua san uaigh é mar tá sé meata,
Is an Pairnéalach sínte laimh leis,
Is gan againn 'na ndiaidh ach draíodar dana
Nach fiú iad a chur go Pairlimint.

If Ó Conaill lived again, in charge,
The days would be safe and pleasant for us,
Because he would lambast every diseased Tory
And we would have Home Rule without dependence,
And my sorrow that he is in the grave because he is fallen,
And Parnell stretched out beside him,
And left for us, after him naught but chancing rogues,
Who aren't worth sending to Parliament.

VII

Tá mo shúilse 'rís le Rí na bhFlaitheas,
Is leis an Mhaighdin mhánla,
Go dtógfaidh Miléiseans arís a gceanna,
D'fhonn bualadh thabhairt do námhdaibh;
Beidh bodaigh an Bhéarla dá ruagairt feasta
As Oileán naofa Phádraig,
Is dlí na nGael arís dá spreagadh–
Sin cabhair ag an Spailpín Fánach.

My eyes are again with the king of Paradise,
And with the gentle Virgin,
That Milenaus again will lift their heads,
With the wish to grant defeat to their enemies,
The English language louts will be chased henceforth
From Patrick's holy isle,
And the laws of the Gaels propagated again,
That is a help to the Wandering Spalpeen.

"Amhran na Mianach"[2]

I

Go hAmericá siar seadh do rachas,
Ag sealg i dtúis mo shaoghail,
Ag cuardach an óir úd fé thalamh
Cois sleasa 'gus ciumhais an tsléibhe;
Cé gur shiubhluigheas gach sráid agus cathair,

Ó Bhostún go Sráid Choiréil,
N'fheacas aon áit mar an baile
D'fhágas ar maidin le fáinne 'n lae.

West to America I went
In my youth to hunt,
Looking for that well-known gold under the ground
On the slopes and edges of the mountain;
Though I walked every village and city,
From Boston to Coral Street,
I never saw a place like the village
I left at the break of day.

II

Is mór atá ráidhte le sealad,
'Ges gach tallaire cnáideach baoth,
Nár bhaoghal duit an ghaoth ann ná 'n fhearthainn,
Ach taitneamh go hárd ón ngréin;
Ach do chonnac-sa go leor ann den tsneachta,
Is clagairt den mbáistigh thréin,
Is coinnleoirí reoidhte ann cois leabtha,
Ar maidin le fáinne 'n lae.

There's been a good deal of talk recently
From every stupid, tiresome scamp
That you won't be bothered there by rain or wind–
There's only bright sunshine;
But I saw plenty of snow there
And heard the rattle of heavy rain,
And frozen candlesticks beside the bed
At the breaking of the day.

III

Siad na ceolta is mo gheobhair id aice–
Ar do leabaidh ní bhfaghaidh tú néall–
Ach mianaigh ah liúighrigh 's ag screadaigh,
Ag cur deataigh go hárd san aer;
Beidh an saoiste ansúd ann 'na sheasamh,

Is i n-aice dho an foreman caol,
Is iad a' fógairt gach stróile chun taisce,
Ar maidin le fáinne 'n lae.

The songs you'll mostly hear–
You'll never get a wink of sleep in bed–
Are the shouts and screams of the miners
As they fill the air with steam;
The bosses will be standing there
And next to him the lean foreman,
Driving all the wretches to work
At the break of day.

IV

Tógfaid siad síos tu go topaidh,
Ó thaithneamh is ó radharc an tsaoghail,
Is tabharfaid duit sluasad nó carra,
Casúr nó tuagh bheag mhaol;
Beidh an maisín go hárd ann ar barra,
Is í ag obair le cómhacht an aeir,
Do shúile le smúit ann dá sndalladh,
Ar maidin fáinne 'n lae.

They'll soon take you down
Where you can't see the light of day,
And give you a shovel or a car
Or a hammer or a blunt little axe;
The machine will be up top
Working by air power,
Your eyes will be blinded by dust
At the break of day.

V

Is mó rud a chífir id aice
Ná feacaís id dhúthaigh féin,
Chífir fear púdar do tharrac
Chun talaimh do bhrúghadh gan bhréig;
Chífir an mhiúil ann 's an capall

Is na carraí 'na ndhiaidh ar hate (.i. Heat),
An cairréaraidhe is lamp' ar a hata,
Ar maidin fáinne 'n lae.

Many things you'll see around
That you never saw at home,
You'll see a man drawing powder
To shake the earth (that's no lie);
You'll see the mule there and the horse
Dragging cars in heat,
The carman with a lamp on his hat
At the break of day.

VI

Is má castar tu choidhche 'na n-aice,
Cuimhneochaidh tú ar bhrí mo scéil;
Gurb aindeis an tslí é chun beathadh
D'aon scaraire rábach tréan;
Mar beidh tú chómh doimhim sin fé thalamh,
Is ná feicfir an ghrian ná an rae,
Ach do choinneal bheag chaoch ar do hata
Ar maidin fáinne 'n lae.

And if you ever come amongst them
You'll remember the point of my story;
It's a miserable way to earn a living
For any strong, vigorous fellow;
For you'll be so deep in the ground
That you'll never see the sun or the moon,
But only the light of the dim little candle in your hat
At the break of day.

"*Mo chiach mar a thána*"[3]

I

Mo chiach mar a thána 'on tír seo riamh,
Is gur fhágas Éire mo stór im dhiaidh,

Ag smaoineamh go dúbhach ar an aimsir fadó,
Mar a raibh agam greann sult agus seó.

Alas that I ever came to this land
And that I left my beloved behind;
I'm thinking sadly of that time long ago
When I had cheer, sport, and play.

II

Fuaireas-sa litir ó bhrathair gaoil,
Dul go tapaidh anon thar toinn,
Go raibh ór go flúirseach le fáil anso,
Is ná feicfinnse choíche lá cruaigh ná bocht.

I got a letter from a relation
Telling me to hasten across the sea,
That gold was to be found in plenty there
And that I'd never have a hard day or a poor one again

III

Is maith is cuimhin liom an mhaidean bhreá úr,
Nuair a d'fhágas slán ag mo mháthair bhocht dubhach,
Muintir an bhaile ag síorshileadh deor–
"Slán leat a Shéamuis, ní fhillfir nios mó."

I well remember that fine fresh morning
When I bade farewell to my poor sad mother;
The people of the village kept shedding tears:
"Farewell, James, you'll never come back."

IV

Thugas liom mála 'na gcuirfinn ann t-ór,
D'fháisceas go daingean córda ar a thóin,
Ar eagla go n-imeodh an t-airgead uaim,
Bhíos chun capall a cheannach dom mháthair is uan.

I brought along a bag to put the gold in
And fastened it tight with a cord around the top
Lest I lose all the money;
I was going to buy my mother a horse, and a lamb as well.

V

Chuas ansan ar bord loinge go baoth,
Mo mhála ar mo ghualainn ag guíochtaint chun Dé,
Mé thabhairt slán og dtí an talamh thar stoirm is gaoth,
Mar a mbeinn im dhuine uasal i gcaitheamh mo shaoil.

Naively I went abroad
With my bag on my shoulder, praying to God
To bring me safe to land through storm and wind,
Where I'd be a gentleman for the rest of my days.

VI

Mo mhairg nuair a dheineas-sa caladh sa tír,
Do ghluaiseas faoin gcathair ag sodar gan mhoill,
N'fheacasa aon ór ar aon chúinne den tsráid
Faraior bhíos im stróille bocht caite ar an bhfán.

Alas, when I landed
I made for the city without delay;
But I never saw gold on the street corners–
Alas, I was a poor aimless person cast adrift.

VII

Sin mar a chaitheas-sa tamall dem shaol,
Ó bhaile go baile gan toinnte ar mo thaobh,
Nuair a thagadh an oíche bhíodh fuar agus fliuch,
Ba mhinic mé sínte ins na coillte amuigh.

That's how I spent part of my life,
Going from place to place, with no company at my side;
When night would come, it was cold and wet;
Often I lay stretched out in the woods.

VIII

Is mór mór go mb'fhearra bheith in Éirinn an ghrinn,
Ag éisteacht le ceolta na n-éanlaithe ró-bhinn,
Ná ag lorg lá oibre ar spriúndlóir beag cam,
Gur dhóigh leis gur asal tú a bhuailfí le feam.

It's far far better to be in Ireland where there's cheer,
Listening to the melodious bird songs,
Than looking for work from a crooked little miser
Who thinks you're only an ass to be beaten with a stick.

IX

Téirigh go hÉirinn a óigchailín chiúin,
A bhuachaill bhig éist liom agus gluais leat anonn,
Mar a mbeidh agaibh réal lá aonaigh is punt
Cead rince go haerach le chéile ar an ndrúcht.

Go back to Ireland, my modest girl;
Listen to me, my little lad, and head for home,
Where you'll have a pound and sixpence on a fair day
And freedom for the carefree dance together on fair dew.

X

Nuair a thiocfaidh an lá is nuair a lasfaidh an spéir,
Go ngluaiseoidh na sluaite faoi bhrat uaithne na nGael,
Go mbeidh sibh an lá úd faoi ghradam sa bhruíghin,
Ag ruagadh an tSasanaigh dhamanta thar toinn.

When the day comes and the sky is alight
And the hosts march under the green banner of the Gael,
That day you'll be glorious in battle
Which will scatter the cursed English across the sea.

"Spancil Hill"[4]

Last night as I lay dreaming, of the pleasant days gone by,
My mind being bent on rambling and to Erin's Isle I did fly.
I stepped on board a vision and sailed out with a will,
'Till I gladly came to anchor at the Cross of Spancil Hill.
Enchanted by the novelty, delighted with the scenes,
Where in my early childhood, I often times have been.
I thought I heard a murmur, I think I hear it still,
'Tis that little stream of water at the Cross of Spancil Hill.

And to amuse my fancy, I lay upon the ground,
Where all my school companions, in crowds assembled 'round.
Some have grown to manhood, while more their graves did fill,
Oh I thought we were all young again, at the Cross of Spancil Hill.
It being on a Sabbath morning, I thought I heard a bell,
O'er hills and vallies sounded, in notes that seemed to tell,
That Father Dan was coming, his duty to fulfill,
At the parish church of Clooney, just one mile from Spancil Hill.
And when our duty did commence, we all knelt down in prayer,
In hopes for to be ready, to climb the Golden Stair.
And when back home returning, we danced with right good will,
To Martin Moilens music, at the Cross of Spancil Hill.
It being on the twenty third of June, the day before the fair,
Sure Erin's sons and daughters, they all assembled there.
The young, the old, the stout and the bold, they came to sport and kill,
What a curious combination, at the Fair of Spancil Hill.
I went into my old home, as every stone can tell,
The old boreen was just the same, and the apple tree over the well,
I miss my sister Ellen, my brothers Pat and Bill,
Sure I only met strange faces at my home in Spancil Hill.
I called to see my neighbors, to hear what they might say,
The old were getting feeble, and the young ones turning grey.
I met with tailor Quigley, he's as brave as ever still,
Sure he always made my breeches when I lived in Spancil Hill.
I paid a flying visit, to my first and only love,
She's as pure as any lily, and as gentle as a dove.
She threw her arms around me, saying Mike I love you still,
She is Mack the Rangers [sic] daughter, the pride of Spancil Hill.
I thought I stooped to kiss her, as I did in days of yore,
Says she Mike you're only joking, as you often were before,
The cock crew on the roost again, he crew both loud and shrill,
And I awoke in California, far far from Spancil Hill.
But when my vision faded, the tears came in my eyes,
In hope to see that dear old spot, some day before I die.
May the Joyous King of Angels, His Choicest Blessings spill,
On that Glorious spot of Nature, the Cross of Spancil Hill.

"Muirsheen Durkin"[5]

In the days I went a courtin' I was never tired resortin'
To an ale-house or a playhouse and many's the house besides
But I told me brother Seamus I'd go off and be right famous
And before I would return again I'd roam the whole world wide
Chorus
Goodbye Muirsheen Durkin sure I'm sick and tired of workin'
No more I'll dig the praties, no longer I'll be fooled
For sure's me name is Carney I'll be off to California
Where instead of diggin' praties I'll be diggin' lumps of gold
I've courted girls in Blarney, in Kanturk and in Killarney
In Passage and in Queenstown, that is the Cobh of Cork
Goodbye to all this pleasure sure I'm off to seek me leisure
And the next time that you hear from me, is a letter from New York
Repeat Chorus
So goodbye all ye boys at home I'm sailing far across the foam
I'm going to make me fortune in far Amerikay
There's gold and money plenty for the poor and for the gentry
And when I return again I never more will stray
Repeat Chorus

Last Words of Jim Moore, Speculator Mine Disaster, June 8, 1917[6]

6-8-17

Dear Pet,

This may be the last message you will get from me. The gas broke about 11:15 p.m. I tried to get all the men out, but the smoke was too strong, I got some of the boys with me in a drift and put in a bulkhead. * * * If anything happens to me you had better sell the house * * * and go to California and live. You will know your Jim died like a man and his last thought was for his wife that I love better than anyone on earth. * * * We will meet again. Tell mother and the boys goodbye.

With love to my pet and may God take care of you.

Your loving Jim,
JAMES D. MOORE

5:00 a.m., 6-9-17

Dear Pet,

Well, we are all waiting for the end. * * * I guess it won't be long. * * * We take turns rapping on the pipe, so if the rescue crew is around they will hear us. Well, my dear little wife, try not [to] worry, I know you will, but trust in God, everything will come out all right. There is a young fellow here, Clarence Marthy. He has a wife and two kiddies. Tell her we done the best we could, but the cards were against us. Goodby little loving wife. It is now 5:10.

7:00 a.m. All alive but air getting bad. One small piece of candle left. Think it is all off.

9:00 a.m. In the dark.

APPENDIX 2

Transcript of Official Oath of the State of Nevada

Note: bold words written in, italics sections typed[1]

Official Oath of the State of Nevada

~~I~~ **We** the undersigned *do solemnly swear* that ~~I~~ **we** *will support, protect and defend the Constitution and Government of the United States, and the Constitution and Government of the State of Nevada, against all enemies, whether domestic or foreign; and that* ~~I~~ **we** *will bear true faith, allegiance and loyalty to the same, any ordinance, resolution or law of any State Convention or Legislature to the contrary notwithstanding; and further, that* ~~I~~ **we** *do this with a full determination, pledge or purpose, without any mental reservation or evasion whatsoever. And* ~~I~~ **we** *do further solemnly swear that* ~~I~~ **we** *have not fought a duel, nor in any manner aided or assisted in such duel, nor been knowingly the bearer of such challenge or acceptance since the adoption of the Constitution of the State of Nevada; and that* ~~I~~ **we** *will not be so engaged or concerned, directly or indirectly, in or about any such duel during my continuance in office; and further, that* ~~I~~ **we** *will well and faithfully perform all the*

https://doi.org/10.5876/9781646422517.c009

duties of the office of Trustees of the Ancient Order of Hibernians Division No. 2 which I am about to enter; so help me God.

Subscribed and sworn before me, this

Felix Boyle

sixteenth *day of* **J. J. O'Reilly** *32877*

April A.D. 1877 **Michael Kelly**

APPENDIX 3

Parentage Percentages and Figures, American West

Table A3.1. US-born miners' parentage by place of birth, 1880[a]

Place of Birth	*Mother %*	*Mother No.*	*Father %*	*Father No.*
US	79.1	32,792	77.7	32,231
Canada	1.2	494	1.2	486
Ireland	8.9	3,692	9.6	3,979
England	3.9	1,610	4.2	1,745
Wales	0.7	296	0.7	300
Scotland	1.6	667	1.7	690
Germany	2.7	1,121	2.8	1,171
France	0.6	241	0.6	253
Mexico	0.4	179	0.4	177

[a] These numbers account for 99.1 percent of the total for the "Mother" columns and 98.9 percent of the total for the "Father" columns. The other results show less than a dozen results for a variety of locations. IPUMS. US-born miners = 41,456.

https://doi.org/10.5876/9781646422517.c010

Table A3.2. Canadian-born miners' parentage by place of birth, 1880[a]

Place of Birth	*Mother %*	*Mother No.*	*Father %*	*Father No.*
Canada	64.8	2,459	61.2	2,321
US	5.1	195	5.5	207
Ireland	12.0	457	13.2	499
England	5.4	205	6.3	238
Wales	0.1	3	0.2	7
Scotland	10.6	403	11.7	445
Germany	0.3	13	0.4	16
France	1.3	51	1.6	60

[a] These numbers account for 99.8 percent of results for the "Mother" columns and 99.9 percent of the results for the "Father" columns for Canadian-born miners. The other results show fewer than a dozen results for a variety of locations. IPUMS.

Notes

Introduction

1. Seamus Ó Muircheartaigh in Seán Ó Dubhda, *Duanaire Duibhneach* (Dublin: Government Publications Office, 1976), 132–133, English translation by Bruce D. Boling, Brown University, in Kerby A,. Miller, *Emigrants and Exiles: Ireland and the Irish Exodus to North America* (New York: Oxford University Press, 1988), xiii.

2. Madame de Bovet, *Three Months' Tour in Ireland* (London: Chapman and Hall, 1891), 280; Arnold Schrier, *Ireland and the American Emigration, 1850–1900* (Chester Springs, PA: Dufour Editions, 1997), 20.

3. Sally Zanjani, *A Mine of Her Own: Women Prospectors in the American West 1850–1950* (Lincoln: University of Nebraska Press, 1997), 27, 37. The contrast between the opening quotes and the story of Nellie Cashman is an intentional effort to draw attention to the relatively few women involved in the occupation and aids in explaining the frequent use of masculine pronouns in this work. Few women were involved in speculative mining, and practically none participated in large-scale industrial mining; thus, the male pronoun is used as the default in later sections. Also see Melanie J. Mayers and Robert N. DeArmond, *Staking Her Claim: The Life of Belinda Mulrooney, Klondike and Alaska Entrepreneur* (Athens: Ohio University Press, 2000).

4. Dashiell Hammett, *Red Harvest* (New York: Alfred A. Knopf, 1929), 3. The first noir novel by Hammett was a fictionalized account of his time as a Pinkerton detective working in the city of Butte, Montana. The first line establishes a direct link between Butte and its fictional stand-in, Personville (nicknamed Poisonville), as well as referencing the unique Irish cultural influence: "I first heard Personville called Poisonville by a red-haired mucker named Hickey Dewey in the Big Ship in Butte. He also called his shirt a shoit." In a fitting coincidence, the chief of police in the corrupt town is Noonan, a figure probably based on the the imposing chief of police in Butte, Jere J. Murphy, nicknamed "Jere the Wise." See the *Montana Standard* (Butte), September 22, 1935.

5. Charles Dickens, *American Notes* (New York: John W. Lovell, 1883), 660.

6. David M. Emmons, *Beyond the American Pale: The Irish in the West, 1845–1910* (Norman: University of Oklahoma Press, 2010), 9; Miller, *Emigrants and Exiles*.

7. Nevada State Mineralogist, *Biennial Report of the State Mineralogist for the State of Nevada 1872–1873* (Carson City: State Printing Office, 1873), 83.

8. Zanjani, *Mine of Her Own*, 37.

9. Zanjani, *Mine of Her Own*, 40.

10. Zanjani, *Mine of Her Own*, 40.

11. There are many cases of children placed into institutional care. Thomas Kyle became depressed after his wife's death and decided to "try his fortunes in the Wild West," so he placed his daughter in a boarding school and left for the goldfields. "Autobiographies and Reminiscences of California Pioneers," vol. 8, manuscript, Society of California Pioneers, Alice Phelan Sullivan Library, San Francisco, CA, 1904, 134. For another important example of a woman mining entrepreneur outside the scope of this work see Mayers and DeArmond, *Staking Her Claim*.

12. Donald Harman Akenson, "Irish Migration to North America, 1800–1920," in Andy Bielenberg, ed., *The Irish Diaspora* (New York: Longman, 2000), 127–133. See also Akenson, *Small Differences: Irish Catholics and Irish Protestants 1815–1922* (Montreal: McGill-Queen's University Press, 1988). This was particularly true for the Scots Irish, who so firmly rejected their association with Irish Catholics that they created a separate heritage to avoid conflation with their brethren. Few Irish Protestants are found mining in the American West. The occupation had little appeal to them, and as a group they were both able and willing to rapidly assimilate into American society and practically disappear within a generation. Emmons argues against any similarity between Irish Catholic and Protestant emigration, calling it a "large difference." Emmons, *Beyond the American Pale*, 1.

13. As Oliver Plunkett wrote in the seventeenth century, "Ulstermen and Leinstermen have never agreed and will not in the future either and the same is true of Munstermen and Connaghtmen. Connaghtmen and Ulstermen will easily agree

as will Munstermen and Leinstermen." John Hanly, ed., *The Letters of Saint Oliver Plunkett, 1625–1681* (Bucks, UK: Colin Smythe Ltd., 1979), 318.

14. W. S. "Shorty" Russell interview, transcribed in Duncan Emrich, *In the Delta Saloon: Conversations with Residents of Virginia City, Nevada, Recorded in 1949 and 1950* (Reno: University of Nevada Oral History Program, 1991), 265. Russell's other job apart from mining was tending bar.

15. There is a certain degree of difficulty establishing who exactly was a miner. To a degree it was contextual, depending on their position in the workforce and the resource mined. The variation in occupational strata is seen in the occasionally confusing distinction between laborers and miners. On one side were those laborers who were conducting physical labor only, and at the other extreme were experienced, skilled miners who had a grasp of engineering and geology through years of experience. And in between? The records detail a range of occupational pride, uneven standards, and privilege that left the skill difference more muddled than most scholars of labor portray.

16. Micí Mac Gabhann, *Rotha mór an tSaoil* (Dublin: National Publications, 1953); Michael MacGowan, *The Hard Road to Klondike*, translated by Valentine Iremonger (Cork: Collins, 2003).

17. For an overview of literacy and fluency, see Aidan Doyle, "Language and Literacy in the Eighteenth and Nineteenth Centuries," in James Kelly, ed., *The Cambridge History of Ireland, vol. 3: 1730–1880* (Cambridge: Cambridge University Press, 2018), 353–379. A nuanced examination of the cultural context is found in Nicholas M. Wolf, *An Irish-Speaking Island: State, Religion, Community, and the Linguistic Landscape in Ireland, 1770–1870* (Madison: University of Wisconsin Press, 2014).

18. "Celtic temperaments clashed when the exhausted Irish laborer found himself face to face with the acquired skills of centuries." Arthur Cecil Todd, *The Cornish Miner in America: The Contribution to the Mining History of the United States by Emigrant Cornish Miners, the Men Called Cousin Jacks* (Glendale, CA: Arthur H. Clark, 1967), 242. A more detailed explanation of the experiences of the Irish in Britain can be found in Alan J. M. Noonan, "Wandering Labourers: The Irish and Mining throughout the United States, 1845–1920," PhD diss., University College Cork, Ireland, 2013, 34–39.

19. Frank A. Crampton, *Deep Enough: A Working Stiff in the Western Mine Camps* (Norman: University of Oklahoma Press, 1982), 42, 264.

20. Crampton, *Deep Enough*, 42.

21. David M. Emmons, *The Butte Irish: Class and Ethnicity in an American Mining Town, 1875–1925* (Urbana: University of Illinois Press, 1989).

22. Frederick Jackson Turner, "The Significance of the Frontier in American History (1893)," in John Mack Faragher, *Rereading Frederick Jackson Turner: "The Significance of the Frontier in American History" and Other Essays* (New Haven, CT: Yale

University Press, 1999), 31–60. Their absence aggravated one author sufficiently to give his book an intentionally adversarial title. Myles Dungan, *How the Irish Won the West* (Dublin: New Island, 2006).

23. *Fifth Report of All Hallows Missionary College* (Dublin: J. F. Fowler, 1852–1853), 25–26.

24. *"fear fabhair nó plandóir nach rabh coinsias ar bith aige."* Mac Gabhann, *Rotha mór an tSaoil*, 104. Emphasis added.

25. MacGowan, *Hard Road to Klondike*, 64–65.

26. Mac Gabhann, *Rotha mór an tSaoil*, 105. At the 2014 American Conference for Irish Studies, Padraig Ó Siadhail questioned the influence of Mac Gabhann's folklorist son-in-law, Seán Ó hEochaidh, to whom the work was dictated and who transcribed it.

27. David Brundage, *The Making of Western Labor Radicalism: Denver's Organized Workers, 1878–1905* (Urbana: University of Illinois Press, 1994); Gunther Peck, *Reinventing Free Labor: Padrones and Immigrant Workers in the North American West, 1880–1930* (Cambridge: Cambridge University Press, 2000).

28. Frederick C. Luebke, ed., *European Immigrants in the American West: Community Histories* (Albuquerque: University of New Mexico Press, 1998), vii.

29. The details of intentionally segregated living conditions are explored in chapter 5, this volume.

30. Emmons, *Beyond the American Pale*, 354–356.

31. The pioneering efforts of Arnold Schrier and Kerby Miller saved many of these precious records from certain loss.

32. "To the Irishmen of Marysville and Vicinity," Yuba County Documents, MS 3A2, Bancroft Library, University of California, Berkeley.

33. Masonic organizations acted as a bedrock for the continuing American hostility toward Catholicism and Irish Catholics in particular. Not all fraternal organizations shared this hostility; for example, the Benevolent and Protective Order of Elks had Irish members in many mining camps.

34. Beyond the American pale is a reference to Emmons's exploration of Irish identity and place in the American West. His dissection of the intellectual struggle of identity stands as a useful thesis in understanding the mind-set of other parts of the Irish diaspora in this period.

Chapter 1: Varied Hues of Green

1. John Brophy, *A Miner's Life: An Autobiography* (Madison: University of Wisconsin Press, 1964), 51.

2. The number of unskilled miscellaneous laborers is difficult to establish, as the census data do not reveal whether they worked directly in the mines as muckers or perhaps in railroad building or construction.

3. One way to mitigate this is to see if a large amount of infrastructural work was ongoing at the time of the census and then rule them in or out, but doing so is not always possible.

4. As stated later, there was a slightly higher rate of female endogamy compared to men. Whether a father's or mother's ethnicity predominated in mixed-ethnicity families would be an interesting study in and of itself. As detailed later, however, even if the mother or father was born in the United States, they were often second-generation Irish-American themselves or Catholic. Since the American Catholic Church at this time was, in effect, an Irish-American Catholic Church, the cultural divide was not as wide as we might perceive.

5. There was an even smaller number of Methodist Cornish miners in Waterford and the Beara Peninsula in Cork. These Cornish were hired to bring their mining talents to Ireland, and the Irish-born children of these small Methodist colonies sometimes followed in their fathers' footsteps as miners. But they did not adopt the label Scots Irish. For example, see William Hall, *Progressive Men of the State of Montana* (Chicago: A. W. Bowen, 1902), 672.

6. Quoted in George Korson, *Coal Dust on the Fiddle: Songs and Stories of the Bituminous Industry* (Hatboro, PA: Folklore Associates, 1965), 442.

7. Korson, *Coal Dust on the Fiddle*, 441.

8. Balch Institute Sheet Music Collection, circa 1824–1945, Collection 3141, Historical Society of Pennsylvania, Philadelphia.

9. Philip J. Mellinger, *Race and Labor in Western Copper: The Fight for Equality, 1896–1918* (Tucson: University of Arizona Press, 1995), 19.

10. As one example, Irish immigrants' understanding of the Irish language ranged from fluent to little or none, which shaped their experiences in the US and their relationships with their own culture. Dividing them into English speaking, white, and Protestant or Catholic is too simplistic.

11. *Miners' Magazine*, October 22, 1903, quoted in Dan Tannacito, "Poetry of the Colorado Miners: 1903–1906," *Radical Teacher* 15 (March 1980): 8. See also Joe Lazure, "Hobo Miner," *Miners' Magazine* 6, no. 94 (March 13, 1905): 13. The Scoto in the title may represent a reference to the earlier Scotch or Scots Irish identity, in which case it represents a clever play on the fact that Scots Irish was used to distinguish one group from another. Here, the author forces that identity together with a separate hyphen to point to the absurdity of the distinctions.

12. See chapter 6, this volume.

13. See chapter 2, this volume.

14. Alan B. Campbell, *The Lanarkshire Miners: A Social History of Their Trade Unions 1775–1974* (Edinburgh: J. Donald, 1979), 182–189; James E. Handley, *The Irish in Modern Scotland* (Cork: Cork University Press, 1947), 44; Donald M. MacRaild, *Culture, Conflict, and Migration: The Irish in Victorian Cumbria* (Liverpool: Liverpool University Press, 1988), 40; Paul O'Leary, "Networking Respectability: Class, Gender and Ethnicity among the Irish in South Wales, 1845–1914," in Donald M. MacRaild and Enda Delaney, eds., *Irish Migration, Networks and Ethnic Identities since 1750* (London: Routledge, 2007), 114; Noonan, "Wandering Labourers," 44–71.

15. While other resources were mined in these states, in particular lead in Dubuque, Iowa, such mines were shallow and did not require hard-rock mining skills. Robert F. Klein, *Dubuque during the California Gold Rush: When the Midwest Went West* (Charleston SC: History Press, 2011), 16–20.

16. This holds true for other states as well, emphasizing New York's importance as a migration hub for the Irish on their way to the American West.

17. The address "James [Séamus] Feiritéar, 200 East 46th St. New York" was written on the inside page of his diary. Diary of Séamus Feiritéar, Butte-Silver Bow Archives, Butte, MT.

18. Depositors, Box 21, Emigrant Savings Bank Records, Manuscripts and Archives Division, New York Public Library, New York, NY. The addresses in New York for the accounts seem to be those of their relatives in the city. Many thanks to Professor Marion Casey for help with this source. There are several dozen listings for "miners"; however, approximately half are actually "minors," children, misspelled in the entry.

19. Marion Casey, "Refractive History: Memory and the Founders of the Emigrant Savings Bank," in Joseph Lee and Marion Casey, eds., *Making the Irish American: History and Heritage of the Irish in the United States* (New York: New York University Press, 2006), 317–318.

20. See Mary C. Waters, *Ethnic Options: Choosing Identities in America* (Berkeley: University of California Press, 1990), 32–37; Ronald M. James, "Erin's Daughters on the Comstock," in Ronald M. James and C. Elizabeth Raymond, eds., *Comstock Women: The Making of a Mining Community* (Reno: University of Nevada Press, 1998), 246–262. For more on identity, see chapter 6, this volume.

21. The two tables in appendix 3 provide full nativity breakdowns for the parents of US-born and Canadian-born miners in the American West, which demonstrate that if a Canadian- or American-born miner had a foreign parent, that parent was most likely Irish. Significant numbers of Canadian-born miners had Scottish parentage, suggesting a miner's migration path from Scotland through Canada and onward to the US. Using the more accurate figures to represent the English, Scottish, Welsh, and German miners shows an underestimation of approximately

1 percent. The Irish percentage increases from 8.9 to 12.5 percent, an increase of 40 percent. This more accurate figure places the Irish-American mining population second numerically to the Chinese in the American West.

22. See Noonan, "Wandering Labourers," 44–45, in particular tables 2.2, 2.3, and 2.4.

23. With further investigation and development of the census data, it is highly likely that Cornwall will not follow this general rule. The hostility in the region toward the Irish will likely produce similar data to those for Irish miners in Wales. See Noonan, "Wandering Labourers," 51–58.

24. Frederick Jackson Turner, *The Frontier in American History* (New York: H. Holt, 1920), 12.

25. David Fitzpatrick, "The Irish in Britain, Settlers or Transients," in Patrick Buckland, and John Belchem, eds., *The Irish in British Labour History: Conference Proceedings in Irish Studies*, no. 1 (Liverpool: Institute of Irish Studies, University of Liverpool with the Society for the Study of Labour History Liverpool, March 1992), 5.

26. Graham Davis and Matthew Goulding, "Irish Hard-Rock Miners in Ireland, Britain and the United States," in Graham Davis and John Fripp, eds., *In Search of a Better Life: British and Irish Migration* (Stroud, Gloucestershire: History Press, 2011), 179–196.

27. Wayland D. Hand, "The Folklore, Customs, and Traditions of the Butte Miner," *California Folklore Quarterly* 5, no. 1 (January 1946): 19–20.

28. Forty-two women with Irish parentage were laborers in California and twenty-two lived in San Francisco. There is a wide range of ages—the youngest is 15, the oldest 78—and both extremes of the range are found in San Francisco. California is the only state in the American West with more than 4 Irish-American women laborers. In contrast, New York, Massachusetts, and Pennsylvania account for more than half of all Irish-American women laborers, with 508, 205, and 190 respectively. IPUMS.

29. Mary R. Stickney, "Mining Women of Colorado," *The Era: An Illustrated Monthly Magazine of Literature and of General Interest* (Philadelphia: Henry T. Coates), 9 (1902): 24–33.

30. Stickney, "Mining Women of Colorado," 29.

31. Stickney, "Mining Women of Colorado," 30.

32. Stickney, "Mining Women of Colorado," 30.

33. Stickney, "Mining Women of Colorado," 30. McCarthy's lifelong curiosity is shown by her journey to the World's Fair in St. Louis for several weeks in 1904. *Intermountain Catholic*, July 2, 1904. McCarthy's birthplace is listed as Indiana (to Irish parents) in the 1900 census but as "Ireland (Eng.)" in the 1910 census. Whether this was a result of her misleading the census inspector or a mistaken entry is not known, and it is unclear what the term "Eng" refers to. Election District 82,

Precinct 5, Denver City, Arapahoe County, Colorado United States Census of 1900, 6B; Election District 90, Precinct 6, Denver City, Arapahoe County, Colorado, United States Census of 1910, 1A; Family Search, familysearch.org.

34. There are hints of a network of women business owners and mine investors in Colorado supporting one another, similar to the Irish-American networks detailed in this book. For example, Delia A. McCarthy worked as a secretary in the Bonacord Company run by Mrs. E. C. Atwood. Further investigation is required beyond their mention in Sandra L. Myres, *Westering Women and the Frontier Experience, 1800–1915* (Albuquerque: University of New Mexico Press, 1999), 264.

35. This book uses the term *sex work* rather than prostitution because of the inherent connotations with criminality and immorality associated with the latter, which is only employed where the cultural context necessitates its use.

36. Anne Seagraves, *Soiled Doves: Prostitution in the Early West* (Hayden, ID: Wesanne, 1994), 109. Mary Welch aka "Chicago Joe" Hensley was a brothel keeper in Helena, Montana. Molly Burdan aka Molly b'Damn, was a brothel keeper in Murray, Idaho.

37. The report combines miners and quarrymen into a single category. This is not a serious deficiency, as quarrymen form a small part—less than 5 percent—of this number. In 1880 Irish-born made up 23 percent and Irish-Americans 31 percent of all quarrymen. IPUMS. United States Congress Joint Immigration Commission, *Report of the Immigration Commission, Occupations of the First and Second Generations of Immigrants in the United States*, vol. 65 (Washington, DC: Government Printing Office, 1911), 6–16.

38. United States Joint Immigration Commission, *Report of the Immigration Commission.*

39. United States Joint Immigration Commission, *Report of the Immigration Commission.*

40. United States Joint Immigration Commission, *Report of the Immigration Commission.*

41. United States Joint Immigration Commission, *Report of the Immigration Commission.*

42. "The Reminiscences of John Brophy," 8, Microfilm, MICRO 56, Columbia Oral History Collection, Elmer Holmes Bobst Library, New York University, New York, NY.

43. One miner, who learned his trade from his Irish friends, barely escaped being crushed by falling rock and later incorrectly credited both Irish and Cornish tradition for saving his life. "I was sure that I owed my life to the tommyknockers, those unseen, wee, small folk, who came over with the Irish and Cousin Jacks, to tap on the rocks and warn mining stiffs when there is some serious underground danger,

as they had warned me." His recollection reveals the popularity of the Cornish superstition among miners. Crampton, *Deep Enough*, 110.

44. Seán Ó Dubhda, ed., *An Duanaire Duibhneach*, 130–131, translation by Bruce D. Boling, Miller Collection, held by Professor A. Kerby Miller, University of Missouri, Columbia [hereafter Miller Collection].

45. Ó Dubhda, *Duanaire Duibhneach*, 132–133.

46. Ó Dubhda, *Duanaire Duibhneach*, 130–131.

47. Frances McCoy, Robert McCoy, and Francis Frederick, interviewed by Paula B. Robke, December 3, 1978, OH 1688, North Idaho College Oral History Program Collection, Idaho Historical Society, Boise [hereafter NICO].

48. Frances McCoy, Robert McCoy, and Francis Frederick interview.

49. Evelyne Stitt Pickett, "Hoboes across the Border: A Comparison of Itinerant Cross-Border Laborers between Montana and Western Canada," *Montana: The Magazine of Western History* 49, no. 1 (Spring 1999): 18–31.

50. E. R. Lewis, "The Ability of the Hobo," *Railway Age Gazette*, June 21, 1912, 1567.

51. E. Keough, "Characteristics of the Hobo," *Railway Age Gazette*, June 21, 1912, 1566. "[A]nd, while they are not necessarily rough, they do not pick their language nor use extra precautions to get along with the men, as do the inexperienced ones." Rough in this case means violent.

52. Mark Wyman, *Hoboes: Bindlestiffs, Fruit Tramps, and the Harvesting of the West* (New York: Hill and Wang, 2010), 268.

53. Lewis, "Ability of the Hobo," 1567.

54. F. D. Calhoon, *49er Irish: One Irish Family in the California Mines* (New York: Exposition, 1977), 88.

55. Calhoon, *49er Irish*, 89.

56. Calhool *49er Irish*. His mother, Mary, died of Spanish influenza a few months later.

57. Songs and Ballads of the Anthracite Miners, recorded and edited by George Korson, AFS LI6, stored at the American Folklife Center, Library of Congress, Washington, DC.

58. Roger Bruns, *Knights of the Road: A Hobo History* (New York: Methuen, 1980), 11.

59. Frederick Rhinaldo Wedge, *Inside the I.W.W. by a Former Member and Official: A Study of the Behavior of the I.W.W., with Reference to Primary Causes* (Berkeley: F. R. Wedge, 1924).

60. Pinkerton Agency Records, Box 60, Manuscript Division, Library of Congress, Washington, DC.

61. TDA to YAMMC, August 31, 1906, Yellow Aster Mining and Milling Company Records, 1898–1918, Manuscript Collection, California History Section, California State Library, Sacramento [hereafter YAMC].

62. Some state governments made further efforts to limit hobos' movements. In Massachusetts, a vagrancy law made riding a freight train *prima facie* evidence of being a tramp, and punishment varied from 90 days of hard labor in New Mexico to being sold into servitude for up to a year in Kentucky. In Missouri, a tramp could be hired out to the highest bidder with cash in hand. In terms of legal definitions, in most states tramps were men. E. R. Lewis, "The Ability of the Hobo," *Railway Age Gazette*, June 21, 1912, 1567.

63. Frances McCoy, Robert McCoy, and Francis Frederick interview, NICO.

64. Wilbur S. Shepperson, *Restless Strangers: Nevada's Immigrants and their Interpreters* (Reno: University of Nevada Press, 1970), 57.

65. Shepperson, *Restless Strangers*, 57; Butte City Ward 3, Election District 109, Silver Bow County, Montana, United States Census of 1900, 69A, Family Search, familysearch.org. Casey was born in August 1855 and could be found in the city prison in the 1900 census. Shepperson's book contain no sources for his interviews, no background details on his interview subjects, and no way for historians to find further details of their lives except through extensive research.

Chapter 2: Digging Lumps of Gold

1. Perhaps as many as half came to San Francisco by way of Australia according to David Noel Doyle, "The Irish in North America, 1776–1845," in Joseph Lee and Marion Casey, eds., *Making the Irish American: History and Heritage of the Irish in the United States* (New York: New York University Press, 2006), 172. Doyle offers no source for this estimate. Given this and the huge numbers of Irish at the gold mining camps in Australia, it seems that the largest proportion of those called the "Sydney Ducks" were Irish. The fact that they and the Chinese were the most viciously targeted by a nativist organization, led by Irishman John Kearney in San Francisco, complicates the simplistic narrative of American nativists supported by the Irish routing foreigners.

2. James Largan to Dr. Forde, in Henry L. Walsh, *Hallowed Were the Gold Dust Trails: The Story of the Pioneer Priests of Northern California* (Santa Clara: University of Santa Clara Press, 1946), 28–30.

3. The newlywed couple emigrated from Belfast, Ireland, in 1847 after they had eloped in County Cork. Calhoon, *49er Irish*, 3–4.

4. Calhoon *49er Irish*, 31–39.

5. Calhoon, *49er Irish*, 31–39.

6. John K. Orr, Sacramento, to his parents, County Down, Ireland, November 1, 1849, Correspondence and Family Papers relating to the Orr and Dunn Families,

1612–1949, D.2908, Public Record of Northern Ireland, Belfast [hereafter Orr Family Letters]. Also accessible at Irish Emigration Database, http://ied.dippam.ac.uk/records/30346.

7. John Orr, Sacramento, to Rev. John Orr, Portaferry, July 13, 1850, Orr Family Letters.

8. John Orr to Rev. John Orr, July 13, 1850.

9. John Orr, California, to Rev. John Orr, Portaferry, October 27, 1850, Orr Family Letters.

10. John Orr to Rev. John Orr, October 27, 1850. The town responded by locally raising $250,000 for the construction of a levee.

11. John Orr to Jane Orr, Portaferry, August 17, 1850, Orr Family Letters.

12. John Orr to Jane Orr, August 17, 1850. The exchange rate at this time was $5 to £1.

13. John Orr to Jane Orr, August 17, 1850.

14. John Orr to Rev. John Orr, October 27, 1850.

15. John Orr to Rev. John Orr, October 27, 1850.

16. E. E. Griggs, Sacramento, to Rev. John Orr, November 8, 1850, Orr Family Letters.

17. E. E. Griggs to Rev. John Orr, November 8, 1850.

18. John Orr to Jane Orr, August 17, 1850.

19. Rev. John Orr, Portaferry, to E. E. Griggs, California, January 22, 1851, Orr Family Letters.

20. Rev. John Orr to E. E. Griggs, January 22, 1851.

21. Rev. John Orr to E. E. Griggs, January 22, 1851.

22. E. E. Griggs, Sacramento, California, to Rev. John Orr, December 12, 1851, Orr Family Letters.

23. E. E. Griggs to Rev. John Orr, December 12, 1851.

24. E. E. Griggs to Rev. John Orr, December 12, 1851.

25. E. E. Griggs to Rev. John Orr, May 15, 1852.

26. E. E. Griggs, Sacramento, to Rev. John Orr, Portaferry, May 15, 1852, Orr Family Letters.

27. William Hayes, Gold Hill, Placer Co. California, to his wife and child, March 15, 1857, Box 4, C-B 547, California Gold Rush Letters, Bancroft Library, University of California, Berkeley.

28. Judith Phelan, Raheen, Queen's Co./Laois, to John Lawlor, St. Louis, July 27, 1852, Teresa Lawlor/Phelan Letters, California Historical Society, San Francisco [hereafter Lawlor Letters].

29. Margaret Mehen, Sonora, California, to sister Teresa Lawlor, St. Louis, July 13, 1854, Lawlor Letters.

30. Judith Phelan, Raheen, Queen's Co./Laois, to Teresa Lawlor Gallagher, St. Louis, July 27, 1852, Lawlor Letters.

31. Margaret Mehen to Teresa Lawlor, July 13, 1854, Lawlor Letters.

32. Judith Phelan, Raheen, Queen's Co./Laois, to Teresa Lawlor, St. Louis, April 16, 1854, Lawlor Letters.

33. Patrick Phelan, Spring Park, near Melbourne, Australia, to Peter Mehen, husband of Margaret Lawlor, Sonora, California, October 15, 1853, Lawlor Letters. Patrick Phelan learned who Margaret had married in California and sent three letters to "Mr. Mihen Cenori, Callifornia." This fourth letter to Peter Mehen did reach him, but it is not stated how he learned the correct spelling of his name, which appeared in various sources as Meehan, Meehen, or Mehan. Patrick Phelan to Peter Mehen, October 15, 1853, Lawlor Letters.

34. An Irish wake is a particular combination of a funeral ceremony and a celebratory gathering held when someone dies. Irish emigrants continued this practice in the American West; see chapter 6, this volume.

35. James Gamble, Canyon City, Wasco Co., Oregon, to his brother, Abel Gamble, Belfast, December 22, 1864, Gamble Family Letters, Miller Collection [hereafter Gamble Family Letters].

36. James Gamble, Jemmy Lind, Calavera Co., California, to his mother, Magherascouse, Co. Down, Ireland, May 8, 1850, Gamble Family Letters.

37. Dudley T. Ross, *Devil on Horseback: A Biography of the "Notorious" Jack Powers* (Fresno: Valley Publishers, 1975), 19–20.

38. *Belfast Newsletter*, January 9, 1849.

39. Franklyn Y. Fitch, *The Life, Travels and Adventures of an American Wanderer: A Truthful Narrative of Events in the Life of Alonzo P. De Milt Containing His Early Adventures among the Indians of Florida; His Life in the Gold Mines of California and Australia* (New York: John W. Lovell, 1883), 115–116.

40. John S. Hittell, *Mining in the Pacific States of North America* (New York: John Wiley, 1862), 34.

41. Census of Ireland 1901, no. 16, Drumcrow, Drumcarbon, Co. Cavan.

42. Census of Ireland 1901, no. 8, Circular, Philipstown, Co. Offaly (King's County). The census does specify some miners as coal miners, but this is less a case of occupational pride than simply a clarification that they were working in nearby Irish mines. In contrast, there are only three references to silver mining as an occupation in the Irish census. One was Mary Sullivan, a thirty-one-year-old "retired silver miner's wife" who returned to Ireland with her children, all of whom are listed as born in "Montana, USA." Another was Ambrose Quilty, a young miner who returned to his family. His occupation lists "Silver Miner (Home on Visit)." The last silver miner, James Brown, was likewise visiting the family home in

Ardagh Co. Kerry. Census of Ireland 1901, no. 17, Urhin, Conlagh, Co. Cork; Census of Ireland 1901, no. 14, Drumclifff North, Carney, Co. Sligo; Census of Ireland 1901, no. 4, Ardagh, Ardoughter, Co. Kerry.

43. Patrick Catherwood, Magherascouse, Co. Down, to his cousin, America, August 12, 1855, Gamble Family Letters.

44. Patrick Catherwood to his cousin, America, August 12, 1855.

45. Seamus P. Lynch, *Short Biography on Pat Magill, Pioneer, Rancher and Miner* (Omagh, Ireland: Centre for Migration Studies, 1939).

46. Lynch, *Short Biography on Pat Magill*. Patrick Magill is found in the 1930 US Census in Clark township in Routt, Colorado. I have used his wife's name as it appears in the census rather than the version in the account by Lynch, which uses Meurl.

47. James Williamson to John Williamson, October 2, 1850, Williamson Papers 1781–ca. 1910, MS T.2680, Public Record Office of Northern Ireland, Belfast [hereafter Williamson Family Letters],

48. *Belfast Commercial Chronicle*, August 18, 1852.

49. John McTurk Gibson, May 22, 1859, diary entry, "Journal of Western Travel," Bancroft Library, University of California, Berkeley [hereafter Gibson diary].

50. May 22, 1859, diary entry, Gibson diary.

51. Irwin Silber, ed., *Songs of the Great American West* (New York: Macmillan, 1967), 116. Also note modifications of other songs: "O California! This is the land for me, / A pick and shovel, and lots of bones! / Who would not come the sight to see, / The golden land of dross and stones. / Oh Susannah, don't you cry for me, / I'm living dead in Californ-nee." Alonzo Delano, *Life on the Plains and among the Diggings* (Buffalo, NY: Miller, Ortan and Mulligan, 1854), 349.

52. *Belfast Commercial Chronicle*, April 15, 1850. This paper was the "Orange" paper in Belfast, that is, pro-union, pro-Protestant, and anti-Catholic. The sneering tone was part of an effort to portray Irish society as dichotomous, pitting Protestant against Catholic, in an effort to meld the deep class and denominational differences of Protestantism (and indeed Catholicism) into one identity. In this image, crafted in Ireland, lay the seeds that would grow into the term *Scots Irish* in America, a continuing effort to distinguish the small differences between them and their perceived adversaries. See Miller, *Emigrants and Exiles*, 78.

53. *Belfast Commercial Chronicle*, April 15, 1850.

54. *Belfast Commercial Chronicle*, April 15, 1850.

55. *Belfast Newsletter*, February 15, 1850.

56. "Lament of the Irish Gold Hunter," American Song Sheets, Rare Books and Special Collections, Library of Congress, Washington, DC. The piece's harsh, mocking tone makes it unlikely that this was an effort at self-deprecation, and so it is assumed that the author is not Irish.

57. Lawrence Taylor, "Bás in Éirinn: Cultural Constructions of Death in Ireland," *Anthropological Quarterly* LXII, no. 4 (1989): 175–188.

58. *Armagh Guardian*, January 15, 1849.

59. John O'Hanlon, *The Irish Emigrant's Guide for the United States* (New York: Arno, 1976), 181.

60. *National Police Gazette*, November 11, 1882. For more on the sexually suggestive publication, see Guy Reel, *The National Police Gazette and the Making of the Modern American Man, 1879–1906* (New York: Palgrave Macmillan, 2006).

61. "Autobiography and Reminiscence of Thomas Kyle," vol. 8, manuscript, Society of California Pioneers, Alice Phelan Sullivan Library, San Francisco, 1904.

62. *Sacramento Daily Union*, April 18, 1859. Bradley's cause of death was established through "A surgical dissection of the brain, by Doctor Simpson, before the Coroner's Jury."

63. "Daley, a very fine young man, aged about thirty, a miner in employ of King & Clark, at Squaw Creek, was fatally injured by the falling in of a slab of dirt in a narrow cut. He was gotten out by his fellow workmen, but had no use of himself and died next morning. His body was conveyed to Marysville, to-day, for burial." *Sacramento Daily Union*, April 18, 1859.

64. May 22, 1859, diary entry, Gibson diary.

65. Francis P. Farquhar, *Up and Down California in 1860–1864: The Journal of William H. Brewer* (Berkeley: University of California Press, 1974), 348.

66. Bill Williamson, California, to Robert Williamson, Ahorey, February 12, 1876, Williamson Family Letters.

67. Bill Williamson to Robert Williamson, Ahorey, February 12, 1876, Williamson Family Letters.

68. Bill Williamson to Robert Williamson, Ahorey, February 12, 1876, Williamson Family Letters.

69. Mann, *After the Gold Rush*, 120.

70. Extracts from Letter—Writer Unknown, *Belfast Commercial Chronicle*, August 18, 1852.

71. Extracts from Letter, August 18, 1852.

72. Herbert O. Lang, *A History of Tuolumne County, California: Compiled from the Most Authentic Records* (San Francisco: B. F. Alley, 1882), 18, 39, 40. Lang writes that it was the "third or fourth within a week." The phrase *Mexican Indians* is unclear and could refer to indigenous peoples from outside of Sonora, probably including Southern California as well as Mexico.

73. Lang, *History of Tuolumne County*, 41.

74. Lang, *History of Tuolumne County*, 41.

75. Lang, *History of Tuolumne County*, 41.

76. Lang, *History of Tuolumne County*, 42.

77. Lang, *History of Tuolumne County*, 43.

78. Lang, *History of Tuolumne County*, 43.

79. Lang, *History of Tuolumne County*, 44.

80. Lang, *History of Tuolumne County*, 45.

81. Lang, *History of Tuolumne County*, 45.

82. Lang, *History of Tuolumne County*, 46.

83. Lang, *History of Tuolumne County*, 46.

84. Lang, *History of Tuolumne County*, 47.

85. Henry V. Huntley, *California: Its Gold and Its Inhabitants*, vol. 2 (London: Thomas Cautley Newby, 1856), 251.

86. Lang, *History of Tuolumne County*, 161.

87. Ross, *Devil on Horseback*, 1–5.

88. Delano, *Life on the Plains and among the Diggings*, 242.

89. Hinton R. Helper, *The Land of Gold: Reality versus Fiction* (Baltimore: Baltimore, Pub., 1855), 96. Helper also wrote *The Impending Crisis of the South*, credited as "the single most important book, in terms of its political impact, in the United States." George M. Fredrickson, *The Arrogance of Race: Historical Perspectives on Slavery, Racism, and Social Inequality* (Middletown, CT: Wesleyan University Press, 1988), 28.

90. Helper, *Land of Gold*, 104. The parallel between Helper's economic disillusionment and the resurgence of white nationalists in the US in the 2010s is obvious.

91. *Sacramento Daily Union*, September 14, 1859; Bryan Anthony Carter, "A Frontier Apart: Identity, Loyalty, and the Coming of the Civil War on the Pacific Coast," PhD diss., Oklahoma State University, Stillwater, 2014, 322–323. The judge's name was John Terry.

92. *Los Angeles Star*, September 13, 1856; Carter, "Frontier Apart," 291. Also, he did not see himself as Irish.

93. There is no historical record of Mehen's religious affiliation. As such, it was unlikely that he was a practicing Catholic, probably a factor in his ascension to the ranks of the "respectable."

94. Lang, *History of Tuolumne County*, 37.

95. Katherine A. White, *A Yankee Trader in the Gold Rush: The Letters of Franklin A. Buck* (Boston: Riverside Press Cambridge, 1930), 110.

96. White, *Yankee Trader in the Gold Rush*, 111.

97. White, *Yankee Trader in the Gold Rush*, 62. Strangely, his faith in lynchings was unaffected. Part of this might have been his romanticized view of the American West. At one point, when a rancher hired fifteen mounted men led by the sheriff to hunt down cattle thieves, the writer armed with rifle and pistol described the event as "romantic and I could imagine myself back in the times of Scott's novels,"

a view that doubtless colored his view of all violence. White, *Yankee Trader in the Gold Rush*, 62.

98. The account does not explain whether Byrne died from his wound or the composition of the people intervening. Huntley, *California*, 213.

99. John B. McGloin and Martin Francis Schwenninger, "A California Gold Rush Padre: New Light on the 'Padre of Paradise Flat,'" *California Historical Society Quarterly* 40 (March 1961): 62.

100. *Seventh Report of All Hallows Missionary College* (Dublin: J. F. Fowler, 1855), 11.

101. *Seventh Report of All Hallows Missionary College*, 12, 19.

102. Record of Internment, Smartsville, Diocese of Sacramento. It is a strange coincidence that William Butler Yeats would write the "Ballad of Father Gilligan," which describes an exhausted priest during the time of the famine who falls asleep and is thereby unable to attend to one of his parishioners to give them his last rites: "'Mavrone, mavrone! the man has died, / While I slept on the chair.' / He roused his horse out of its sleep, / And rode with little care." The priest then arrives to the man's house where the woman greets him, saying, "'Father! You come again.' The priest is deeply upset before he realized the woman mentioned him appearing, and so he thanks God for his mercy, 'He who hath made the night of stars / For souls who tire and bleed, / Sent one of His great angels down / To help me in my need.'" As one priest wrote, such was the determination to administer last rites: "It is no ordinary thing to start at eight or ten o'clock at night on a sick call, sometimes ten, twenty, or thirty miles over dreary and lonely plains." *Annals of All Hallows Missionary College* (Dublin: J. F. Fowler, 1863), 91. Emphasis added.

103. Denis Hurley to John Hurley, May 12, 1922, Hurley Family Emigrant Letters, MS U170 [hereafter Hurley Letters], Cork City Archives, Cork, Ireland. A few decades earlier he wrote, "There is nothing true but heaven." Denis Hurley, Carson City, to his mother, March 12, 1894, Hurley Letters.

104. See Mullany Letters, chapter 5, this volume.

105. James P. Shannon, *The Catholic Colonization of the Western Frontier* (New Haven, CT: Yale University Press, 1957), 28, 71.

106. O'Hanlon, *Irish Emigrant's Guide*, 138.

107. A book that focuses on these misunderstandings and cultural differences is Emmons, *Beyond the American Pale*.

108. Clustering appears to have been a factor for the Irish in many rural settlements throughout the American West. See Shannon, *Catholic Colonization*, 152; Alan J. M. Noonan, "From Ireland to Montana: A Study of the Frontier 1860–1900," M.Phil. thesis, University College Cork, Ireland, 2008, 55–63.

109. Rollo Ogden, ed., *Life and Letters of Edwin Lawrence Godkin*, vol. 1 (New York: Macmillan, 1907), 183.

110. McGloin and Schwenninger, "California Gold Rush Padre," 58.

111. McGloin and Schwenninger, "California Gold Rush Padre," 64.

112. McGloin and Schwenninger, "California Gold Rush Padre," 59–60: "the United States, Canada, Mexico, Chile, Peru, Brazil, Ireland, England, France, Spain, Portugal, Italy, Switzerland, Germany, Austria, Poland, China, Australia, and the Sandwich Islands . . . all understand the Holy Mass and take part in the One Sacrifice." Reverend Florian's letter was written on January 29, 1859. McGloin and Schwenninger, "California Gold Rush Padre," 64.

113. Calhoon, *49er Irish*, 75. See chapter 6, this volume.

114. McGloin and Schwenninger, "California Gold Rush Padre," 58.

115. Susan Johnson, *Roaring Camp: The Social World of the Californian Gold Rush* (New York: W. W. Norton, 2001), 150. Nellie Cashman, who chose to live in remote Alaska following her last gold rush, remained a devout Catholic and even when enfeebled and bed-bound would habitually pray the rosary.

116. *Belfast Newsletter*, September 25, 1866. It is difficult to confirm the veracity of this case of concealed cannibal clergy. It is possible that it was a popular horror story recounted in newspapers. Or not.

117. Lynn I. Perrigo, "Law and Order in Early Colorado Mining Camps," *Mississippi Valley Historical Review* 28, no. 1 (June 1941); 50. See chapter 6, this volume.

118. James Gamble, Canyon City, Wasco Co., Oregon, to his brother, Abel Gamble, December 22, 1864, Gamble Family Letters.

119. "A Home in the Golden West," *Belfast Telegraph*, June 1, 1928.

120. Richard White, *"It's Your Misfortune and None of My Own": A History of the American West* (Norman: University of Oklahoma Press, 1991), 199.

121. James Gamble to Abel Gamble, December 22, 1864, Gamble Family Letters.

122. James Gamble to Abel Gamble, December 22, 1864, Gamble Family Letters.

123. James Gamble to Abel Gamble, December 22, 1864, Gamble Family Letters. Somewhat strangely, it was six years before his brother learned of his death.

124. McGloin and Schwenninger, "California Gold Rush Padre," 61. Whether he was referring to the quality of the food or the lodging was not clear.

125. McGloin and Schwenninger, "California Gold Rush Padre," 61.

126. Charles Canning to Lizzie McSparron, January 20, 1878, T.2743, McSparron (McSparran) Papers, 1860–1916 [hereafter McSparron Letters], Public Record Office of Northern Ireland, Belfast.

127. Charles Canning and Lizzie McSparron, January 20, 1878.

128. See Alan J. M. Noonan, "'Oh Those Long Months without a Word from Home,' Migrant Letters from Mining Frontiers," *The Boolean* (2011): 135–142. Walla Walla, Idaho Territory, was within what would later become the state of Washington.

129. James Mullany to his sister Mary, November 28, 1861, MSS 2417, Mullany Letters, Oregon Historical Society, Portland [hereafter Mullany Letters].

130. James Mullany to his sister Mary, November 28, 1861.

131. Denis Hurley to his brother Tim, April 1, 1895, Hurley Letters.

132. Paschal L. Mack letter to his sister, San Francisco, August 20, 1853, ALS, 1852–1853, MSS C-B 547, California Gold Rush Letters, 1848–1859, Bancroft Library, University of California, Berkeley.

133. "The reason of me not writing sooner, I have been living far away from any post office." James Gamble, Canyon City, Wasco Co., Oregon, to his brother Abel, Ireland, July 24, 1864, Gamble Family Letters.

134. James Gamble to his brother Abel, July 24, 1864.

135. White, *"It's Your Misfortune and None of My Own,"* 303.

136. John Orr to Jane Orr, Portaferry, August 17, 1850, Orr Family Letters.

137. Paschal L. Mack letters to sister, San Francisco, December 13, 1852, ALS, 1852–1853, MSS C-B 547, California Gold Rush Letters, 1848–1859, Bancroft Library, University of California, Berkeley.

138. McGloin and Schwenninger, "California Gold Rush Padre," 62.

139. McGloin and Schwenninger, "California Gold Rush Padre," 62.

140. Denis Hurley, Carson City, to his parents, Clonakilty, March 16, 1874, Hurley Letters.

141. McGloin and Schwenninger, "California Gold Rush Padre," 62.

142. William Breault, *The Miner Was a Bishop: The Pioneer Years of Patrick Manogue in California and Nevada 1854–1895* (Rancho Cordova, CA: Landmark Enterprises, 1988), 4.

143. Breault, *The Miner Was a Bishop*, 14.

144. Breault, *The Miner Was a Bishop*, 30–40.

145. The Diary of Robert Williamson of Ahorey, 19 [unpublished manuscript], Miller Collection. He traveled from Ireland to visit his relatives who lived in California in 1874.

146. Noonan, "From Ireland to Montana," 120–125.

147. *Daily Free Press*, December 23, 1880.

148. Some suggested paths of migration include the fact that the majority of those born in Canada who immigrated to California in 1880 were of Irish parentage.

149. *Puck Magazine*, May 15, 1878, Library of Congress, Print and Photographs Division, Washington, DC.

150. *Puck Magazine*, May 15, 1878.

151. Perhaps it was inevitable there would be some cultural friction when four Irishmen were hired by a company called the English Mining Company. Huntley, *California*, 160–161.

152. White, *Yankee Trader*, 144. Original emphasis. Unlike most nativists, Buck shows some nuance in his understanding of Indians, distinguishing those who lived in the mountains with a "meaner race" in the valley. He also commented on the treaty between the townspeople and Indians: "I don't think it will be their fault if the Treaty is broken, as the wrong in most all cases is first on the sides of the whites." White, *Yankee Trader*, 109–110. His father was Rufus Buck, elected to the House of Representatives in 1835, 1844, and 1855. Accessed January 20, 2017, legislature.maine.gov. The state website offers only Republican, Democratic, and Whig Party affiliations, obscuring its legislators' bigoted and nativist history.

153. White, *Yankee Trader*, 144.

154. We do not know the form of the oath or how exactly he conducted himself in this position, but given his account, he was likely more aggressive and insulting than he recounts. In his accounts, Buck also fetishized foreign-born women. While traveling to California he spoke to a "superlatively green" woman from Belfast whom he fantasized about later that night. Elsewhere he details a ball, saying: "I am considerable of a Know Nothing but I rather took to the Spanish set, especially as Mrs. Davidson happens to be the prettiest specimen of the Mexican people I have ever seen. I wish you could see her waltz." White, *Yankee Trader*, 19, 145.

155. Benjamin B. Richards, *California Gold Rush Merchant: The Journal of Stephen Chaplin Davis* (San Marino, CA: Huntington Library, 1959), 59.

156. Huntley, *California Gold Rush Merchant*, 138–139.

157. Richards, *California*, 61–62.

158. Record of Internment, Smartsville, Diocese of Sacramento, CA [hereafter Record of Internment].

159. William T. Ellis, *Memories: My Seventy-Two Years in the Romantic County of Yuba, California* (Eugene: University of Oregon, 1939), 44.

160. Noonan, "Wandering Labourers," 43–69.

161. Huntley, *California*, 91

162. Calhoon, *49er Irish*, 74; Mann, *After the Gold Rush*, 2.

163. *Daily Alta California*, January 17, 1860.

164. Twogood Diary, October 12, 1856, Special Collections, California Historical Society, San Francisco.

165. Township No. 4, Placer County, California, 1860 United States Census, 180, Family Search, familysearch.org. In the census, Township No. 4 included Colfax and Dutch Flats, another miner settlement twelve miles northeast of Colfax.

166. Township No. 4, familysearch.org. White and Rice were the only two with any personal cash, $100 and $200, respectively.

167. Terry Lee Anderson and Peter Jensen Hill, eds., *The Not So Wild, Wild West: Property Rights on the Frontier* (Stanford: Stanford Economics and Finance, 2004), 3–5, 29, 151–152.

168. *Sacramento Daily Union*, August 26, 1856.

169. *Sacramento Daily Union*, August 26, 1856.

170. *Marysville Daily Herald*, August 22, 1856. A fascinating comparison to this "gentle" whipping is the graphic detail given by one pro-vigilante writer after a gang of Australian miners entrusted a gold nugget to an assayer who "lost it" and was then whipped by the "Sydney Ducks" and his store ransacked. The following day the writer helped whip the crowd into a righteous frenzy, but they were unable to find any Australian to punish. Thomas N. Wand, "Early Days in the Mountains," *San Francisco Call*, July 21, 1889, in Marvin Lewis, ed., *The Mining Frontier: Contemporary Accounts from the American West in the Nineteenth Century* (Norman: University of Oklahoma Press, 1967), 22–27.

171. *Marysville Daily Herald*, August 22, 1856; *Sacramento Daily Union*, August 26, 1856.

172. See Noonan, "From Ireland to Montana," 125–130.

173. *Marysville Daily Herald*, August 22, 1856. The *Sacramento Daily Union* editor, commenting on the case, considered this statement "too personal to be entirely pertinent to the subject." *Sacramento Daily Union*, August 26, 1856.

174. "To the Irishmen of Marysville and Vicinity," Yuba County Documents, MS 3A2, Bancroft Library, University of California, Berkeley. Original spelling; emphasis added.

175. Nationalists sometimes referred to Ireland as the fatherland in their speeches. See Timothy D. Sullivan, *Speeches from the Dock: or, Protests of Irish Patriotism* (Dublin: M. H. Gill and Son, Ltd. 1945); Calhoon, *49er Irish*, 129.

176. Thomas N. Brown, *Irish American Nationalism, 1870–1890* (Philadelphia: J. B. Lippincott), 1966.

177. James Gamble, Auburn, Baker Co., Oregon, to his mother, December 28, 1871, Gamble Family Letters.

178. James Gamble to his mother, December 28, 1871, Gamble Family Letters.

179. *Sacramento Daily Union*, April 21, 1866.

180. *Sacramento Daily Union*, April 21, 1866. Also see Duane A. Smith, *Mining America: The Industry and the Environment, 1800–1920* (Lawrence: University Press of Kansas, 1987).

181. *Sacramento Daily Union*, February 5, 1877.

182. *Sacramento Daily Union*, February 5, 1877.

183. *Sacramento Daily Union*, April 28, 1887.

184. *Sacramento Daily Union*, April 28, 1887.

185. *Sacramento Daily Union*, April 28, 1887.

186. *Daily Alta California*, February 17, 1886.

187. *Daily Alta California*, August 11, 1889.

188. Kathleen Smith and Lane Parker, *Smartsville and Timbuctoo* (Charleston, SC: Arcadia, 2008), 95. McGanney also owned a large house situated next to the Smartsville graveyard.

189. Smith and Parker, *Smartsville and Timbuctoo*, 95.

190. *Sacramento Daily Union*, May 4, 1866.

191. *Sacramento Daily Union*, May 4, 1866. The proposal to create an organization of Irishmen may have been in response to this organization.

192. *Daily Alta California*, October 4, 1884.

193. Lawrence Austin, Colfax, California, to his sister in New York, July 15, 1885, Mary Ann Landy Letters, Miller Collection.

194. *Bodie Standard*, March 8, 1882.

195. Patrick Reddy was a larger-than-life figure who probably deserves his own book. Conceived in Ireland and born in Rhode Island in 1839, he came to California in 1863 and visited many mining camps, where he earned a reputation as something of a wild figure. He lost his arm after he was shot by an unknown assailant in Virginia City but retained his reckless reputation, becoming a famous lawyer in Bodie with powerful rhetorical skills and his "lilting Irish brogue." In Bodie he founded a local branch of the Land League, as well as becoming a state senator. At age fifty-eight he foiled a post office robbery in Sacramento by tackling a gunman and using his thumb to jam the gun's hammer and cylinder. After the police arrived, he had his thumb bandaged and continued his day as normal. See the full account in Roger D. McGrath, *Gunfighters, Highwaymen and Vigilantes: Violence on the Frontier* (Berkeley: University of California Press, 1984), 121.

196. *Bodie Standard*, May 24, 1882.

197. *Sacramento Daily Union*, February 5, 1877.

198. Collected Records of Museum of St. Mary of the Mountain, Virginia City, NV. Daniel O'Sullivan was ordained in All Hallow on June 24, 1871, was assistant at Crescent City in 1872, left Smartsville in 1878 to serve as assistant in Virginia City, and succeeded Reverend Manogue as pastor. He served as pastor in Mendocino, California, from 1883 to 1886, as founding pastor of Redwood City from 1887 to 1895, and as pastor of All Hallows in San Francisco from 1898 to 1928. He died on February 3, 1928, at age eighty-two. Some of those in the internment book were buried in consecrated ground near Browne's Valley, where mass was frequently celebrated for the miners who worked in more remote areas of the county. The Irish-American community as a whole made up over 80 percent of the names recorded, and half of those Irish-Americans were born in Ireland. A total of 149 males and 112 females are listed, reflecting a more mature society in contrast to the earlier male-dominated

population typical of mining regions. The average age of death was fifty-four. One entry lists a name "Bryass," Irish, but without details about gender or age.

199. Collected Records of Museum of St. Mary of the Mountain.

200. Kevin Kenny, *Making Sense of the Molly Maguires* (New York: Oxford University Press, 1998), 212, 278–279; Fergus Macdonald, *The Catholic Church and the Secret Societies in the United States* (New York: United States Catholic Historical Society, 1946), 86.

201. *Coeur d'Alene Press* 1, no. 5 (March 19, 1892).

202. Record of Internment, Smartsville, Diocese of Sacramento, CA.

203. Record of Internment, Smartsville.

204. The priest faced other difficult situations and remarked on the death of Johanna Bohan, a sixty-four-year-old Irish woman: "This was a case of suicide by strangulation hanging which did not result in imminent death. The woman herself sent for me requesting absolution and I attended her." Record of Internment, Smartsville.

205. Record of Internment, Smartsville.

206. Record of Internment, Smartsville.

207. D. Robertson PDA to M. Donohue, Anaconda Copper Mining Company, January 17, 1899, Box 9, Anaconda Copper Mining Company Records, Collection no. 169, Montana Historical Society Archives, Helena; *Anaconda Standard*, August 12, 1898. Also see Neill to Hauser, October 4, 1898, Box 26, Samuel T. Hauser Papers, 1864–1914, Montana Historical Society, Helena.

208. D. Robertson to M. Donohue, January 17, 1899.

209. D. Robertson to M. Donohue, January 17, 1899.

210. D. Robertson to M. Donohue, January 17, 1899.

211. D. Robertson to M. Donohue, January 17, 1899.

212. *Marysville (CA) Daily Appeal*, June 22, 1905.

213. *Marysville (CA) Daily Appeal*, January 17, 1907.

214. Others applications included that of C. T. McMahon, thirty-eight, American miner, spouse Annie McMahon, worked in "Atolia Mining County, Woodsman [fraternity Woodsman of the World] #134 'Have' Idaho, KP [a likely abbreviation for the fraternity Knights of Pythias] #324 Globe, Arizona." Irishman James Duffy was hired as a blacksmith for Yellow Aster. He was single, age thirty-six, and spent nine years in the US and worked at "Russell+Green Joel County Baldwin Park, California." His contact was listed as Mr. James Duffy, Knockhather, Keltulla, Athenry, County Galway, Ireland. Thomas Donohue, age forty-nine, was also hired as a blacksmith and had been in the US for twelve years; his contact was listed as "J. J. Scanlon, Gorminotou Aust." Application for Employment, Box 2297, YAMC.

215. Application for Employment, YAMC.

216. Thiel Detective Agency [TDA] to Yellow Aster Mining and Milling Company [YA], March 19, 1904, Box 2299, YAMC. The detectives were not always successful at infiltrating the miners' unions, for a variety of reasons. A TDA official wrote to the YA: "When you asked us to send you a foreman I brought this man from Northern California at considerable expense to us so as to give you one of our best operatives in this line, with the result that after his arrival in Randsburg he was put to mining for several days and then put to mucking, and then discharged." R. C. Ackley to YA, January 26, 1905, YAMC.

217. TDA to YA, March 19, 1904, YAMC.

218. TDA to YA, March 19, 1904, YAMC.

219. TDA to YA, March 20, 1904, YAMC.

220. TDA to YA, March 26, 1904, YAMC.

221. TDA to YA, April 20, 1904, YAMC.

222. TDA to YA, October 19, 1904, YAMC.

223. TDA to YA, October 19, 1904.

224. *Annals of All Hallows Missionary College* (Dublin: J. F. Fowler, 1903), 99.

225. *Annals of All Hallows Missionary College*, 100.

226. TDA to YA, March 17, 1904, YAMC. The size of the Irish community is reflected in this anecdote. When Bishop Thomas Conaty, a native of Kilnaleck, County Cavan, visited the small Catholic community in Randsburg for the church's first confirmation, five of the eight children confirmed were McCarthy's. None of the other children were Irish-American. C. John Di Pol, *A History of the Parish of Randsburg and St. Ann Ridgecrest* (Ridgecrest, CA: C. John Di Pol, 1996), 8. As had Father Twomey, discussed earlier, another Irish-born priest, Father John J. Crowley, died while traveling across his parish in 1940. See Joan Brooks, *Desert Padre: The Life and Writings of Father John J. Crowley, 1891–1940* (Desert Hot Springs, CA: Mesquite, 1997).

227. TDA to YA, November 11, 1904, YAMC.

228. TDA to YA, November 12, 1904, YAMC.

229. TDA to YA, November 12, 1904.

230. TDA to YA, November 12, 1904. "Tom Atkinson came up from Los Angeles in an automobile to vote for McCarthy, but he was too late by one hour and ten minutes." TDA to YA, November 12, 1904.

231. TDA to YA, November 12, 1904.

232. TDA to YA, December 22, 1904, YAMC. Even in mining's early days, various communities often targeted brothels, particularly scandalous sex workers or madams, for exclusion or expulsion. Later movements sought to segregate "respectable" society from these moral deviants. See Perrigo, "Law and Order in Early Colorado

Mining Camps," 56–57. Marguerite was more commonly known as "French" Marguerite Roberts.

233. TDA to YA, December 22, 1904, YAMC.

234. TDA to YA, December 22, 1904.

235. TDA to YA, December 22, 1904. Marguerite was also trying to get the case tried in Mojave or Tehachapi rather than in Bakersfield. TDA to YA, December 22, 1904.

236. TDA to YA, January 3, 1905, YAMC.

237. TDA to YA, January 3, 1905.

238. TDA to YA, January 3, 1905.

239. TDA to YA, January 3, 1905.

240. TDA to YA, January 16, 1905, YAMC.

241. TDA to YA, January 16, 1905.

242. TDA to YA, January 16, 1905.

243. Some made racial arguments that poorer southern Italians were not white. See Thomas A. Guglielmo, *White on Arrival: Italians, Race, Color, and Power in Chicago, 1890–1945* (New York: Oxford University Press, 2004).

244. TDA to YA, March 17, 1905, YAMC.

245. TDA to YA, March 17, 1905.

246. TDA to YA, February 24, 1905, YAMC.

247. TDA to YA, December 22, 1906, YAMC.

248. Ed Larsh and Robert Nichols, *Leadville, U.S.A.* (Boulder: Johnston Books, 1992), 216.

249. TDA to YA, November 21, 1905, YAMC.

250. TDA to YA, December 21, 1906, YAMC.

251. TDA to YA, January 17, 1907, YAMC. John O'Leary was named in detective reports on January 16 and 23, 1907, YAMC. On the ethnic mixture in the mines of Randsburg, the detective noted that "[Frank Johnson] is a Finlander by birth and all Finlanders in Bingham usually join the union; in fact many of them join before going to work. In other words, that is about their first official act on arriving in Bingham from the old country." TDA to YA, January 31, 1906. On another occasion the detective wrote "On Feb 7th the operative noticed some words inscribed on the wall of the change room as follows: *Levator lubii superioris alaeque nasi*. He was later told by an Italian miner that they meant 'The boss has gone up to the first level.' " However, this was Latin, not Italian. It is possible that an Irish miner who knew Latin used it to communicate with other Italian miners, simultaneously using the language as a form of coded communication. TDA to YA, February 12 1907, YAMC.

252. TDA to YA, May 21, 1907, YAMC.

253. TDA to YA, May 21, 1907.
254. TDA to YA, May 21, 1907.
255. TDA to YA, May 21, 1907.
256. TDA to YA, May 21, 1907.

Chapter 3: Mirages in the Desert

1. Grant H. Smith, *The History of the Comstock Lode 1850–1920* (Reno: University of Nevada and Nevada State Bureau of Mines, 1943), 7.

2. William Hickman Dolman, *Before the Comstock, 1857–1858: Memoirs of William Hickman Dolman* (Reno: University of Nevada, 1947), 36. He wrote earlier of a few miners he remembered, stating that Peter O'Reilly (the different spelling might indicate that he meant a different person than the Peter O'Riley in the above quote) was "half-witted and 'half-cracked,' lazy and stupid" and Henry Comstock was "an industrious visionary prospector, thought little more than half-witted." Dolman, *Before the Comstock*, 6.

3. Following in the footsteps of many unsuccessful miners, O'Riley and McLaughlin sold their mine leases for a pittance. The former continued to work odd jobs until his death, while the latter ended his days in a mental institution in California.

4. Ronald M. James, "Erin's Daughters on the Comstock," in James and Raymond, eds., *Comstock Women*, 259.

5. Virgil A. Bucchianeri, *Nevada's Bonanza Church, Saint Mary's in the Mountains* (Virginia City, MT: Saint Mary in the Mountains Parish, 2009), 14. Also see John T. Dwyer, *Condemned to the Mines: The Life of Eugene O'Connell, 1815–1891* (New York: Vantage, 1976). James is incorrect in stating, "It is not possible to equate church activities with the Irish community exclusively or even predominantly." James, "Erin's Daughters on the Comstock," 257.

6. 1880 United States Census, IPUMS.

7. 1880 United States Census.

8. Michael Hurley, Cedar Pass, to his parents, Tawnies, Clonakilty, August 26, 1871, Hurley Letters.

9. J. N. Flint, "The Comstock in the Early Days," *San Francisco Call*, July 28, 1889, in Lewis, *Mining Frontier*, 168.

10. Women constituted the overwhelming majority in this category, with 4 Irish-born widows for each widower, again highlighting the dangers of mining as an occupation. There were 53 Irish-born "male" widowers and 196 Irish-born "female" widows. IPUMS. The mean age of Irish-born widows was forty-seven; the youngest

was twenty-three and the oldest was eighty-one. Approximately 74 percent of these widows were fifty or younger. The mean age for widowers was forty-seven.

11. Wells Drury, *An Editor on the Comstock Lode* (Oakland: Pacific Books, 1948), 71.

12. Drury, *Editor on the Comstock*, 90, 194; *Gold Hill News*, July 26, 1864.

13. *Territorial Enterprise* (Virginia City, NV), March 12, 1871.

14. Alfred Doten, *Territorial Enterprise*, March 12, 1871.

15. Walter Van Tilburg Clark, ed., *The Journals of Alfred Doten, 1849–1903*, vol. 1 (Reno: University of Nevada Press, 1973), 1133.

16. Clark, *Journals of Alfred Doten*, 1133.

17. Clark, *Journals of Alfred Doten*, vol. 1, 1218. The date of the talk was sometime between January 18 and February 6, 1874. Alfred Doten wrote for the *Virginia Daily Union* (Virginia City, NV), the *Territorial Enterprise*, and the *Gold Hill News* at different times during his life—although there were a number of nationalistic priests throughout the mining camps of the West and one, Fr. Thomas J. Hagerty, who was very influential in the establishment of the Industrial Workers of the World. See chapter 6, this volume. Sheehy had joined Clan na Gael in 1872. For more details on the life of Fr. Eugene Sheehy, see Celia I. Knox, "The Patriot Priest—Father Eugene Sheehy: His Life, Work, and Influence," PhD diss., University of Sussex, England, 1998.

18. Knox, "Patriot Priest," 13.

19. *Irish World*, April 1, 1872.

20. Brown, *Irish American Nationalism*, 110.

21. Emmons, *Butte Irish*, 52.

22. Michael Davitt, *The Fall of Feudalism in Ireland, or the Story of the Land League Revolution* (New York: Harper and Brothers, 1904), 254–255. Another interesting insight on Irish-American views on emigration is present in the following letter: "From New York I went to Boston for the purpose of endeavoring to enlist the coöperation of the editor of the *Boston Pilot* in the promotion of Irish emigration but that gentleman for whom independent of his political opinions I have the greatest respect and admiration frankly informed me that he would give all the opposition in his power and that I might feel assured the rest of the Irish-American and Irish Press would do the same because it was felt that there was a rooted desire and design in the mind of the British Government and Imperial Parliament to banish the Irish people from Ireland in order to substitute English colonists in their place. I endeavored to persuade my friend O'Reilly that such a theory was arrant nonsense and reasoned with him as to the cruelty of putting difficulties in the way of assistance to enable poor people to emigrate from the barren and notoriously overcrowded districts of the West of Ireland to the unoccupied and practically unlimited fertile

regions of the United States and Canada but to no purpose. So I returned by the next train to New York and [have] appealed for co-operation of the Irish-American people through the friendly medium of the *New York Herald, New York Tribune, New York Times*." Letter by Vere Foster, Relating a Journey across Canada and America, Foster Papers 1529–1980, D.3618, Public Record of Northern Ireland, Belfast.

23. Davitt, *Fall of Feudalism*, 255.

24. *Gold Hill News*, November 1, 1865.

25. *Gold Hill News*, November 1, 1865.

26. Drury, *Editor on the Comstock*, 56.

27. Emrich, *In the Delta Saloon*, 152–153.

28. Clark, *Journals of Alfred Doten*, vol. 1, 1133.

29. Michael Hurley, Cedar Pass, to his parents, Tawnies, Clonakilty, August 26, 1871, Hurley Letters. Michael MacGowan had a violent encounter with an Orange man in the Alaska goldfields on St. Patrick's Day. MacGowan, *Hard Road to Klondike*, 120–126.

30. Clark, *Journals of Alfred Doten*, vol. 1, 1192.

31. Clark, *Journals of Alfred Doten*, vol. 1, 1193.

32. J. N. Flint, "The Comstock in the Early Days," *San Francisco Call*, July 28, 1889, in Lewis, *Mining Frontier*, 169.

33. Clark, *Journals of Alfred Doten*, vol. 3, 1621.

34. Clark, *Journals of Alfred Doten*, vol. 3, 1621. This could also be evidence of the impact of John Devoy's New Departure in 1879.

35. Clark, *Journals of Alfred Doten*, vol. 3, 1621.

36. Clark, *Journals of Alfred Doten*, vol. 3, 2062.

37. Clark, *Journals of Alfred Doten*, vol. 3, 2062. Interestingly, Thomas Brick notes that there was friction between the Irish and the English on the transatlantic route for the opposite reason: "The sentiment among the English against the Irish at that time was much in evidence because of the number of Irishmen who volunteered and fought against them in the Boor [*sic*] War with the Boors in the Transvaal and Orange Free State South Africa." Thomas W. Brick, *Memoirs of an Emigrant to USA, Spring, 1902*, 31, Miller Collection.

38. Donal P. McCracken, *The Irish Pro-Boers, 1877–1902* (Johannesburg: Perskor, 1989), 123–127; Michael Davitt, *The Boer Fight for Freedom* (London: Funk and Wagnalls, 1902), 325–327.

39. Anne M. Butler, "Mission in the Mountains: The Daughters of Charity in Virginia City," in James and Raymond, eds., *Comstock Women*, 144.

40. They were also commonly known as the Sisters of Charity.

41. Drury, *Editor on the Comstock*, 13.

42. Butler, "Mission in the Mountains," 146.

43. Meredith B. Linn, "From Typhus to Tuberculosis and Fractures in Between: A Visceral Historical Archaeology of Irish Immigrant Life in New York City 1845–1870," PhD diss., Columbia University, New York, NY, 2008, 388. Medical discrimination in New York prompted the establishment of Catholic medical services in the city, and it remains a strong possibility that this was another reason for doing so in other parts of the US.

44. *Territorial Enterprise*, February 15, 1877; Charles Collins, *Mercantile Guide and Directory for Virginia City, Gold Hill, Silver City and American City* (San Francisco: Agnew and Deffebach, 1865), 29.

45. Annals of St. Mary's in the Mountains, cited in Butler, "Mission in the Mountains," 148.

46. Hospital Logbook of St. Mary, Museum of St. Mary's in the Mountains, Virginia City, NV.

47. The hospital was named after Mary Louise Bryant Mackay, the wife of the silver magnate who donated the three-story building to the nuns, but was shortened to St. Mary's.

48. Hospital Logbook of St. Mary, Museum of St. Mary's in the Mountains.

49. Hospital Logbook of St. Mary, Museum of St. Mary's in the Mountains.

50. Annals of St. Mary's in the Mountains, cited in Butler, "Mission in the Mountains," 149.

51. Further analysis of those from Cork listed in the logbook and cross-referenced with headstones from the Catholic graveyard just outside Virginia City reveals that many of these Cork-born Irish were not from the mining region of the Beara Peninsula.

52. Hospital Logbook of St. Mary, Museum of St. Mary's in the Mountains.

53. Drury, *Editor on the Comstock*, 14.

54. J. N. Flint, "The Comstock in the Early Days," *San Francisco Call*, July 28, 1889, in Lewis, *The Mining Frontier*, 169.

55. Flint, "Comstock in the Early Days."

56. Butler, "Mission in the Mountains," 155. For an overview of the history of the Daughters of Charity, see Daniel Hannefin, *Daughters of the Church: A Popular History of the Daughters of Charity in the United States, 1809–1987* (Brooklyn: New City Press, 1990).

57. Storey County, Nevada, 1870 United States Census, IPUMS. There were eight "colored" people in the entire state of Nevada.

58. *Territorial Enterprise*, March 5, 1867; Butler, "Mission in the Mountains," 155.

59. *Territorial Enterprise*, February 21, 1873; Butler, "Mission in the Mountains," 155–156.

60. Denis Hurley, Carson City, to his parents, Clonakilty, March 16, 1874, Hurley Letters.

61. *Carson City News*, November 5, 1908; December 31, 1908.

62. Denis Hurley, Carson City, to John Hurley, Clonakilty, July 12, 1912, Hurley Letters.

63. Ancient Order of Hibernians, April 16, 1877, NHR MS File, Nevada Historical Society, Reno.

64. Ancient Order of Hibernians, April 16, 1877.

65. Senate Bill 81, Nevada Legislative Senate, *The Journal of the Senate of the Sixth Session of the Legislature of the State of Nevada, 1873* (Carson City, NV: Charles A. V. Putnam, State Printer, 1873), 228–229.

66. Assembly Bill 54, Nevada Legislative Assembly, *The Journal of the Assembly of the Seventh Session of the Legislature of the State of Nevada, 1875* (Carson City, NV: John J. Hill, State Printer, 1875). 98.

67. Anne M. Butler, *Daughters of Joy, Sisters of Misery: Prostitutes in the American West, 1865–90* (Urbana: University of Illinois Press, 1985), 5. Butler asserts that they were castigated for their poverty. While class remained an element of their exclusion, equally important was their transgression against the social mores of society. Few women on the early frontier were sex workers; more often, they were independent entrepreneurs. Sally Zanjani, *Goldfield: The Last Gold Rush on the Western Frontier* (Athens: Ohio University Press, 1992), 102–108.

68. Marion S. Goldman, *Gold Diggers and Silver Miners: Prostitution and Social Life on the Comstock Lode* (Ann Arbor: University of Michigan Press, 1981), 142.

69. Don Toms, *Tenderloin Tales: Prostitution, Gambling and Opium on the Gold Belt of the Northern Black Hills, 1867–1915* (Pierre, SD: State Publishing Company, 1997), 5.

70. Ronald M. James and Kenneth H. Feiss, "Women of the Mining West, Virginia City Revisited," in James and Raymond, eds., *Comstock Women*, 21–32; Goldman, *Gold Diggers*, 64–70, 92.

71. Goldman, *Gold Diggers*, 102–103.

72. The Irish had a similarly low proportional representation in sex work in New York. See Timothy J. Gilfoyle, *City of Eros* (New York: W. W. Norton, 1994), 346n16.

73. Peggy Pascoe, *Relations of Rescue: The Search for Female Moral Authority in the American West, 1874–1939* (New York: Oxford University Press, 1993).

74. George Williams, *Rosa May: The Search for a Mining Camp Legend* (Riverside: Tree by the River Publications), 1980.

75. Michael Rutter, *Upstairs Girls: Prostitution in the American West* (Helena, MT: Farcountry, 2005), 196.

76. Thomas Harold Kinnersley, *Nevada, 1859–1881* (Ann Arbor, MI: University Microfilms, 1978), 54.

77. Robert D. Armstrong, *Nevada Printing History: A Bibliography of Imprints and Publications, 1881–1890* (Reno: University of Nevada Press, 1991), 118; Goldman, *Gold Diggers*, 68.

78. Donald R. Abbe, *Austin and the Reese River Mining District: Nevada's Forgotten Frontier* (Reno: University of Nevada Press, 1985), 2–3.

79. Abbe, *Austin*, 4–5.

80. Abbe, *Austin*, 10–12.

81. Rodney H. Smith, *Austin, Nevada, 1862–1888* (Reno: University of Nevada Press, 1963), 70.

82. Abbe, *Austin*, 13. See also 89–90 for a list of the districts.

83. This compared to the Chinese at 10.1 percent, Indians at 6.2 percent, and the German population at 4.1 percent. After the Irish, these were the largest immigrant groups in 1880.

84. *Reese River Reveille*, August 5, 1865.

85. See appendix 1. Emmons, *Beyond the American Pale*, 356; *Reese River Reveille*, March 16, 1865; Abbe, *Austin*, 59.

86. Golden Family Records, MS NC29, Special Collections, University of Nevada, Reno.

87. Bernice Maher Mooney, *Salt of the Earth: The History of the Catholic Church in Utah, 1776–2007, 3rd ed.* (Salt Lake City: University of Utah Press, 2008), 320.

88. Myron Angel, *History of Nevada* (Oakland: Thompson and West, 1881), 205–207.

89. *Annals of All Hallows Missionary College*, 1903, All Hallows College Archives, Dublin, 61. Born September 8, 1839, in Knockahaw, Rathdowney, County Laois, he immigrated to California in 1857 and engaged in gold mining in Nevada County, California, for nine years. Collected Records of Museum of St. Mary of the Mountain, Virginia City, NV. Rev. Phelan became Chaplain of St. Vincent's Orphanage, San Rafael, California, from 1894 until his death on February 15, 1903. He is buried in Holy Cross Cemetery, Colma, CA.

90. Dwyer, *Condemned to the Mines*, 183. In 1880, 566 people in Lander County, Nevada, had Irish-born parents on both sides. The place of birth for 326 of that number—57.6 percent—was Ireland. The next most popular place of birth was Nevada, with 67 people—11.8 percent—most of whom were children of the first-generation Irish mining population. Most Irish occupations were in the laborer category, but they probably worked in the mines as muckers, since railroad workers had their own category and very little construction was taking place in Lander County by 1880. Since the Irish endogamy rates were so high, with 92.3 percent of

Irish-born males marrying Irish women and 87.5 percent of Irish women marrying Irish-born males, the trend of Irish women having lower rates of endogamy vis-à-vis Irish men is in line with the rest of the American West. Irish women in California and Nevada were slightly more likely to marry someone who was not Irish. The majority of these women were Catholic, which again shows the strong Irish preference for marrying other Catholics. For similar evidence on Montana, see Noonan, "From Ireland to Montana," 164–165.

91. *Annals of All Hallows Missionary College*, 1903, 62. Phelan claimed to have baptized many non-Catholics, Lutherans, Mormons, and "one Jew" in his time at Austin and detailed a story about one skeptic who converted after the priest said "remember that you do not know everything. If all that you do not know was written in a book you could not lift it." *Annals of All Hallows Missionary College*, 1903, 63.

92. *Report of the Director of the Mint upon the Production of the Precious Metals in the United States* (Washington, DC: US Government Printing Office, 1883), 159.

93. James Stewart, San Francisco, to Mr. Coulter, Belfast, September 22, 1884, T.3399/1, James Stewart Emigrant Documents, 1884–1885, Public Record Office of Northern Ireland, Belfast.

94. Denis Hurley, Carson City, to his parents, Clonakilty, May 25, 1873, Hurley Letters.

95. Denis Hurley, Carson City, Ormsby County, to his mother, Clonakilty, September 14, 1893, Hurley Letters.

96. Denis Hurley to his cousin Denis Ryan, Clonakilty, May 25, 1873, Hurley Letters.

97. Denis Hurley to Denis Ryan, May 25, 1873.

98. Denis Hurley to Denis Ryan, May 25, 1873.

99. Denis Hurley to his parents, May 25, 1873.

100. Denis Hurley, Carson City, to his parents, Clonakilty, July 21, 1875, Hurley Letters.

101. Denis Hurley, Carson City, to his parents, Clonakilty, January 6, 1876, Hurley Letters.

102. Dennis Hurley to his parents, January 6, 1876.

103. Denis Hurley, Carson City, to his parents, Clonakilty, May 26, 1877, Hurley Letters.

104. Denis Hurley to his parents, May 26, 1877. Emphasis added.

105. Denis Hurley, Carson City, to his parents, Clonakilty, August 21,1877, Hurley letters.

106. Denis Hurley to his parents, August 21, 1877.

107. Michael Hurley, Carson City, to his parents, Clonakilty, December 1, 1877, Hurley Letters.

108. Denis Hurley to his mother, February 24, 1899, Hurley Letters.

109. Denis Hurley to his mother, December 5, 1891, Hurley Letters.

110. Michael Hurley, Spokane Falls, Washington, to his mother, January 13, 1891, Hurley Letters.

111. Denis Hurley to Mary Hurley, Clonakilty, January 6, 1924, Hurley Letters.

112. Denis Hurley, Carson City, to Mary Hurley, Clonakilty, April 8, 1925, Hurley Letters. "My investments are bringing me a fair income but I suffered with the failure of the banks." Denis Hurley, Carson City, to his niece Mrs. Dan Deasy, January 21, 1937, Hurley Letters.

113. Denis Hurley to John Hurley, June 27, 1907, Hurley Letters.

114. Denis Hurley to John Hurley, December 11, 1913, Hurley Letters. In later letters to his niece, Denis writes of his hope to return to Ireland and visit his home during the Eucharistic Congress of Dublin in 1932, an unfulfilled dream because of an unexpected drop in the value of his investments.

115. Denis Hurley to John Hurley, August 1, 1901, Hurley Letters.

116. Denis Hurley to John Hurley, January 30, 1905, Hurley Letters. Emphasis added.

117. Denis Hurley to his mother, July 3, 1896, Hurley Letters.

118. W. L. Kennedy to James Gilmore, February 2, 1879, Kennedy Family Emigrant Papers, 1869–1912, MS T.3152, Public Record Office of Northern Ireland, Belfast [hereafter Kennedy Family Letters].

119. W. L. Kennedy to James Gilmore, February 2, 1879.

120. W. L. Kennedy to James Gilmore, February 2, 1879.

121. Patrick Dunny to his parents, December 30, 1856, Arnold Schrier Papers, UA-16-07, Archives and Rare Books Library, University of Cincinnati, Cincinnati, OH.

122. Thomas Higgins, Los Angeles, California, to Cecilia Keefe, Mosinee, Wisconsin, May 16, 1916, Higgins Family Letters, MS 22442 [hereafter Higgins Collection], Wisconsin Historical Society, Madison.

123. Smith, "History of the Comstock," 192.

Chapter 4: Mollies in the Mountains

1. Rodman W. Paul, *Mining Frontiers of the Far West, 1848–1880* (Austin: Holt, Rinehart and Winston, 1963), 113.

2. Paul, *Mining Frontiers*, 119.

3. Paul, *Mining Frontiers*, 125.

4. That was 11.2 percent of all miners with Welsh parentage in Colorado, totaling 54 miners. IPUMS. This turns out to have been an isolated group, as there was

not a single Welshman in Las Animas County, where 21 Irishmen and 24 Englishmen made up a sizable portion of the 109 miners and the remainder were Americans and Scots.

5. IPUMS.

6. *Denver Tribune*, January 28, 1884. See Thomas G. Andrews, *Killing for Coal: America's Deadliest Labor War* (Cambridge, MA: Harvard University Press, 2008), 123.

7. *Leadville Herald*, January 26, 1884.

8. *Denver Tribune*, January 27, 1884. In the *Leadville Herald* the same incident was reported differently: "A party of men were coming up from town to lynch him." *Leadville Herald*, January 27, 1884.

9. David A. Wolff, *Industrializing the Rockies: Growth, Competition, and Turmoil in the Coalfields of Colorado and Wyoming, 1868–1914* (Boulder: University Press of Colorado, 2003), 43.

10. Andrews, *Killing for Coal*, 155.

11. Andrews, *Killing for Coal*, 156.

12. 1890 United States Census, IPUMS.

13. Andrews, *Killing for Coal*, 104. Andrews (p. 89) further noted the ethnic divisions: "The Welsh and Scots despised the Irish, the French bore a grudge against the Germans, and the Germans claimed superiority over the Poles, who could not forgive the Austrians, who despised the African Americans, who distrusted Yankees, who saw the Hispanos as dirty lazy and primitive."

14. James R. Dow, Roger L. Welsch, and Susan D. Dow, eds., *Wyoming Folklore: Reminisces, Folktales, Beliefs, Customs, and Folk Speech* (Lincoln: University of Nebraska Press, 2010), 97.

15. Paul, *Mining Frontiers of the Far West*, 129.

16. Thomas F. Dawson, "The Old-Time Prospector," *Colorado Magazine* 1 (March 1924): 104–105.

17. Although a legitimate argument can be made for including the children of the English, Scottish, and Welsh who married native-born Americans in the native-born American category, that number would still fall below the figure of 16,470.

18. IPUMS. The majority of Irish-born (53.2 percent) and those of Irish parentage (53.3 percent) were miners, while another 13.8 and 12.3 percent, respectively, were laborers. The next two largest occupations were keeping house (9.1 percent of Irish-born and 9.0 percent of those with Irish parentage) and domestic service (3.5 percent and 3.3 percent, respectively)—most of them single women. A sizable number of Irish were also saloon keepers, with 52 of Irish parentage on both sides.

19. Carlyle Channing Davis, *Olden Days in Colorado* (Los Angeles: Phillips, 1916), 118.

20. Gretchen Scanlon, *A History of Leadville Theatre: Opera Houses, Variety Acts and Burlesque Shows* (Charleston: History Press, 2012).

21. IPUMS.

22. IPUMS. Butte did still manage to have 738 more Irish-born living there when this 1890 census was taken.

23. IPUMS.

24. [Life story of Hugh O'Daly, recounted by Quillen], 6, Hannon, Brosnan, Daly Collection, Butte-Silver Bow Archives, Butte, MT.

25. Michael McGee to his parents, Keeldrum, Co. Donegal, May 9, 1873, McGee Letters, Miller Collection [hereafter McGee Letters]. He possibly worked in Ohio on his journey west. Michael also mentioned relatives and friends in Scotland, highlighting the transnational nature of the diaspora for Irish emigrants.

26. Michael McGee to his parents, May 9, 1873.

27. Michael McGee to his parents, February 23, 1878, McGee Letters.

28. Michael McGee, Leadville, Colorado, to his parents, February 11, 1880, McGee Letters.

29. Michael McGee to his parents, February 11, 1880.

30. Michael McGee to his parents, February 11, 1880; Dennis McGee, Leadville, Colorado, to his parents, March 1, 1882, both in McGee Letters.

31. Michael McGee, Leadville, Colorado, to his parents, December 30, 1880, McGee Letters.

32. Michael McGee, Springfield, Illinois, to his parents, March 17, 1879; Michael McGee to his parents, n.d. possibly March 1, 1882; both in McGee Letters.

33. Michael McGee, Leadville, Colorado, to his parents, August 2, 1880, McGee Letters.

34. Michael McGee, Leadville, Colorado, to his parents, December 30, 1880, McGee Letters.

35. Dennis McGee, Leadville, Colorado, to his parents, March 1, 1882, McGee Letters.

36. R. G. Dill, "History of Lake County," in *History of the Arkansas Valley, Colorado* (Chicago: O. L. Baskin, 1881), 308.

37. Michael McGee, Leadville, Colorado, to his parents, February 14, 1883, McGee Letters.

38. Michael McGee to his parents, February 14, 1883.

39. Michael McGee to his parents, February 14, 1883.

40. Michael McGee to his parents, February 14, 1883.

41. Dennis McGee, Leadville, Colorado, to his parents, March 14, 1884, McGee letters.

42. Aunt Grace Gallagher, Marshall, Illinois, to Mr. and Mrs. McGee, n.d., McGee letters.

43. Dennis McGee, Leadville, Colorado, to his parents, February 29, 1884, McGee Letters.

44. Grace Gallagher, Marshall, Illinois, to her nephew, Keeldrum, Co. Donegal, March 24, 1891; Dennis McGee to his parents, February 29, 1884, March 14, 1884; both in McGee Letters.

45. Dennis McGee to his parents, February 29, 1884, McGee Letters.

46. Grace Gallagher, Marshall, Illinois, to her nephew, Keeldrum, Co. Donegal, March 24, 1891, McGee Letters.

47. Dennis and Dan McGee, Leadville, Colorado, to their mother, April 20, 1888, McGee Letters.

48. Dennis and Dan McGee to their mother, April 20, 1888.

49. Dennis and Dan McGee to their mother, April 20, 1888.

50. Patrick Mullins, Leadville, Colorado, to Patrick McGee, Keeldrum, Co. Donegal, n.d., McGee Letters.

51. G. F. Willison, *Here They Dug the Gold* (New York: Brentanos, 1931), 236–237.

52. James Patrick Walsh, "Michael Mooney and the Leadville Irish: Respectability and Resistance at 10,200 Feet, 1875–1900," PhD diss., University of Colorado, Boulder, 2010, 131.

53. Dill, *"History of Lake County,"* 250.

54. Walsh, "Michael Mooney and the Leadville Irish," 101–102. Walsh's dissertation is the most thorough study of Leadville's Irish community. His sustained study of an Irish mining community is matched only by Emmons's work on Butte.

55. Dill, *"History of Lake County,"* 239; Walsh, "Michael Mooney and the Leadville Irish," 172.

56. *Leadville Democrat*, May 27, 1880, in Walsh, "Michael Mooney and the Leadville Irish," 104.

57. Walsh, "Michael Mooney and the Leadville Irish," 104.

58. Don L. Griswold, *History of Leadville and Lake County, Colorado: From Mountain Solitude to Metropolis* (Denver: Colorado Historical Society, 1996), 413.

59. Walsh, "Michael Mooney and the Leadville Irish," 115.

60. Walsh, "Michael Mooney and the Leadville Irish," 115.

61. Walsh, "Michael Mooney and the Leadville Irish," 196.

62. *Leadville Evening Chronicle*, June 3, 1879. The reporter asked a question after this statement from the nun: "But you Sisters are all members of the Catholic Church?" The nun gave the witty retort, "Yes, that is true, but that does not prevent us from trying to do good." See *Leadville Evening Chronicle*, June 3, 1879.

63. Cited in Carol K. Coburn and Martha Smith, "'Pray for Your Wanderers': Women, Religious on the Colorado Mining Frontier, 1877–1917," *Frontiers: A Journal of Women Studies* 15, no. 3 (1995): 38.

64. Cited in Coburn and Smith, "Pray for Your Wanderers," 38.

65. Coburn and Smith, "Pray for Your Wanderers," 49.

66. See Carroll Smith-Rosenberg, *Disorderly Conduct: Visions of Gender in Victorian America* (New York: Alfred A. Knopf, 1985), 180.

67. Sister Julia, *Annunciation Parish: Leadville, Colorado, A History* (Leadville, CO: n.p., 1953), 15–17.

68. Walsh, "Michael Mooney and the Leadville Irish," 186.

69. Walsh, "Michael Mooney and the Leadville Irish," 197. One man who had been shot was cared for by the nuns. As Walsh notes dryly, "The nuns had a reputation for never turning anyone away, regardless of circumstance." Walsh, "Michael Mooney and the Leadville Irish," 197.

70. Sister Julia, *Annunciation Parish*, 12–13.

71. Walsh, "Michael Mooney and the Leadville Irish," 203.

72. *Leadville Chronicle*, June 15, 1880; Walsh, "Michael Mooney and the Leadville Irish," 79.

73. *Leadville Chronicle*, September 15, 1882.

74. An anti-Chinese song "Since the Chinese Ruint the Thrade" does appear that plays on a supposed conflict between the two groups in an unspecified place. The first and third verses are included below. Note the link between the danger of Mike's labor, politics, and possibly crime and the reference to the daughters' descent into sex work:

From me shanty down on Sixth Street,
It's meself have jist kim down;
I've lived there this eighteen year—
It's in phat they call Cork Town.
I'm on my way to the City Hall
To get a little aid;
It's meself that has to ax it now
Since the Chinese ruint the thrade.

I'm a widdy woman, I'd have ye know—
Poor Mike was kilt at work.
He got a fall from the City Hall,
For he was a mason's clerk,
An' me daughter Ellen is gone this year
Wid a Frinch bally troupe, ther jade,

So I find it hard to get along
Since the Chinese ruint the thrade.

G. W. Greene, *The "Poor Little Man" and the Man in the Moon Is Looking, Love, Songster* (San Francisco: G. W. Greene, 1871), 11, quoted in in Philip Foner, *American Labor Songs of the Nineteenth Century* (Urbana: University of Illinois Press, 1975), 109–110.

75. It seems that the Chinese population of the American West largely avoided Leadville after several early instances of anti-Chinese violence in the region in the 1870s. By 1880, the population of Gilpin County was heavily immigrant—a trait it maintained into the 1880s. At the time of the 1880 census, Gilpin County was the only county in Colorado with a sizable number of Chinese miners, 113 of the total 163 in the state, and was one of only three counties with a Chinese population over 100 (the others were Arapahoe and Park Counties). The conversely low percentage of native-born American miners (those for whom both parents were born in the US) in Gilpin County in 1880, 1.4 percent—which was well below the total statewide average of 5.4 percent—might at first appear to be solely considered a side effect of the increasing mechanization of mines. But another trend emerges with the data from other counties. Where the percentage of native-born miners was above average, as in Gunnison and Custer Counties, no Chinese miners are found; as such, these figures appear to reflect a virulent strain of nativism among native-born miners.

76. *Leadville Democrat*, June 17, 1880; Walsh, "Michael Mooney and the Leadville Irish," 99.

77. Dill, *"History of Lake County,"* 240; Walsh, "Michael Mooney and the Leadville Irish," 107.

78. *Leadville Weekly Herald*, May 29, 1880; Walsh, "Michael Mooney and the Leadville Irish," 159.

79. *Denver Tribune*, May 29, 1880; Walsh, "Michael Mooney and the Leadville Irish," 159. See also Griswold, *History of Leadville*, 581.

80. Dill, *"History of Lake County,"* 321.

81. *Rocky Mountain News*, June 18, 1880; Walsh, "Michael Mooney and the Leadville Irish," 127.

82. *Denver Tribune*, June 18, 1880; Walsh, "Michael Mooney and the Leadville Irish," 165. This was a different Workingmen's Party than the Workingmen's Party of California (WPC), the viciously anti-Chinese organization led by Denis Kearney.

83. *Denver Tribune*, June 18, 1880; Walsh, "Michael Mooney and the Leadville Irish," 161. Daly was a wrecker and was responsible for scuttling an earlier

agreement in May between the miners and businessmen: "Business men talked compromise and seemed on the point of triumph when George Daly, superintendent of the Little Chief and mouthpiece of the mine owners, made a most unconciliatory speech. Prophesying violence and bloodshed, he walked out of the meeting." Willison, *Here They Dug the Gold*, 237.

84. Rowland T. Berthoff, "The 'Freedom to Control' in American Business History," in David H. Pinkney and Theodore Ropp, eds., *A Festschrift for Frederick B. Artz* (Durham, NC: Duke University Press, 1964), 158–160; Perry K. Blatz, *Democratic Miners: Work and Labor Relations in the Anthracite Coal Industry, 1875–1925* (New York: State University of New York Press, 1994), 35, 37.

85. *Leadville Weekly Herald*, June 5, 1880.

86. *Leadville Weekly Herald*, June 5, 1880.

87. See also Lynn Perrigo, "A Social History of Central City, Colorado, 1859–1900," PhD diss., University of Colorado, Boulder, 1936.

88. *Leadville Weekly Democrat*, Septeber 3, 1881.

89. *Bodie Standard*, September 7, 1881.

90. *Homer Mining Index*, July 17, 1880.

91. *Leadville Weekly Democrat*, September 3, 1881.

92. US Bureau of Labor, *Report on Labor Disturbances in the State of Colorado, from 1880 to 1904* (Washington, DC: US Government Printing Office, 1905), 71. Emphasis added.

93. Davis, *Olden Days in Colorado*, 249.

94. Davis, *Olden Days in Colorado*, 256.

95. Davis, *Olden Days in Colorado*, 256.

96. Richard M. Brown, *Strain of Violence: Historical Studies of American Violence and Vigilantism* (New York: Oxford University Press, 1975), 134.

97. Brown, *Strain of Violence*, 255–256.

98. Davis, *Olden Days in Colorado*, 251–252. Emphasis added.

99. Dill, *"History of Lake County,"* 244.

100. David J. Cook, *Hands Up, or, Thirty-Five Years of Detective Life in the Mountains and on the Plains* (Denver: W. F. Robinson, 1897), 347.

101. Walsh, "Michael Mooney and the Leadville Irish," 143.

102. Cook, *Hands Up*, 348; Ruby G. Williamson, *Gold, God, the Devil, and Silver: Leadville, Colorado, 1878–1978* (Gunnison, CO: B&B Printers, 1977), 48.

103. *Leadville Democrat*, June 17, 1880, quoted in Walsh, "Michael Mooney and the Leadville Irish," 151.

104. Walsh, "Michael Mooney and the Leadville Irish," 151.

105. Walsh, "Michael Mooney and the Leadville Irish," 151.

106. Walsh, "Michael Mooney and the Leadville Irish," 151–152.

107. Thiel Detective Reports, August 8, 1898, John F. Campion Papers, MS 1099, Archives and Special Collections, University of Colorado, Boulder [hereafter Campion Papers]. Emphasis added.

108. District 199, Danville, Montour County, Pennsylvania, 1880 United States Census, 180A; Ward 2, District 47, Leadville City, Lake County, Colorado, 1900 United States Census, 1A; both from familysearch.org. Knuckey is listed as "Knuckley" in the Thiel reports.

109. Thiel Detective Reports, June 14, 1898, Campion Papers.

110. Thiel Detective Reports, August 6, 1898, Campion Papers. A John, Michael, and Patrick Ahern, all miners, are living at the same address on E. 532 4th Street. *Ballenger and Richards' Annual Leadville City Directory, 1898* (Leadville, CO: Ballenger and Richards, 1898), 62. The Ahern in the reports is probably one of these men.

111. Walsh, "Michael Mooney and the Leadville Irish," 129.

112. Walsh, "Michael Mooney and the Leadville Irish," 129.

113. Walsh, "Michael Mooney and the Leadville Irish," 129.

114. Operative F.M.G., July 28, 1896, Leadville Strike Reports 1896–1898, MSS 334, History Colorado, Denver; Walsh, "Michael Mooney and the Leadville Irish," 277.

115. Operative A.C.H., August 1, 1896, LSR; Walsh, "Michael Mooney and the Leadville Irish," 276.

116. Thiel Detective Reports, May 17, 1899, Campion Papers.

117. Operative J. N., August 4, 1896, LSR; Walsh, "Michael Mooney and the Leadville Irish," 276. It is unlikely that this is the same Phil Conley who appears in the conflict between the union and YA in Randsburg, California, discussed in chapter 2, this volume.

118. Thiel Detective Reports, August 10, 1898, Campion Papers.

119. Thiel Detective Reports, August 10, 1898.

120. Thiel Detective Reports, August 10, 1898.

121. Thiel Detective Reports, August 3, 1898, Campion Papers.

122. The public notices in newspapers detailed membership turnout for the funerals of Irish miners and fundraising events such as dances. *Intermountain Catholic*, February 17, 1900; March 14, 1903.

123. Thiel Detective Reports, August 3 and 8, 1898, Campion Papers.

124. Thiel Detective Reports, August 3, 1898.

125. Richard E. Lingenfelter, *The Hardrock Miners: A History of the Mining Labor Movement in the American West, 1863–1893* (Berkeley: University of California Press, 1974), 227.

126. Thiel Detective Reports, August 8, 1898, Campion Papers. It is also possible that the Thiel detective manager, W. E. Giese, who forwarded the reports to the mine owner, could have been responsible for spicing up dull reports.

127. Thiel Detective Reports, August 6 and 7, 1898, Campion Papers. McGowen's name is spelled James MacGowan in the city directory, and he lived at 811 E. 6th St. *Leadville City Directory, 1898*, 188.

128. Thiel Detective Reports, June 8, 1898, Campion Papers. Cross-referenced with local census reports. 1900 United States Census, familysearch.org.

129. Thiel Detective Reports, August 8, 1898, Campion Papers.

130. Thiel Detective Reports, August 3 and 10, 1898, Campion Papers.

131. Thiel Detective Reports, August 3 and 10, 1898, Campion Papers.

132. Thiel Detective Reports, August 10, 1898, Campion Papers.

133. Thiel Detective Reports, August 14, 1898, Campion Papers.

134. Thiel Detective Reports, August 13, 1898, Campion Papers.

135. Thiel Detective Reports, August 13, 1898, Campion Papers; *Leadville City Directory, 1898*, 114.

136. Thiel Detective Reports, August 13, 1898, Campion Papers.

137. Thiel Detective Reports, August 13, 1898, Campion Papers.

138. Thiel Detective Reports, August 13, 1898, Campion Papers.

139. Thiel Detective Reports, August 14, 1898, Campion Papers.

140. Thiel Detective Reports, August 14, 1898, Campion Papers.

141. Thiel Detective Reports, August 14, 1898, Campion Papers; *Leadville City Directory, 1898*, 213, 310. In the directory the saloon is called Newman and Rogers, on Harrison Avenue.

142. Thiel Detective Reports, August 15, 1898, Campion Papers.

143. Thiel Detective Reports, August 15, 1898, Campion Papers.

144. Thiel Detective Reports, August 15, 1898, Campion Papers. Again, this might have been an attempt by the Thiel outfit to scare Campion.

145. Thiel Detective Reports, August 15, 1898, Campion Papers; *Leadville City Directory, 1898*, 213. Ernest Nicholas is "Earnest Nichols" in the Campion Papers.

146. Thiel Detective Reports, August 15, 1898, Campion Papers; *Leadville City Directory, 1898*, 131. Two of the three John Gallaghers of Leadville specified their occupation as "miner" in the directory.

147. Thiel Detective Reports, August 15, 1898, Campion Papers.

148. Thiel Detective Reports, August 15, 1898, Campion Papers.

149. Thiel Detective Reports, June 5, 1899, Campion Papers. This may have been a ruse by the men, using counterintelligence against the company.

150. Thiel Detective Reports, June 5, 1898, Campion Papers. See chapter 5, this volume.

151. Thiel Detective Reports, June 8, 1899, Campion Papers.

152. Thiel Detective Reports, June 8, 1899, Campion Papers.

153. Thiel Detective Reports, June 12, 1899, Campion Papers.
154. Thiel Detective Reports, June 12, 1899, Campion Papers.
155. Thiel Detective Reports, July 28, 1899, Campion Papers.
156. Thiel Detective Reports, July 28, 1899.
157. Thiel Detective Reports, July 28, 1899.
158. Thiel Detective Reports, July 28, 1899.
159. Card no. 6112, Box 69, Portland Gold Mining Co. Records, University of Colorado, Boulder [hereafter PGM Co.].
160. Card no. 7267, PGM Co.
161. Card no. 7175, PGM Co.
162. Card no. 4561, PGM Co.
163. Card no. 6566, PGM Co.
164. Card no. 6566, PGM Co.
165. Card no. 8011, PGM Co.
166. Card no. 402, PGM Co.
167. Card no. 8338, PGM Co.
168. Card no. 2855, PGM Co.
169. Card no. 1356, PGM Co.
170. Card no. 8162, PGM Co. Note that it did not say he was a good worker.
171. Card no. 13180, 7096, PGM Co. Both men were Irish-born.
172. Card no. 5470, PGM Co.
173. Walsh, "Michael Mooney and the Leadville Irish," 335–336.
174. Ward no. 3, Butte City, Silver Bow Township, 1900 United States Census, 329, familysearch.org.

Chapter 5: In Search of Respect

1. Thomas Higgins, Los Angeles, California, to Cecilia Keefe, Mosinee, Wisconsin, May 16, 1916, Higgins Collection.

2. William W. Staley, "Gold in Idaho," Pamphlet no. 68, *Idaho Bureau of Mines and Geology* (Moscow: University of Idaho Press, 1946), 26–28. Hard-rock miners were also known as quartz miners.

3. Table 30, US Census Office, *Census of Population, 1870, vol. I: The Statistics of the Population of the United States* (Washington, DC: US Government Printing Office, 1872), 730.

4. Thomas Higgins to Cecilia Keefe, May 16, 1916, Higgins Collection.

5. Hyman Palais, "Black Hills Miners' Folklore," *California Folklore Quarterly* 4, no. 3 (1945): 266.

6. Mining proved a tempting prospect for many, particularly naive investors, both near and far. Indeed, Democratic Boise County treasurer Alfred Slocum refused to turn over the taxes to the territorial treasury for the year 1865, ostensibly because the treasury was in Republican hands. By the end of the year it had become clear that he had used the taxes in mine speculation to buy 410 shares in the Elkhorn mines, which were subsequently sold to try to recoup the taxes.

7. James Mullany, McMinnville, Yamhill County, to his sister Mary, September 11, 1858, Mullany Letters.

8. James Mullany to his sister Mary, September 11, 1858.

9. James Mullany to his sister Mary, September 11, 1858.

10. James Mullany to his sister Mary, September 11, 1858.

11. James Mullany to his sister Mary, September 11, 1858.

12. James Mullany, McMinnville, to his sister Mary, August 5, 1860, Mullany Letters. Mullany used "their" instead of "there," which is replaced in the quoted material to aid legibility. The absence of a Catholic Church was often given as a reason Irish women would not venture into the frontier, even with their husbands. For example, James Grayston's wife refused to stay in Kansas without a Catholic Church nearby. Charlotte Erickson, *Invisible Immigrants: The Adaptation of English and Scottish Immigrants in Nineteenth-Century America* (Leicester: Leicester University Press, 1972), 73. The Irish clannishness was matched in this regard by another Catholic immigrant group, the Polish. See William Thomas and Florian Znaniecki, *The Polish Peasant in Europe and America* (Urbana: University of Illinois Press, 1996), 286–287.

13. Edward G. Bourne, "The Legend of Marcus Whitman," *American Historical Review* 6 (January 1901): 283–287. See in particular note 13. Catholic prelates reporting from the region were not concerned solely with the American perception of the events but also with the Indians. One wrote back to Ireland in 1854, "The murder of Dr. Whitman may have injured our cause . . . but it killed Protestantism among the Indians." *Seventh Report of All Hallows Missionary College*, 63.

14. James Mullany to his sister Mary, August 5, 1860, Mullany Letters.

15. James Mullany to his sister Mary, August 5, 1860. "There" is substituted for "their" in the letter for legibility.

16. James Mullany to his sister Mary, August 5, 1860.

17. James Mullany to his sister Mary, August 5, 1860.

18. James Mullany, Walla Walla, to his sister Mary, November 5, 1860, Mullany Letters.

19. James Mullany to his sister Mary, November 5, 1860.

20. James Mullany, Oro Fina, to his sister Mary, August 18, 1861, Mullany Letters.

21. James Mullany to his sister Mary, August 18, 1861.

22. James Mullany to his sister Mary, August 18, 1861.

23. James Mullany, McMinnville, to his sister Mary, September 11, 1858, Mullany Letters.

24. James Mullany to his sister Mary, September 11, 1858; O'Hanlon, *Irish Emigrant's Guide*, 257.

25. James Mullany to his sister Mary, September 11, 1858.

26. James Mullany to his sister Mary, September 11, 1858.

27. James Mullany to his sister Mary, September 11, 1858. Such sentiments were repeated regularly in many prospectors' letters—Irish as well as Americans: "I cannot think of returning home from the Land of Gold with Nothing, notwithstanding the chords of Fraternal, Conjugal and Paternal Love are drawing me towards home." Levi C. Hillman Letters, Historical Society of Minnesota, St. Paul, in Malcolm J. Rohrbough and the American Council of Learned Societies, *Days of Gold: The California Gold Rush and the American Nation* (Berkeley: University of California Press, 1997), 263. Many failed and wrote back in letters despairing at their failure, but it was possible that some were simply trying to extend their excursion in the West.

28. IPUMS.

29. Judy Yung, Gordon Chang, and Him Mark Lai, eds., *Chinese American Voices: From the Gold Rush to the Present* (Berkeley: University of California Press, 2006); Judy Yung, *Unbound Feet: A Social History of Chinese Women in San Francisco* (Berkeley: University of California Press, 1995).

30. Elliott West, "Five Idaho Mining Towns: A Computer Profile," *Pacific Northwest Quarterly* 73 (1982): 112.

31. West, "Five Idaho Mining Towns," 112. These figures only include the Irish-born because the 1870 census does not offer parentage nativity statistics, unlike the 1880 census. The figures used are conservative estimates since the number of Irish-Americans in Idaho at this time was higher than the statistics might indicate, with first- and second-generation Irish-Americans making up a third of the population of southern Idaho Territory as it went through boom to bust phases of mine rushes.

32. *Idaho Daily Statesman*, September 18, 1932.

33. Idaho State Historical Reference Series, *Rocky Bar Mines*, no. 199 (Boise, ID: State Historical Society, 1996), 5.

34. *Illustrated History of North Idaho: Embracing Nez Perces, Idaho, Latah, Kootenai and Shoshone Counties, State of Idaho* (Racine, WI: Western Historical Publishing Company, 1903), 909.

35. *Owyhee Avalanche*, February 1, 1868.

36. Hubert H. Bancroft, Bancroft Scraps 1849, volume 111: Idaho Miscellany, Bancroft Library, University of California, Berkeley;; Liping Zhu, *A Chinaman's Chance:*

The Chinese on the Rocky Mountain Mining Frontier (Boulder: University Press of Colorado, 2000), 133.

37. Only 5.2 percent of the tax on Chinese miners was collected, while the tax on all foreign nationals targeting the Chinese similarly failed. Zhu, *Chinaman's Chance*, 122, 136.

38. For a detailed history of Chinese women in the West, see Huping Ling, *Surviving on the Gold Mountain: A History of Chinese American Women and Their Lives* (Albany: State University of New York Press, 1998).

39. Zhu, *Chinaman's Chance*, 161.

40. When I first came across the phrase *Burke Irish* in oral history records, it seemed as though the Irish fled to Burke because the owners of the Tiger Mine, the major employer there, were sympathetic to the Irish and the unions. However, the letter, written in the 1880s, indicates that Burke was an Irish town long before the Bunker Hill and Sullivan troubles. A. A. Albinola, oral history conducted by W. A. Albinola, December 14, 1973, OH 1511; Mary Achord, oral history conducted by Connie Ringel, December 11, 1978, OH 1516; both in NICO; Cataldo Mission Letters, Coeur d'Alene's Old Mission State Park, Coeur d'Alene, Idaho [hereafter Cataldo Mission Letters].

41. William D. Haywood, *Bill Haywood's Book: The Autobiography of William D. Haywood* (New York: International Publishers, 1929), 56. She is listed as Maggie in the 1900 census, a widow with Irish boarders that included several Irish miners and waitresses, an Irish-American miner born in Pennsylvania, and an Irish-American waitress born in Iowa. She had a similar mix of Irish in 1910. Electoral District 98, Gem, Burke Precinct, Shoshone County, 1900 United States Census, 20, familysearch.org; Electoral District 249, Burke Precinct, Shoshone County, 1910 United States Census, 7, familysearch.org.

42. Cataldo Mission Letters.

43. Cataldo Mission Letters.

44. Cataldo Mission Letters.

45. Cataldo Mission Letters; Idaho State Historical Reference Series, *Placer Mining in Southern Idaho*, no. 166 (Boise, ID: State Historical Society, 1980), 3.

46. Efforts to enforce "respectable" practices in Irish religiosity have had varying success.

47. Katherine G. Aiken, *Idaho's Bunker Hill: The Rise and Fall of a Great Mining Company, 1885–1981* (Norman: University of Oklahoma Press, 2005), 6.

48. Alfred D. Chandler, *The Visible Hand: The Managerial Revolution in American Business* (Cambridge, MA: Belknap, 1977), 3. See also Katherine Aiken, "'Not Long Ago a Smoking Chimney Was a Sign of Prosperity': Corporate and Community

Response to Pollution at the Bunker Hill Smelter in Kellogg, Idaho," *Environmental History Review* 18 (Summer 1994): 67–86.

49. Jerome Karabel, *The Chosen: The Hidden History of Admission and Exclusion at Harvard, Yale, and Princeton* (Boston: Houghton Mifflin, 2005), 24–25.

50. Chandler, *Visible Hand*, 7, 90.

51. Robert Wayne Smith, *The Coeur d'Alene Mining War of 1892: A Case Study of an Industrial Dispute, 2nd ed.* (Gloucester, MA: P. Smith, 1961), 10–20.

52. Smith, *Coeur d'Alene Mining War*. While the MOA probably formed in response to the unionization of the miners, it first flexed its strength not against the workers but rather against the railroad companies, which it claimed charged excessive rates.

53. John Fahey, *The Ballyhoo Bonanza; Charles Sweeny and the Idaho Mines* (Seattle: University of Washington Press, 1971), 15–24.

54. Fahey, *Ballyhoo Bonanza*, 46.

55. Fahey, *Ballyhoo Bonanza*, 46.

56. Fahey, *Ballyhoo Bonanza*, 60.

57. Smith, *The Coeur d'Alene*, 20–22; Aiken, *Idaho's Bunker Hill*, 30–45.

58. Fahey, *Ballyhoo Bonanza*, 75.

59. F. W. Bradley to W. B. Harris, December 6, 1894, Bunker Hill Company Records 1887–1984, Box 7, MG 367, Special Collections and Archives, University of Idaho, Moscow [hereafter BHM].

60. F. W. Bradley to W. B. Harris, December 6, 1894.

61. *Coeur d'Alene Press*, April 10, 1892. See also Mark Wyman, *Hard-Rock Epic: Western Miners and the Industrial Revolution, 1860–1910* (Berkeley: University of California Press, 1979), 114–140.

62. Fahey, *Ballyhoo Bonanza*, 76–77.

63. V. Clement to H. Hammond, 11, 12, 15 July 11, 12, 15, 1892, BHM.

64. F. W. Bradley to N. H. Harris, November 9, 1894, BHM.

65. F. W. Bradley to N. H. Harris, November 9, 1894.

66. Ibid.

67. F.W. Bradley, to N. H. Harris, August 10, 1894, BHM. ". . . 18 Swedes; 15 Englishmen; 15 Welshmen; 13 Scotchmen; 11 Austrians; 8 Finlanders; 6 Frenchmen; 5 Danes; 5 Norwegians; 2 Swiss; 1 Icelander; 1 Portuguese." While the list is notable for its ethnic diversity, the Chinese are conspicuous by their absence. Only 100 Chinese remained in the state, a shadow of the 3,853 Chinese who lived in Idaho in 1870. Most Chinese miners had returned to China, and the Chinese Exclusion Act of 1882 had halted further immigration. See Zhu, *Chinaman's Chance*, 154–155; Fern Coble Trull, "The History of the Chinese in Idaho from 1864 to 1910," master's thesis, University of Oregon, Eugene, 1949.

68. F. W. Bradley to F. R. Moore, Spokane Washington, November 25, 1894, BHM.
69. F. W. Bradley to N. H. Harris, November 23, 1894, BHM.
70. F. W. Bradley to N. H. Harris, November 23, 1894.
71. F. W. Bradley to N. H. Harris, November 23, 1894.
72. F. W. Bradley to N. H. Harris, November 28, 1894, BHM.
73. F. W. Bradley to N. H. Harris, November 28, 1894.
74. F. W. Bradley to N. H. Harris, November 28, 1894.
75. F. W. Bradley to F. R. Moore, Spokane, Washington, November 25, 1894, BHM.
76. W. S. Haskins to F. W. Bradley, November 28, 1894, BHM.
77. Emmons, *Beyond the American Pale*, 316.
78. C. G. Griffith to F. W. Bradley, December 3, 1894, BHM. Emphasis added.
79. F. W. Bradley to N. H. Harris, December 2, 1894, BHM.
80. F. W. Bradley to W. J. McConnell, December 17, 1894, BHM.
81. F. W. Bradley to W. J. McConnell, December 17, 1894.
82. F. W. Bradley to W. J. McConnell, December 17, 1894.
83. F. W. Bradley to W. J. McConnell, December 17, 1894.
84. F. W. Bradley to W. J. McConnell, December 17, 1894.
85. Richard Thomas, Wardner, Idaho, to his family, St. Just, Cornwall, November 22, 1904, in Arthur Cecil Todd, "Cousin Jack in Idaho," *Idaho Yesterdays* (Winter 1964): 9. The letters were printed from a private collection in the journal.
86. Richard Thomas, Mace, Idaho, to family, February 13, 1905, in Todd, "Cousin Jack in Idaho," 9.
87. Richard Thomas to family, February 13, 1905.
88. Richard Thomas, Burke, Idaho, to family, March 13, 1905, in Todd, "Cousin Jack in Idaho," 10.
89. Todd, "Cousin Jack in Idaho," 6.
90. Todd, "Cousin Jack in Idaho," 11.
91. F. W. Bradley to N. H. Harris, January 9, 1895, BHM.
92. F. W. Bradley to N. H. Harris, January 9, 1895.
93. F. W. Bradley to N. H. Harris, January 22, 1895, BHM.
94. F. W. Bradley to N. H. Harris, January 22, 1895. Emphasis added.
95. F. Burbidge to F. W. Bradley, April 4, 1895, BHM.
96. F. W. Bradley to N. H. Harris, January 17, 1895, BHM.
97. Fahey, *Ballyhoo Bonanza*, 71
98. F. W. Bradley to N. H. Harris, February 8, 1895, BHM.
99. F. W. Bradley to N. H. Harris, February 8, 1895.
100. F. W. Bradley to N. H. Harris, February 8, 1895.
101. F. W. Bradley to N. H. Harris, March 8, 1895, BHM.
102. F. Burbidge to F. W. Bradley, April 8 and 22, 1895, BHM.

103. F. Burbidge to F. W. Bradley, April 4, 1895, BHM.

104. F. Burbidge to F. W. Bradley, April 4, 1895.

105. F. Burbidge to F. W. Bradley, April 4, 1895.

106. F. Burbidge to F. W. Bradley, April 4, 1895.

107. F. W. Bradley to N. H. Harris, June 25, 1895, Box 8, BHM.

108. Frederick Burbidge to F. W. Bradley, May 5, 1895, Box 8, BHM.

109. Frederick Burbidge to F. W. Bradley, May 5, 1895; *San Francisco Call*, May 8, 1895.

110. Frederick Burbidge to F. W. Bradley, May 5, 1895.

111. Merle Travis, *Sixteen Tons*, 1947, vinyl record.

112. F. W. Bradley to N. H. Harris, February 11, 1895; Frederick Burbidge to F. W. Bradley, April 3, 1895; both in Box 7, BHM.

113. N. H. Harris to F. W. Bradley, May 15, 1896, Box 8, BHM.

114. F. W. Bradley to N. H. Harris, May 15, 1896, Box 8, BHM.

115. Frederick Burbidge to F. W. Bradley, May 16, 1896, Box 8, BHM.

116. Frederick Burbidge to F. W. Bradley, May 18, 1895, Box 8, BHM.

117. Frederick Burbidge to F. W. Bradley, May 18, 1895.

118. McConnell wrote admiringly of the vigilantes in his history of Idaho. See William John McConnell, *Early History of Idaho* (Boise, ID: Caxton, 1912), 206, 230–253.

119. See Noonan, "From Ireland to Montana," 128–131.

120. J. Anthony Lukas, *Big Trouble: A Murder in a Small Western Town Sets off a Struggle for the Soul of America* (New York: Simon and Schuster, 1997), 142–144.

121. US Industrial Commission, *Report of the Industrial Commission on the Relations of Capital and Labor in the Mining Industry*, vol. 12 (Washington, DC: US Government Printing Office, 1901), xix.

122. Mary Purcell Achord interview, OH 1511, NICO.

123. A. A. Albinola interview, OH 1516, NICO.

124. Aiken, *Idaho's Bunker Hill*, 70–75.

125. Quoted in John Fahey, *The Days of the Hercules* (Moscow: University Press of Idaho, 1978), 182.

126. Fahey, *Days of the Hercules*, 178.

127. "Wallace, Idaho—Feb. 6," *Engineering and Mining Journal* 103, no. 7 (February 17, 1917): 323.

128. The prohibition of alcohol in 1916 was another move that would have encouraged miners to join unions as a fraternal alternative to saloons. A. W. Muir, the vice president of the American Federation of Labor, had found only sixty-two active members in the Coeur d'Alene mining district, indicating that men fell in and out of union membership at times. Within a few months they organized more fully in the region and opened branches in Burke, Mullan, Wallace, and Gem.

129. *San Francisco Call*, June 12, 1904.

130. Lukas, *Big Trouble*, 9–10.

131. Edward Boyce, "Miners' Union Day at Butte, Montana," *Miners' Magazine* (April 1900): 39–42.

132. Fahey, *Ballyhoo Bonanza*, 77–78. The two states specifically mentioned from which scabs were hired were California and Colorado. Meanwhile, the wife of Victor Clement, a mine owner, claimed hyperbolically that talking to the miners was "like putting his head in the lion's mouth," but she was probably correct that, because of the action, "was looked upon more or less as a hero." Fahey, *Ballyhoo Bonanza*, 77–78.

Chapter 6: Oro y cobre, Gold and Copper

1. AFS 8893, Wayland Hand Collection of Irish Songs Recorded in Butte, Montana, 1945, AFC 1948/045, Archive of Folk Culture, American Folklife Center. Library of Congress, Washington, DC.

2. Hall, *Progressive Men of Montana*, 708. As the final gold rushes played out by 1875, the territory witnessed an increase in farm acreage, from 15,827 acres in 1870 to 41,969 in 1880. S. J. Coons, "Influence of Gold Camps on the Economic Development of Western Montana," *Journal of Political Economy* 38, no. 5 (October 1930): 593. For information on more Irish who transitioned between occupations, see Alan J. M. Noonan, "Progressive Men of the State of Montana: A Kaleidoscope of Irish Montanans," *Journal of the Trinity Postgraduate Seminar Series* 1 (2008): 12–19.

3. *Montana Post*, November 5, 1864.

4. Robert A. Burchell, "Irish Property Holding in the West in 1870," *Journal of the West* 31 (Summer 1992): 10–11. The next-richest Irishman was Edward Gallagher, a thirty-three-year-old stock dealer worth $10,000.

5. Clyde Francis Murphy Papers, MSS 285, Section C, 2, Folder 1, Series 1, Archives and Special Collections, Maureen and Mike Mansfield Library, University of Montana, Missoula [hereafter Murphy Papers]. It is often written as Drum Lummon as well.

6. John Darlington, *Report: The Drum Lummon Gold and Silver Mine, Montana* (London: Waterlow and Sons Limited, 1882), 6, 15, in Box 4, Thomas Cruse Papers, 1841–1956, MC 36, Montana Historical Society Archives, Helena [hereafter Cruse Papers].

7. Darlington, *Report*. The stamp mill employed seven men, with a smith and a "wagoner" rounding out Drumlummon's workforce

8. Section D, 2, Folder 1, Series 1, Murphy Papers; "The Drum Lummon," *River Press* (Fort Benton, MT), March 7, 1883.

9. *Mountaineer* (Maryville, MT), March 14, 1895, Murphy Papers.

10. *Mountaineer*, March 14, 1895.

11. Hall, *Progressive Men of Montana*, 40.

12. Section C, 4, Folder 1, Series 1, Murphy Papers; Table 8, US Census Office, *Census of Population, 1890: vol. 1, Report on Population of the United States, Part 1* (Washington, DC: US Government Printing Office, 1895), 385. Unfortunately, the 1890 census collated tables do not list foreign-born in small towns, and with the destruction of the census returns in a fire, there is no way of knowing the town's ethnic makeup. However, the Catholic Church does hint at an Irish presence in the town.

13. Section C, 4, Folder 1, Series 1, Murphy Papers.

14. Pinkerton Detective Agency [PDA] to George H. Robinson, the Montana Mining Company [MMC], May 10, 1892, Box 5, MSS 142, Montana Mining Company Records, Archives and Special Collections, Maureen and Mike Mansfield Library, University of Montana, Missoula [hereafter MMC Papers].

15. PDA to MMC, May 12, 1892, MMC Papers.

16. PDA to MMC, May 14, 1892, MMC Papers.

17. PDA to MMC, May 15, 1892, MMC Papers.

18. PDA to MMC, May 15, 1892.

19. PDA to MMC, May 15, 1892.

20. PDA to MMC, May 16, 1892, MMC Papers.

21. PDA to MMC, May 16, 1892.

22. PDA to MMC, May 16, 1892.

23. PDA to MMC, May 17, 1892, MMC Papers.

24. PDA to MMC, May 17, 1892.

25. PDA to MMC, May 17, 1892.

26. PDA to MMC, May 17, 1892.

27. PDA to MMC, May 17, 1892.

28. John Higham, *Strangers in the Land: Patterns of American Nativism, 1860–1925* (New Brunswick, NJ: Rutgers University Press, 2002), 57, 58.

29. Hall *Progressive Men of Montana*, 1426.

30. David Harry Bennett, *The Party of Fear: From Nativist Movements to the New Right in American History* (Chapel Hill: University of North Carolina Press, 1988), 171.

31. Emmons, *Butte Irish*, 101.

32. H. L. Simmons, Cornwall to R. J. Bayliss, Marysville, July 2, 1888; James Harris, London, on behalf of his son in Cornwall, Harold J. Harris to R. J. Bayliss, Marysville, July 10, 1888; both in Box 5, MMC Papers.

33. Joseph Kinsey Howard, *Montana: High, Wide, and Handsome* (Lincoln: University of Nebraska Press, 1983), 40.

34. Merrill G. Burlingame, "The Mining Frontier in Montana," in Michael P. Malone and Richard B. Roeder, eds., *Montana's Past: Selected Essays* (Missoula: University of Montana, 1969), 93.

35. Quoted in Clark C. Spence, ed., *The American West* (New York: Crowell, 1966), 407.

36. Patrick Kearney to John Kearney, December 21, 1890, in Séamus De Búrca, ed., *Soldier's Song: The Story of Peadar Kearney* (Dublin: P. J. Bourke, 1957), 251.

37. Frank T. Gilbert, *Resources, Business, and Business Men of Montana* (Walla Walla, WA: Historic Publishing Company, 1888), 60; Section R, Folder 1, Series 1, Murphy Papers. Early positions in the bank were President Thomas Cruse; Vice President Frank H. Cruse; and his nephew, Treasurer W. J. Sweeney. Joaquin Miller, *An Illustrated History of the State of Montana* (Chicago: Lewis, 1894), 417; Section 1, Folder 1, Series 1, Murphy Papers.

38. Hall, *Progressive Men of Montana*, 41.

39. A man named Wilbur Fisk Sanders suspiciously arrived in town the day Meagher died. Notably, the memorial association resolved to accept all donations except those from one man: Sanders.

40. David Emmons, "Orange and the Green in Montana: A Reconsideration of the Clark-Daly Feud," *Arizona and the West* 28 (Autumn 1986): 245.

41. Emmons, *Beyond the American Pale*, 330–332.

42. Translated by Patrick J. Barrett and used with permission.

43. F. E. Richter, "The Copper-Mining Industry in the United States, 1845–1925," *Quarterly Journal of Economics* 41, no. 2 (February 1927): 242; F. E. Richter, "The Copper-Mining Industry in the United States, 1845–1925," *Quarterly Journal of Economics* 41, no. 4 (August 1927): 686.

44. Richter, "Copper-Mining Industry in the United States" (February 1927): 238.

45. Richter, "Copper-Mining Industry in the United States" (February 1927): 238.

46. Howard, *Montana: High, Wide and Handsome*, 85.

47. Emmons, *Butte Irish*, 13.

48. Emmons, *Butte Irish*, 17.

49. "Board off Eerin," Miscellaneous Members and Correspondences, Butte Irish Collection, Archives and Special Collections, Maureen and Mike Mansfield Library, University of Montana, Missoula. Original spellings.

50. Emmons, *Butte Irish*, 64.

51. Hall, *Progressive Men of Montana*, 16.

52. Hall, *Progressive Men of Montana*, 18.

53. Hall, *Progressive Men of Montana*, 15, 17, 661.

54. Richter, "Copper-Mining Industry in the United States" (August 1927): 715. When he died on November 12, 1900, the *New York Times* estimated that his

property holdings in Montana alone were worth $25 million and his one-fourth holding in the Amalgamated Copper Company was worth $35 million. *New York Times*, November 13, 1900.

55. Quoted in Emmons, *Butte Irish*, 20.

56. Emmons, *Butte Irish*, 102.

57. Emmons, *Butte Irish*.

58. Emmons, *Butte Irish*, 157.

59. An Craoibín Aoibhinn, aka Douglas Hyde, *Mo Thurus go hAmerice: no imeasg na nGaedheal ins an Oilean Ur* (Dublin: Government Publications Office, 1937), 129–130. "Cathair Éireannach í seo, beagnach. Is Éireannaigh iad an chuid is mó de na daoinibh atá innti. Is Éireannach an Maor, agus tá striúrú gach rud ar láimh na nÉireannach." Author translation.

60. Aoibhinn, *Mo Thurus go hAmerice*, 130. "Mar sin féin ní abróchainn gur sásamhail í staid na Éireannach annso. Tá a seasamh ar fad ar na mianachaibh." Author translation.

61. Aoibhinn, *Mo Thurus go hAmerice*, 130. "Annsin táinig brosgán de phuncánaidh ó Mhissouri, agus chuireadar fúta ins an ngleann, an nidh nach ndéanfadh na hÉireannaigh, agus ní fhéadhá an glean so do cheannach indiu ar dháchad milliúin dollar!" Author translation.

62. Emmons, *Butte Irish*, 15; see also Michael O'Connell, "Emigration from the Berehaven Copper Mining District to the United States of America, 1840–1900," MA thesis, University College Cork, Ireland, 2007, 66. O'Connell shows that 30.5 percent of Kilnamanagh parish went to Butte, whereas Massachusetts was more popular for the Irish from Killaconeagh parish. Tabulated using a sample total of 200 for Kilnamanagh and 209 for Killaconeagh from Riobard O'Dwyer, *Who Were My Ancestors? Genealogy (Family Trees) of the Allihies (Copper Mines) Parish, County Cork, Ireland* (Self-published, 1976); Riobard O'Dwyer, *Who Were My Ancestors? Genealogy (Family Trees) of the Bere Island Parish, County Cork, Ireland* (Self-published, 1976).

63. Report of the Board of Trade, Sea Fisheries Act, 1868, Main Papers 531, Journal Office, Parliament Office, Records of the House of Lords, Parliamentary Archives, UK Parliament, London, 25; O'Connell, "Emigration from the Berehaven Copper Mining District," 68.

64. *Helena Independent*, June 30, 1878.

65. This was the same event during which the BMS manager, Bradley, decided to send two "missionaries" to Butte to recruit strikebreakers. See chapter 5, this volume. F. W. Bradley to N. H. Harris, June 25, 1895, Box 8, BHM; H. Minar Shoebotham, *Anaconda: Life of Marcus Daly, the Copper King* (Harrisburg, PA: Stackpole, 1956), 78–80.

66. Donald Lorenzo Kemmerer and C. Clyde Jones, *American Economic History* (New York: McGraw-Hill, 1959), 191; Richter, "Copper Mining Industry in the United States" (February 1927): 263–264.

67. O'Connell, "Emigration from the Berehaven Copper Mining District," 71.

68. Mary Murphy, "A Place of Greater Opportunity: Irish Women's Search for Home, Family and Leisure in Butte, Montana," *Journal of the West* 31, no. 2 (1992): 78.

69. Murphy, "Place of Greater Opportunity," 78.

70. Emmons, *Butte Irish*, 68.

71. Laurie Mercier, " 'We Are Women Irish': Gender, Class, Religious, and Ethnic Identity in Anaconda," in Elizabeth Jameson and Susan Armatige, eds., *Writing the Range: Race, Class, and Culture in the Woman's West* (Norman: University of Oklahoma Press, 1997), 311.

72. Mercier, "We Are Women Irish," 314.

73. Debbie Bowman Shea, *Irish Butte* (Mount Pleasant, SC: Arcadia, 2011), 18. See also Janet L. Finn and Ellen Crain, *Motherlode: Legacies of Women's Lives and Labors in Butte, Montana* (Butte: Clark City Press, 2005).

74. Mary Murphy, *Mining Cultures: Men, Women, and Leisure in Butte, 1914–41* (Chicago: University of Illinois Press, 1998), 35.

75. Emmons, *Butte Irish*, 25.

76. J. J. Lee, "The Irish Diaspora," in Laurence M. Geary and Margaret Kelleher, eds., *Nineteenth-Century Ireland: A Guide to Recent Research* (Dublin: University College Dublin Press, 2005), 217.

77. Emmons, *Butte Irish*, 67.

78. Emmons, *Butte Irish*, 137.

79. Emmons, *Butte Irish*, 26.

80. Emmons, *Butte Irish*, 408.

81. Peter Buckingham, *Red Tom Hickey: The Uncrowned King of Texas Socialism* (College Station: Texas A&M University Press, 2019).

82. *Butte Miner*, June 14, 1917.

83. Census of Ireland, 1911, no. 4, Thornhill, Curryglass, Co. Cork. Accessible at census.nationalarchives.ie.

84. Murphy, *Mining Cultures*, 47–48.

85. Ray Calkins, *Looking Back from the Hill: Recollections of Butte People* (Butte: Butte Historical Society, 1982), 71.

86. Murphy, *Mining Cultures*, 158. "The threat of unemployment loomed over the working class and encouraged people to acquire or improve only what they could take with them." Murphy, *Mining Cultures*, 225.

87. George Prescott, 1948, Montana Folklore Survey Project, 1979, AFS 1981/005, Folder 41, Box 2, Archive of Folk Culture, American Folklife Center, Library of Congress, Washington, DC.

88. Emmons, *Butte Irish*, 17.

89. Emmons, *Butte Irish*, 17.

90. Emmons, *Butte Irish*, 17.

91. Emmons, *Butte Irish*, 17.

92. Noonan, "Wandering Labourers," 104–106, 131–134.

93. Haywood, *Bill Haywood's Book*, 52. Despite the high praise, Haywood also wrote that "the tremendous power of this big union was not always used to the best advantage; it nearly always allowed itself to be divided on questions of strategy, tactics, and political issues." Haywood, *Bill Haywood's Book*, 52.

94. Emmons, *Butte Irish*, 184.

95. Timothy J. Sarbaugh, "The Irish in the West: An Ethnic Tradition of Enterprise and Innovation, 1848–1991," *Journal of the West* 31, no. 2 (1992): 6.

96. Murphy, *Mining Cultures*, 18.

97. Murphy, *Mining Cultures*, 18.

98. Emmons, *Butte Irish*, 148.

99. AFS 8893, Wayland Hand Collection of Irish Songs Recorded in Butte, Montana, 1945, AFC 1948/045, Archive of Folk Cuture, American Folklife Center, Library of Congress, Washington, DC.

100. MacGowan, *Hard Road to Klondike*, 73.

101. Murphy, *Mining Cultures*, 22.

102. *Butte Mining Journal*, September 21, 1887, quoted in Emmons, *Butte Irish*, 150.

103. Aoibhinn, *Mo Thurus go hAmerice*, 129–130. "Tá mé cinnte gurab í an áit is gránna d' á bhfaca mé riamh. Níl crann, níl sgeach, níl luibh, níl oiread agus tráithnín féir innti ná i n-aice léi. Tá na mílte agus na mílte spáis ar gach taobh di gan fás ann. Tá gach rud dóighte agus ithte ag an deatach nimhneach thagann ó na similéaraidh móra in a leaghann Mac Uí Chléirigh an Seanadóir ó Mhontana a chuid copair." Author translation.

104. Emmons, *Butte Irish*, 149.

105. Emmons, *Butte Irish*, 157.

106. Emmons, *Butte Irish*, 147.

107. Emmons, *Butte Irish*, 147.

108. SR 292, AFS #10,506A, Sam Eskins Collection, 1939–1969, AFC 1999/004, Archive of Folk Culture, American Folklife Center, Library of Congress, Washington, DC.

109. Ó Dubhda, *Duanaire Duibhneach*, 132–133.

110. Quoted in Archie Green, *Wobblies, Pile Butts, and Other Heroes: Laborlore Explorations* (Urbana: University of Illinois Press, 1993), 191. It was also an implicit endorsement of labor rights and the eight-hour day.

111. See Green, *Wobblies, Pile Butts, and Other Heroes*, 180.

112. Green's book contains an extended analysis of "Marcus Daly Goes to Heaven." Green, *Wobblies, Pile Butts, and Other Heroes*, 182–183.

113. Kevin Shannon, 1945, Montana Folklore Survey Project, 1979, AFS 1981/005, Folder 41, Box 2, Archive of Folk Culture, American Folklife Center, Library of Congress, Washington, DC. It is possible that this was intended as a jab at foreign shareholders or at John D. Ryan.

114. Oscar A. Dingman, 1948, Montana Folklore Survey Project, 1979, AFS 1981/005, Folder 41, Box 2, Archive of Folk Culture, American Folklife Center, Library of Congress, Washington, DC.

115. Uncredited, Montana Folklore Survey Project, 1979, AFS 1981/005, Folder 41, Box 2, Archive of Folk Culture, American Folklife Center, Library of Congress, Washington, DC. The Cornish referred to the Irish as "savage" when the Parrott Mine in Butte was purchased by Daly's Anaconda Company: "Goodbye, birdie, savage got thee, no more place for we." For details of the more lighthearted pranks and rivalries between the Irish and Cornish, see Hand, "Folklore, Customs, and Traditions," 6–10.

116. AFS 8892B, Wayland Hand Collection of Montana Silver Miners' Songs, AFC 1950/004, Archive of Folk Culture, American Folklife Center, Library of Congress, Washington, DC.

117. Green, *Wobblies, Pile Butts, and Other Heroes*, 192.

118. *Miners' Magazine*, November 10, 1910.

119. Emmons, *Butte Irish*, 268.

120. Emmons, *Butte Irish*, 268.

121. Emmons, *Butte Irish*, 269.

122. Emmons, *Butte Irish*, 278; Alan J. M. Noonan, "Real Irish Patriots Would Scorn to Recognise the Likes of You," in David Convery, ed., *Locked Out: A Century of Working Class Life* (Dublin: Irish Academic Press, 2013), 57–73.

123. Emmons, *Butte Irish*, 341. For example, Roger Casement's efforts in Germany to form an Irish brigade and the reference to "our gallant allies in Europe" in the Proclamation of the Irish Republic, 1916. See Noonan, "Real Irish Patriots," 68–69.

124. *Anaconda Standard*, March 18, 1915; Emmons, *Butte Irish*, 348.

125. Brosnan to his father, February 18, 1917, Brosnan Letters, Miller Collection; Emmons, *Butte Irish*, 257.

126. If it caused difficulty for Irish-America, the harm to German ethnic identity during the war was disastrous. The cultural cleansing of German-American

identity, judiciously begun in a display of American patriotism during World War I, was rendered total during World War II.

127. Noonan, "Real Irish Patriots," 57–73.

128. *Montana Socialist*, October 9, 1915. See Emmons, *Butte Irish*, 352–354.

129. *Montana Socialist*, July 22, 1916.

130. Noonan, "Real Irish Patriots," 70–71.

131. *Montana Socialist*, July 22, 1916.

132. Emmons offers an educated guess as to the people Larkin was referring to: Judge/Jeremiah J. Lynch, lawyer/Walter Breen, editor/James B. Mulcahy, and priest/Rev. Michael Hannan. Emmons, *Butte Irish*, 358. In the case of the priest, Emmons is probably incorrect in thinking it was a direct reference to the revolutionary, socialist-sympathizing Hannan. More likely it was a reference to clergy in general who were, on the whole, opposed to both nationalist and socialist organizations. If it was a reference to a local prelate, then it may have been a reference to Bishop John Carroll and other like-minded Irish-American clergy. See Noonan, "Real Irish Patriots," 57–73.

133. AFS 9736, Wayland Hand Collection of Irish Songs Recorded in Butte, Montana, 1945, AFC 1948/045, Archive of Folk Culture, American Folklife Center, Library of Congress, Washington, DC.

134. *Montana Socialist*, January 27, 1917.

135. RELA Minutes, June 25, 1896, Irish Collection, Butte-Silver Bow Archives, Butte, MT.

136. RELA Minutes, June 25, 1896, 360.

137. AOH Minutes Book, Division III, July 16, 1916, Irish Collection, Butte-Silver Bow Archives, Butte, MT.

138. Emmons, *Butte Irish*, 359.

139. Emmons, *Butte Irish*, 360.

140. Emmons, *Butte Irish*, 376. Irish Republican leader Liam Mellows described him as a "tower of strength to the movement." Emmons, *Butte Irish*, 360.

141. Emmons, *Butte Irish*, 268.

142. Emmons, *Butte Irish*, 364–365.

143. Emmons, *Butte Irish*, 364–365.

144. *New York Times*, August 2, 1917. The number is on the badges of Montana state troopers to this day, signaling the early influence of these figures in government and a coded threat directed at the public. The reasons for the numbers themselves vary, from their representation being the time in hours, minutes, and seconds for someone to leave to being the unusual measurements of a overly shallow, long grave. Masonic organizations relished the use of obscure and coded messages in an effort to extend the idea that they were holders of the truth, in opposition to

other groups. In reality, the invented origin story is irrelevant because the meaning behind its use is the salient point: it was a threat.

145. *Anaconda Standard*, August 6, 1917.

146. *Butte Daily Post*, July 28, 1917; *Butte Miner*, July 20, 1917.

147. Will Roscoe, *"The Murder of Frank Little: An Injury to One Is an Injury to All"* [unpublished manuscript] (Helena: Montana Historical Society, 1973), 37.

148. Murphy, *Mining Cultures*, 50.

149. Murphy, *Mining Cultures*, 51.

150. AOH Minutes Book, Division III, July 16, 1906, 216, Irish Collection, Butte-Silver Bow Archives, Butte, MT.

151. Emmons, *Butte Irish*, 243.

152. AOH Minutes Book, Division III, January 10, 1910, 448, Irish Collection, Butte-Silver Bow Archives, Butte, MT.

153. AOH Minutes Book, Division III, November 30, 1908, 373, Irish Collection, Butte-Silver Bow Archives, Butte, MT. Father Hannan strongly disagreed, stating that "his ideal Irishman was the noblest type of manhood and hence the best American." AOH Minutes Book, November 30, 1908, 373.

154. *Anaconda Standard*, May 28, 1918.

155. Dennis L. Swibold, *Copper Chorus: Mining, Politics, and the Montana Press, 1889–1959* (Helena: Montana Historical Society Press, 2006), 26–29.

156. AOH Minutes Book, Division III, July 6, 1908, 353, Irish Collection, Butte-Silver Bow Archives, Butte, MT.

157. AOH Minutes Book, July 6, 1908.

158. Emmons, *Butte Irish*, 398.

159. Thomas J. Hagerty emerged as the remarkable Catholic clerical figure who tried to reconcile Christianity and socialism. He argued: "Socialism has no more to do with religion than astronomy or biology." Thomas J. Hagerty, *Economic Discontent and Its Remedy* (Cincinnati: Standard Publishing Company, 1902), 44. He was suspended as a priest following his efforts to recruit miners for the ALU and the WFM in Colorado. Robert E. Doherty, "Thomas J. Hagerty, the Church, and Socialism," *Labor History* 3, no. 1 (Winter 1962): 39–56.

160. Quoted in the *Butte Daily Bulletin*, July 29, 1919. Jack Carney claimed that de Valera directly insulted Larkin in Butte, calling him "an English man," and said he sent Irish children to England during the lockout "for the purpose of undermining their faith," though it was unlikely that such outrageous attacks would have gone unmentioned in Butte or by Larkin himself. Gerard Watts, "James Larkin and the British, American and Irish Free State Intelligence Services: 1914–1924," PhD diss., National University of Ireland, Galway, 2016, 190–191.

161. Lawrence J. McCaffrey, "Diasporic Comparisons and Irish American Uniqueness," in Charles Fanning, ed., *New Perspectives on the Irish Diaspora* (Carbondale: Southern Illinois University Press, 2000), 23–24.

162. Murphy, *Mining Cultures*, 33.

163. Jerry W. Calvert, *The Gibraltar: Socialism and Labor in Butte, Montana, 1895–1920* (Helena: Montana Historical Society Press, 1988), 120–124.

164. Emmons, *Butte Irish*, 157.

165. Emmons, *Butte Irish*, 95–96. Big Bill Haywood seems to have misunderstood a phrase used to describe those living in Dublin Gulch. He wrote it was the "home of the 'Paddy-come-latelies,' the Irish 'big-wheelers' direct from the old sod." Haywood, *Bill Haywood's Book*, 52. The Irish meant that they were recent arrivals, not that they were well-to-do, but what he writes does reinforce the growing divide within the Irish community in Butte.

166. Ó Dubhda, *Duanaire Duibhneach*, 132–133.

167. AFS 8893, Wayland Hand Collection of Irish Songs Recorded in Butte, Montana, 1945, AFC 1948/045, Archive of Folk Culture, American Folklife Center, Library of Congress, Washington, DC.

168. MacGowan, *Hard Road to Klondike*, 76.

169. As Butte's ability to offer secure employment faded, migration bolstered other smaller Irish communities in Montana. Noonan, "From Ireland to Montana," 120–150.

170. Emmons, *Butte Irish*, 409.

Conclusion

1. Martin R. Ansell, *Oil Baron of the Southwest: Edward L. Doheny and the Development of the Petroleum Industry in California and Mexico* (Columbus: Ohio State University Press, 1998). See also Margaret L. Davis, *Dark Side of Fortune: Triumph and Scandal in the Life of Oil Tycoon Edward L. Doheny* (Berkeley: University of California Press, 1998).

2. Ansell, *Oil Baron of the Southwest*, 11.

3. Ansell, *Oil Baron of the Southwest*, 23–29.

4. Ansell, *Oil Baron of the Southwest*, 180–187, 243–244.

5. Bertie C. Forbes, *Men Who Are Making the West* (New York: B. C. Forbes, 1923), 101–106; Ansell, *Oil Baron of the Southwest*, 183. It is possible that Doheny left home at such a young age because of family difficulty, but if this was the case, it did not affect his attachment to his ethnic identity.

6. Ellen Wogan, September 2, 1870, Arnold Schrier Papers, UA-16-07, Archives and Rare Books Library, University of Cincinnati, Cincinnati, OH.

7. William L. Kennedy, February 3, 1870, Kennedy Family Letters.

8. William L. Kennedy, February 3, 1870.

9. Evalyn Walsh Mclean, *Father Struck It Rich* (New York: Little, Brown, 1936), 56.

10. Mclean, *Father Struck It Rich*, 57. Another interesting episode occurred at the end of the family visit to Ireland when a woman fell off the train at the station and cursed her father: "'Tom Walsh, I curse Evalyn. I want you before you're dead to see her in the gutter, to see her worse from drink than I am' . . . An Irish curse on me . . . [Father's] face was wet with tears. The night was full of evil . . . Of course I was just a child, and this should have touched me no more than a bad dream; perhaps it did no more—and yet I wish, because of the Irish in me, that curse had never been spoken." Evalyn understood the cultural meaning behind such an act, even as a child. Later in her book she speaks of her struggles with addiction to alcohol and morphine, linked to this earlier curse. She would also ask a priest to bless an item and remove a curse. Mclean, *Father Struck It Rich*, 178.

11. Thomas Higgins, Los Angeles, California, to Cecilia Keefe, Mosinee, Wisconsin, May 16, 1916, Higgins Collection.

12. Thomas Higgins to Cecilia Keefe, May 16, 1916. Denis Hurley, who lived in neighboring Nevada, noted: "Persons who have lived in a dry country like this for several years do not take kindly to the rain—too dreary and forlorn." Denis Hurley to his mother, March 9, 1896, Hurley Letters.

13. Thomas Higgins to Cecilia Keefe, May 16, 1916.

14. Thomas Higgins to Cecilia Keefe, May 16, 1916.

15. Thomas Higgins to Cecilia Keefe, May 16, 1916. Contrast a statement in the *Cork Examiner*, May 26, 1890, which argued that remittances kept people from being evicted and in effect stopped them from emigrating to a better life.

16. For the most comprehensive research on remittances, see Schrier, *Ireland and the American Emigration*. Elderly parents probably did become dependent on them when they were unable to work.

17. Thomas Higgins to Cecilia Keefe, May 16, 1916.

18. Thomas Higgins to Cecilia Keefe, May 16, 1916.

19. Ó Dubhda, *Duanaire Duibhneach*, 133. In his book *Special Sorrows*, Jacobson discusses the desire of Irish migrants to maintain a connection to the home country despite their migration. Matthew Frye Jacobson, *Special Sorrows: The Diasporic Imagination of Irish, Polish, and Jewish Immigrants in the United States* (Berkeley: University of California Press, 2010).

20. Thadeus W. Healy, performed in Virginia City, March 17, 1874, Box 114, Collection of Ron Bammarito, Douglas County Historical Society, Gardnerville, NV;

SR 292, AFS #10,506A, Sam Eskins Collection, 1939–1969, Archive of Folk Culture, American Folklife Center, Library of Congress, Washington, DC.

21. *Miners' Magazine*, January 28, 1904, in Tannacito, "Poetry of the Colorado Miners," 6.

22. John S. Kelly, *The Bodyke Evictions* (Scariff, Ireland: Fossabeg Press, 1987), 91. The term *Yank* is used by the Irish to describe any Irish-American or American-born person.

23. Noonan, "Wandering Labourers," 55–59.

24. Ó Dubhda, *Duanaire Duibhneach*, 132.

25. Lara Vapnek, *Elizabeth Gurley Flynn: Modern American Revolutionary* (Boulder: Westview, 2015), 13.

26. Once again, the exceptional gold rushers and early placer mining gamblers are not the groups referred to here. See the introduction and chapter 2, this volume.

27. Mann, *After the Gold Rush*, 218.

28. Patricia Nelson Limerick, *The Legacy of Conquest: The Unbroken Past of the American West* (New York: W. W. Norton, 1987), 99.

29. Fitzpatrick, "The Irish in Britain," 7.

30. Rowland T. Berthoff, *British Immigrants in Industrial America, 1790–1950* (Cambridge, MA: Harvard University Press, 1953), 210.

31. As pointed out succinctly by Barbara J. Fields, "Whiteness, Racism, and Identity," *International Labor and Working-Class History* 60 (Fall 2001): 48–56. Quite simply, "Whiteness leads to no conclusions that it does not begin with assumptions." Fields, "Whiteness, Racism, and Identity," 53. Jacobson, though not in full agreement with her, makes a similar point: "White ethnics may indeed be white, but their history is severed from the broader structural history of whiteness in America." Matthew Frye Jacobson, *Whiteness of a Different Color: European Immigrants and the Alchemy of Race* (Cambridge, MA: Harvard University Press, 1998), 278.

32. See Christian G. Samito, *Becoming American under Fire: Irish Americans, African Americans, and the Politics of Citizenship during the Civil War Era* (Ithaca, NY: Cornell University Press, 2009), 43, 103, 217.

33. Efforts to chart a historical "high point" in anti-Irish rhetoric are as misplaced as similar efforts in British historiography to establish the high point in anti-Irish disturbances. L. Perry Curtis Jr., *Anglo-Saxons and Celts: A Study of Anti-Irish Prejudice in Victorian England* (Bridgeport, CT: University of Bridgeport, 1968), 90. A succinct criticism of this approach can be found in Louise Miskell, "Custom, Conflict and Community: A Study of the Irish in South Wales and Cornwall, 1861–1891," PhD diss., University of Wales, Cardiff, 1996, 7–9.

34. Dale T. Knobel, *Paddy and the Republic, Ethnicity and Nationality in Antebellum America* (Middletown, CT: Wesleyan University Press, 1986), 180–181.

35. Woodrow Wilson, *A History of the American People*, vol. 5 (New York: Harper Brothers, 1902), 168.

36. Wilson, *History of the American People*, vol. 5, 168.

37. Wilson, *History of the American People*, vol. 5, 168.

38. Wilson, *History of the American People*, vol. 5, 168.

39. Barbara Miller Soloman, *Ancestors and Immigrants: A Changing New England Tradition* (Cambridge, MA: Harvard University Press, 1956), 75, 116–117; United States Congress, Immigration Commission, *Reports of the Immigration Commission: Abstracts of Reports of the Immigration Commission with Conclusions and Recommendations and Views of the Minority*, vol. 1 (Washington, DC: US Government Printing Offices, 1911), 248. Others conflated the Irish with Anglo-Saxons for similar a-historical reasons. As Frank Sullivan asserted to the Society of California Pioneers, "May the Californian cherish the memory of the Celtic-Saxon pioneer." Frank J. Sullivan, "*Autobiography and Reminiscence of Frank J. Sullivan*" (unpublished manuscript, Society of California Pioneers, 1904), 15.

40. United States Congress, Immigration Commission Report, vol. 1, 83, 120. "High-strung" could be a reference to their Catholicism or to the Polish workers' tendency to agitate for better pay, or it might be trying to link poverty to a nervous disposition.

41. United States Congress, Immigration Commission Report, vol. 1, 130.

42. United States Congress, Immigration Commission Report, vol. 1, 494; Edward Prince Hutchinson, *Immigrants and Their Children, 1850–1950*, Census Monograph Series (New York: Wiley, 1956), 65.

43. Timothy M. O'Neil, "Miners in Migration: The Case of Nineteenth Century Irish and Irish American Copper Miners," *Éire-Ireland: A Journal of Irish Studies* 36, nos. 1 and 2 (Spring–Summer 2001): 133.

44. Joshua L. Rosenbloom, "Looking for Work, Searching for Workers: U.S. Labor Markets after the Civil War," *Social Science History* 18 (Autumn 1994): 379.

45. Ronald M. James, "Defining the Group: Nineteenth Century Cornish on the North American Mining Frontier," *Cornish Studies* 2 (1994): 41–42.

46. While Dykstra does not mention this distinction, he easily debunks the myth of casual violence in the American West. Robert R. Dykstra, "Overdosing on Dodge City," *Western Historical Quarterly* 27 (Winter 1996): 505–514. See also Murphy, *Mining Cultures*, 47.

47. Berthoff, *British Immigrants in Industrial America*, 208.

48. Noonan, "Wandering Labourers," 52–59.

49. Thomas Colley Grattan, *Civilized America*, vol. 2 (London: Bradbury and Evans, 1859), 28.

50. Roger Daniels, *Asian America: Chinese and Japanese in the United States since 1850* (Seattle: University of Washington Press, 1988), 4; Zhu, *Chinaman's Chance*, 2. It could be said that whiteness studies focuses on the Irish only in terms of who they oppressed.

51. Ó Dubhda, *Duanaire Duibhneach*, 127–129.

52. AFS 9732, N-2, "Wanderlust" reading by W. L. Davis, Wayland Hand Collection of Irish Songs Recorded in Butte, Montana, 1945, AFC 1948/045, Archive of Folk Culture, American Folklife Center, Library of Congress, Washington, DC.

53. AFS 9730, L-1, sung by Matthew Hanafin, Butte, Montana, July 28, 1948, Wayland Hand Collection of Irish Songs Recorded in Butte, Montana, 1945, AFC 1948/045, Archive of Folk Culture, American Folklife Center, Library of Congress, Washington, DC.

54. Emmons, *Beyond the American Pale*, 351–356.

55. Mary Harris Jones, *Autobiography of Mother Jones, American Labor: From Conspiracy to Collective Bargaining* (New York: Arno, 1969 [1925]).

56. *Parkersburg Morning News*, June 23, 1902, in Philip Foner, ed., *Mother Jones Speaks* (New York: Pathfinder, 2003), 484.

57. Foner, *Mother Jones Speaks*, 183.

58. Foner, *Mother Jones Speaks*, 182.

59. Elliott J. Gorn, *Mother Jones: The Most Dangerous Woman in America* (New York: Farrar, Straus and Giroux, 2015), 303.

60. US Industrial Commission, *Report of the Industrial Commission*, 41.

61. US Industrial Commission, *Report of the Industrial Commission*, 32.

62. US Industrial Commission, *Report of the Industrial Commission*, 33–46.

63. "Final Address in Support of the League of Nations," delivered September 25, 1919, in Pueblo, Colorado, in Arthur S. Link, ed., *The Papers of Woodrow Wilson*, vol. 63 (Princeton, NJ: Princeton University Press, 1994), 500–501.

64. *Daily Tribune (Salt Lake City)*, March 16, 1890; William Fox, "Patrick Edward Connor, 'Father' of Utah Mining," master's thesis, Brigham Young University, Provo, UT, 1966.

65. Mac Gabhann, *Rotha mór an tSaoil*, 222.

66. MacGowan, *Hard Road to Klondike*, 144.

67. Terry Coleman, *Passage to America: A History of Emigrants from Great Britain and Ireland to America in the Mid-Nineteenth Century* (London: Pimlico, 1992), 248.

68. John McCue Voyage Poem, May 12, 1856, Miller Collection.

69. Patrick Kearney to John Kearney, December 21, 1890; De Búrca, *Soldier's Song*, 251.

70. *Final Report of the Commission on Industrial Relations* (Washington, DC: US Government Printing Office, 1916), 9.

Appendix 1: Irish Poems, Songs, and Notes about Mining

1. Seán Ó Dubhda, ed., *Duanaire Duibhneach* (Dublin: Government Publications Office, 1976), 127–129. Translation by Patrick J. Barrett.

2. Ó Dubhda, *Duanaire Duibhneach*, 130–131, translation by Dr. Bruce D. Boling, Miller Collection.

3. Ó Dubhda, *Duanaire Duibhneach*, 132–133, translation by Dr. Bruce D. Boling, in Miller, *Emigrants and Exiles*, xiii.

4. These are the original lyrics written by Michael Considine in a letter that was kept by his nephew John Considine. See the Carroll Mackenzie Collection, Clare County Library, http://www.clarelibrary.ie/eolas/coclare/songs/cmc/spancil_hill_msflanaghan.htm.

5. This copy of this song is taken from Robert Gogan, *130 Great Irish Ballads* (Dublin: Music Ireland, 2006), 94. The labor song "On Johnny Mitchell's Train" uses the same air, first verse, and chorus.

6. *Butte Miner*, June 14, 1917.

Appendix 2: Transcript of Official Oath of the State of Nevada

1. Ancient Order of Hibernians, April 16, 1877, NHR MS File, Nevada Historical Society, Reno.

Bibliography

Primary Sources

All Hallows College Archives, Dublin, Ireland

Annals of All Hallows Missionary College. Dublin: J. F. Fowler, 1860, 1861, 1862, 1863, 1897, 1899, 1902, 1903, 1911–1912, 1914–1915, 1922.

Fifth Report of All Hallows Missionary College. Dublin: J. F. Fowler, 1852–1853.

Seventh Report of All Hallows Missionary College. Dublin: J. F. Fowler, 1855.

Bancroft Library, University of California, Berkeley

California Gold Rush Letters 1848–1859, MSS C-B 547

Hubert H. Bancroft Scraps 1849, volume 111: Idaho Miscellany

John McTurk Gibson, "Journal of Western Travel"

Yuba County Documents, MS 3A2

Butte-Silver Bow Archives, Butte, MT

Diary of Séamus Feiritéar

Hannon, Brosnan, Daly Collection

Irish Collection

https://doi.org/10.5876/9781646422517.c011

California Historical Society, San Francisco
Teresa Lawlor/Phelan Letters
Twogood Diary
California State Library, Sacramento
Yellow Aster Mining and Milling Company Records, 1898–1918, Manuscript Collection
Centre for Migration Studies, Omagh, Ireland
Lynch, Seamus P. Short Biography on Pat Magill, Pioneer, Rancher and Miner. 1939.
Clare County Library, Ennis, Ireland
Carroll Mackenzie Collection
Coeur d'Alene's Old Mission State Park, Coeur d'Alene, ID
Cataldo Mission Letters
Cork City Archives, Cork, Ireland
Hurley Family Emigrant Letters, MS U170
Diocese of Sacramento, CA
Record of Internment, Smartsville
Douglas County Historical Society, Gardnerville, NV
Collection of Ron Bammarito
Elmer Holmes Bobst Library, New York University, New York, NY
"The Reminiscences of John Brophy." Microfilm. MICRO 56, Columbia Oral History Collection.
Historical Society of Pennsylvania, Philadelphia
Balch Institute Sheet Music Collection, circa 1824–1945, Collection 3141
History Colorado, Denver
Leadville Strike Reports 1896–1898, MSS 334
Idaho Historical Society, Boise
Idaho State Historical Reference Series
North Idaho College Oral History Program Collection
Library of Congress, Washington, DC
American Song Sheets, Rare Books and Special Collections
Montana Folklore Survey Project, 1979, AFS 1981/005, Archive of Folk Culture, American Folklife Center
Pinkerton Agency Records, Manuscript Division
Sam Eskins Collection, 1939–1969, AFC 1999/004, Archive of Folk Culture, American Folklife Center
Wayland Hand Collection of Irish songs recorded in Butte, Montana, 1945, AFC 1948/045, Archive of Folk Culture, American Folklife Center

Wayland Hand Collection of Montana Silver Miners' Songs, AFC 1950/004, Archive of Folk Culture, American Folklife Center

Maureen and Mike Mansfield Library, University of Montana, Missoula

Butte Irish Collection, MS 122

Clyde Francis Murphy Papers, MS 285

Montana Mining Company Records, MS 142

Miller Collection (copies held by Kerby A. Miller, University of Missouri, Columbia)

Diary of Robert Williamson of Ahorey

Gamble Family Letters

John McCue Voyage Poem

Journal kept by James J. Mitchell, commenced on leaving Ahascragh, County Galway, Ireland, for the United States of America, March 16, 1853

Mary Ann Landy Letters

McGee Letters

Thomas W. Brick, Memoirs of an Emigrant to USA, Spring 1902, Typescript, 1970

Translation of poems by Dr. Bruce D. Boling from Seán Ó Dubhda, ed., Duanaire Duibhneach (Dublin: Government Publications Office, 1976)

Montana Historical Society Archives, Helena

Anaconda Copper Mining Company Records, Collection no. 169

Roscoe, Will. "The Murder of Frank Little: An Injury to One Is an Injury to All." Unpublished Manuscript.

Samuel T. Hauser Papers, 1864–1914

Thomas Cruse Papers, 1841–1956, MC 36

Museum of St. Mary of the Mountain, Virginia City, NV

Collected Records of the Museum of St. Mary of the Mountain

Hospital Logbook of St. Mary

National Library of Ireland, Manuscripts Collection, Dublin

Letters from Irish Emigrants in the USA to Relatives and Friends in Ireland, ca. 400 Items, 19th and 20th Centuries. Arranged in Alphabetical Order of State etc. of Origin, MSS 22441–22442

Nevada Historical Society, Reno

Ancient Order of Hibernians, NHR MS File

New York Public Library, New York, NY

Emigrant Savings Bank Records

Oregon Historical Society, Portland

James Mullany Letters, MS 2417

Public Record Office of Northern Ireland, Belfast

Correspondence and Family Papers relating to the Orr and Dunn Families 1612–1941, D.2908

Foster Papers 1529–1980, D.3618

James Stewart Emigrant Documents 1884–1885, T.3399/1

Kennedy Family Emigrant Papers 1869–1912, MS T.3152

McSparron (McSparran) Papers 1860–1916, T.2743

Williamson Papers 1781–ca.1910, MS T.2680

Society of California Pioneers, Alice Phelan Sullivan Library, San Francisco, CA

Autobiography and Reminiscence of Frank J. Sullivan, 1904

Autobiography and Reminiscence of Thomas Kyle, San Francisco, 1904

University of Cincinnati, Archives and Rare Books Library, Cincinnati, OH

Arnold Schrier Papers, UA-16–07

University of Colorado, Boulder

John F. Campion Papers, MS 1099

Portland Gold Mining Co. Records

University of Idaho, Moscow

Bunker Hill Company Records 1887–1984, MG 367

University of Nevada, Reno

Golden Family Records, MS NC29

Nevada Legislature Records, 1875, MS 01/2B

William H. Hannon Library, Loyola Marymount University, Los Angeles, CA

Emmet Guard march in military regalia in Virginia City, Nevada, on St. Patrick's Day, circa 1870s, Department of Archives and Special Collections

Wisconsin Historical Society, Madison

Higgins Family Letters, MS 22442

Newspapers

Ireland

Armagh Guardian

Belfast Commercial Chronicle

Belfast Newsletter

Belfast Telegraph

Cork Examiner

United States

Anaconda (MT) Standard
Arizona Silver Belt (Globe)
Bodie (CA) Free Press
Bodie (CA) Standard
Butte (MT) Bystander
Butte (MT) Daily Bulletin
Butte (MT) Daily Post
Butte (MT) Miner
Butte (MT) Mining Journal
Carson City (NV) News
Coeur d'Alene (ID) Press
Daily Alta California (San Francisco)
Daily Free Press (Bodie, CA)
Daily Tribune (Salt Lake City)
Denver Tribune
Engineering and Mining Journal (New York)
Gold Hill (NV) News
Helena (MT) Independent
Homer Mining Index (Lundy, CA)
Idaho Daily Statesman (Boise)
Intermountain Catholic (Salt Lake City)
Irish World (New York)
Leadville (CO) Chronicle
Leadville (CO) Democrat
Leadville (CO) Evening Chronicle
Leadville (CO) Herald
Leadville (CO) Herald Democrat
Leadville (CO) Weekly Democrat
Leadville (CO) Weekly Herald
Los Angeles Star
Marysville (CA) Daily Appeal
Marysville (CA) Herald
Miners' Magazine (Denver)
Montana Post (Virginia City)
Montana Socialist (Butte)

Mountaineer (Marysville, MT)
National Police Gazette (New York)
New York Times
Owyhee Avalanche (Silver City, ID)
Pioche (NV) Weekly Record
Puck Magazine (New York)
Radical Teacher (Pittsburgh)
Railway Age Gazette (New York)
Reese River Reveille (Austin, NV)
River Press (Fort Benton, MT)
Rocky Mountain News (Denver)
Sacramento Daily Union
San Francisco Call
Territorial Enterprise (Virginia City, NV)
Virginia Daily Union (Virginia City, NV)

Secondary Sources

Abbe, Donald R. *Austin and the Reese River Mining District: Nevada's Forgotten Frontier.* Reno: University of Nevada Press, 1985.

Aiken, Katherine G. *Idaho's Bunker Hill: The Rise and Fall of a Great Mining Company, 1885–1981.* Norman: University of Oklahoma Press, 2005.

Aiken, Katherine G. " 'Not Long Ago a Smoking Chimney Was a Sign of Prosperity': Corporate and Community Response to Pollution at the Bunker Hill Smelter in Kellogg, Idaho." *Environmental History Review* 18 (1994): 67–86.

Akenson, Donald Harman. "Irish Migration to North America, 1800–1920." In *The Irish Diaspora,* edited by Andy Bielenberg. New York: Longman, 2000, 111–138.

Akenson, Donald Harman. *Small Differences: Irish Catholics and Irish Protestants, 1815–1922.* Montreal: McGill–Queens University Press, 1988.

Allen, Theodore. *The Invention of the White Race: The Haymarket Series.* New York: Verso, 1994.

Allyn, Cambell Loosley. "Foreign Born Populations of California 1848–1920." Master's thesis, University of California, Berkeley, 1927.

Almaguer, Tomas. *Racial Fault Lines: The Historical Origins of White Supremacy in California.* Berkeley: University of California Press, 2008.

Anbinder, Tyler Gregory. *Nativism and Slavery: The Northern Know Nothings and the Politics of the 1850's.* New York: Oxford University Press, 1992.

Anderson, Terry Lee, and Peter Jensen Hill. *The Not So Wild, Wild West: Property Rights on the Frontier.* Stanford: Stanford Economics and Finance, 2004.

Andrews, Thomas G. *Killing for Coal: America's Deadliest Labor War*. Cambridge, MA: Harvard University Press, 2008.

Angel, Myron. *History of Nevada*. Oakland: Thompson and West, 1881.

Ansell, Martin R. *Oil Baron of the Southwest: Edward L. Doheny and the Development of the Petroleum Industry in California and Mexico*. Columbus: Ohio State University Press, 1998.

Armstrong, Robert D. *Nevada Printing History: A Bibliography of Imprints and Publications, 1881–1890*. Reno: University of Nevada Press, 1991.

Ballenger and Richards' Annual Leadville City Directory, 1888. Leadville, CO: Ballenger and Richards, 1888.

Ballenger and Richards' Annual Leadville City Directory, 1898. Leadville, CO: Ballenger and Richards, 1898.

Bennett, David Harry. *The Party of Fear: From Nativist Movements to the New Right in American History*. Chapel Hill: University of North Carolina Press, 1988.

Berlanstein, Lenard R. *Rethinking Labor History: Essays on Discourse and Class Analysis*. Urbana: University of Illinois Press, 1993.

Berthoff, Rowland T. *British Immigrants in Industrial America, 1790–1950*. Cambridge, MA: Harvard University Press, 1953.

Berthoff, Rowland T. "The 'Freedom to Control' in American Business History." In *A Festschrift for Frederick B. Artz*, edited by David H. Pinkney and Theodore Ropp. Durham, NC: Duke University Press, 1964, 158–175.

Bielenberg, Andy. *Ireland and the Industrial Revolution: The Impact of the Industrial Revolution on Irish Industry, 1801–1922*. New York: Routledge, 2009.

Billington, Ray Allen. *The Protestant Crusade, 1800–1860: A Study of the Origins of American Nativism*. Chicago: Quadrangle Books, 1964.

Blatz, Perry K. *Democratic Miners: Work and Labor Relations in the Anthracite Coal Industry, 1875–1925*. New York: State University of New York Press, 1994.

Bourne, Edward G. "The Legend of Marcus Whitman." *American Historical Review* 6 (January 1901): 276–300.

Boyce, Edward. "Miners' Union Day at Butte, Montana." *Miners' Magazine* (April 1900): 39–42.

Breault, William. *The Miner Was a Bishop: The Pioneer Years of Patrick Manogue in California and Nevada 1854–1895*. Rancho Cordova, CA: Landmark Enterprises, 1988.

Brooks, Joan. *Desert Padre: The Life and Writings of Father John J. Crowley, 1891–1940*. Desert Hot Springs, CA: Mesquite, 1997.

Brophy, John. *A Miner's Life: An Autobiography*. Madison: University of Wisconsin Press, 1964.

Brown, Richard M. *Strain of Violence: Historical Studies of American Violence and Vigilantism*. New York: Oxford University Press, 1975.

Brown, Thomas N. *Irish American Nationalism, 1870–1890*. Philadelphia: J. B. Lippincott, 1966.

Brundage, David Thomas. *The Making of Western Labor Radicalism: Denver's Organized Workers, 1878–1905*. Urbana: University of Illinois Press, 1994.

Bruns, Roger. *Knights of the Road: A Hobo History*. New York: Methuen, 1980.

Bucchianeri, Virgil A. *Nevada's Bonanza Church, Saint Mary's in the Mountains*. Virginia City, NV: Saint Mary in the Mountains Parish, 2009.

Buckingham, Peter. *Red Tom Hickey: The Uncrowned King of Texas Socialism*. College Station: Texas A&M University Press, 2019.

Burchell, Robert A. "Irish Property Holding in the West in 1870." *Journal of the West* 31 (Summer 1992): 9–16.

Burlingame, Merrill G. "The Mining Frontier in Montana." In *Montana's Past: Selected Essays*, edited by Michael P. Malone and Richard B. Roeder. Missoula: University of Montana, 1969.

Butler, Anne M. *Daughters of Joy, Sisters of Misery: Prostitutes in the American West, 1865–90*. Urbana: University of Illinois Press, 1985.

Calhoon, F. D. *49er Irish: One Irish Family in the California Mines*. New York: Exposition, 1977.

Calkins, Ray. *Looking Back from the Hill: Recollections of Butte People*. Butte, MT: Butte Historical Society, 1982.

Calvert, Jerry W. *The Gibraltar: Socialism and Labor in Butte, Montana, 1895–1920*. Helena: Montana Historical Society Press, 1988.

Campbell, Alan B. *The Lanarkshire Miners: A Social History of Their Trade Unions 1775–1974*. Edinburgh: J. Donald, 1979.

Carter, Bryan Anthony. "A Frontier Apart: Identity, Loyalty, and the Coming of the Civil War on the Pacific Coast." PhD diss., Oklahoma State University, Stillwater, 2014.

Census of Ireland, 1901, 1911. census.nationalarchives.ie.

Chandler, Alfred D. *The Visible Hand: The Managerial Revolution in American Business*. Cambridge, MA: Belknap, 1977.

Chaput, Don. *Nellie Cashman and the North American Mining Frontier*. Tucson, AZ: Westernlore, 1995.

Coble Trull, Fern. "The History of the Chinese in Idaho, from 1864 to 1910." Master's thesis, University of Oregon, Eugene, 1949.

Coburn, Carol K., and Martha Smith. "'Pray for Your Wanderers': Women, Religious on the Colorado Mining Frontier, 1877–1917." *Frontiers: A Journal of Women Studies* 15, no. 3 (1995): 27–52.

Coleman, Terry. *Passage to America: A History of Emigrants from Great Britain and Ireland to America in the Mid-Nineteenth Century*. London: Pimlico, 1992.

Collins, Charles. *Mercantile Guide and Directory for Virginia City, Gold Hill, Silver City and American City*. San Francisco: Agnew and Deffebach, 1865.

Convery, David, ed. *Locked Out: A Century of Working Class Life*. Dublin: Irish Academic Press, 2013.

Cook, David J. *Hands Up, or, Thirty-Five Years of Detective Life in the Mountains and on the Plains*. Denver: W. F. Robinson, 1897.

Coons, S. J. "Influence of Gold Camps on the Economic Development of Western Montana." *Journal of Political Economy* 38, no. 5 (October 1930): 580–599.

Cowman, Des. "Life and Labour in Three Mining Communities C. 1840." *Saothar* 9 (1983): 10–19.

Cowman, Des. *The Making and Breaking of a Mining Community*. Dublin: Mining Heritage Trust of Ireland, 2006.

Cowman, Des. "Survival, Statistics, and Structures: Knockmahon Copper Mines, 1850–1878." *Decies* 46 (1992): 10–20.

An Craoibín Aoibhinn, aka Douglas Hyde. *Mo Thurus go hAmerice: no imeasg na nGaedheal ins an Oilean Ur*. Dublin: Government Publications Office, 1937.

Crampton, Frank A. *Deep Enough: A Working Stiff in the Western Mine Camps*. Norman: University of Oklahoma Press, 1982.

Culver, William, and Thomas C. Greaves. *Miners and Mining in the Americas*. Dover, NH: Manchester University Press, 1985.

Curtis, L. Perry, Jr. *Anglo-Saxons and Celts: A Study of Anti-Irish Prejudice in Victorian England*. Bridgeport, CT: University of Bridgeport, 1968.

Daniels, Roger. *Asian America: Chinese and Japanese in the United States since 1850*. Seattle: University of Washington Press, 1988.

Davis, Carlyle Channing. *Olden Days in Colorado*. Los Angeles: Phillips, 1916.

Davis, Graham, and John Fripp, eds. *In Search of a Better Life: British and Irish Migration*. Stroud, Gloucestershire: History Press, 2011.

Davis, Margaret L. *Dark Side of Fortune: Triumph and Scandal in the Life of Oil Tycoon Edward L. Doheny*. Berkeley: University of California Press, 1998.

Davitt, Michael. *The Boer Fight for Freedom*. London: Funk and Wagnalls, 1902.

Davitt, Michael. *The Fall of Feudalism in Ireland, or, the Story of the Land League Revolution*. London: Harper and Brothers, 1904.

Dawson, Thomas F. "The Old-Time Prospector." *Colorado Magazine* 1 (March 1924): 103–108.

de Bovet, Madame. *Three Months' Tour in Ireland*. London: Chapman and Hall, 1891.

De Búrca, Séamus. *The Soldier's Song; The Story of Peadar O'Cearnaigh*. Dublin: P. J. Bourke, 1957.

Delano, Alonzo. *Life on the Plains and among the Diggings*. Buffalo, NY: Miller, Ortan and Mulligan, 1854.

Deutsch, Sarah. *No Separate Refuge: Culture, Class, and Gender on an Anglo-Hispanic Frontier in the American Southwest, 1880–1940*. New York: Oxford University Press, 1987.

Dewey, Squire P. *The Bonanza Mines of Nevada: Gross Frauds in the Management Exposed, Reply of S. P. Dewey to the Misrepresentations of the Bonanza Firm in Their Libelous Publication of May 25th, 1878*. San Francisco: [no publisher], 1878.

Dickens, Charles. *American Notes*. New York: John W. Lovell, 1883.

Dill, R. G. "History of Lake County." In *History of the Arkansas Valley, Colorado*. Chicago: O. L. Baskin, 1881, 207–388.

Di Pol, C. John. *A History of the Parish of Randsburg and St. Ann Ridgecrest*. Ridgecrest, CA: C. John Di Pol, 1996.

Doherty, Robert E. "Thomas J. Hagerty, the Church, and Socialism." *Labor History* 3, no. 1 (Winter 1962): 39–56.

Dolman, William Hickman. *Before the Comstock, 1857–1858: Memoirs of William Hickman Dolman*. Reno: University of Nevada, 1947.

Dow, James R., Roger L. Welsch, and Sudan D. Dow, eds. *Wyoming Folklore: Reminisces, Folktales, Beliefs, Customs, and Folk Speech*. Lincoln: University of Nebraska Press, 2010.

Drury, Wells. *An Editor on the Comstock Lode*. Oakland: Pacific Books, 1948.

Dungan, Myles. *How the Irish Won the West*. Dublin: New Island, 2006.

Dwyer, John T. *Condemned to the Mines: The Life of Eugene O'Connell, 1815–1891*. New York: Vantage, 1976.

Dykstra, Robert R. "Overdosing on Dodge City." *Western Historical Quarterly* 27 (Winter 1996): 505–514.

Ellis, William T. *Memories: My Seventy-Two Years in the Romantic County of Yuba, California*. Eugene: University of Oregon, 1939.

Emmons, David. "An Aristocracy of Labor: The Irish Miners of Butte, 1880–1914." *Labor History* 28 (1987): 275–306.

Emmons, David M. *Beyond the American Pale: The Irish in the West, 1845–1910*. Norman: University of Oklahoma Press, 2010.

Emmons, David M. *The Butte Irish: Class and Ethnicity in an American Mining Town, 1875–1925*. Urbana: University of Illinois Press, 1989.

Emmons, David M. "Orange and the Green in Montana: A Reconsideration of the Clark-Daly Feud." *Arizona and the West* 28 (Autumn 1986): 225–245.

Emrich, Duncan. *In the Delta Saloon: Conversations with Residents of Virginia City, Nevada, Recorded in 1949 and 1950*. Reno: University of Nevada Oral History Program, 1991.

Erickson, Charlotte. *Invisible Immigrants: The Adaptation of English and Scottish Immigrants in Nineteenth-Century America*. Leicester: Leicester University Press, 1972.

Fahey, John. *The Ballyhoo Bonanza: Charles Sweeny and the Idaho Mines*. Seattle: University of Washington Press, 1971.

Fahey, John. *The Days of the Hercules*. Moscow: University Press of Idaho, 1978.

Fallows, Marjorie R. *Irish Americans: Identity and Assimilation*. Upper Saddle River, NJ: Prentice Hall, 1979.

Fanning, Charles, ed. *New Perspectives on the Irish Diaspora*. Carbondale: Southern Illinois University Press, 2000.

Faragher, John Mack. *Rereading Frederick Jackson Turner: "The Significance of the Frontier in American History" and Other Essays*. New Haven, CT: Yale University Press, 1999.

Farquhar, Francis P. *Up and Down California in 1860–1864: The Journal of William H. Brewer*. Berkeley: University of California Press, 1974.

Ferguson, Charles D. *The Experiences of a Forty-Niner during Thirty-Four Years' Residence in California and Australia*. Cleveland: Williams, 1888.

Fields, Barbara J. "Whiteness, Racism, and Identity." *International Labor and Working-Class History* 60 (Fall 2001): 48–56.

Final Report of the Commission on Industrial Relations. Washington, DC: US Government Printing Office, 1916.

Finn, Janet L., and Ellen Crain. *Motherlode: Legacies of Women's Lies and Labors in Butte, Montana*. Butte, MT: Clark City Press, 2005.

Fitch, Franklyn Y. *The Life, Travels and Adventures of an American Wanderer: A Truthful Narrative of Events in the Life of Alonzo P. De Milt, Containing His Early Adventures among the Indians of Florida; His Life in the Gold Mines of California and Australia*. New York: John W. Lovell, 1883.

Fitzpatrick, David. *Irish Emigration 1801–1921*. Dublin: Economic and Social History Society of Ireland, 1984.

Fitzpatrick, David. "The Irish in Britain, Settlers or Transients." In *The Irish in British Labour History: Conference Proceedings in Irish Studies*, no. 1, edited by Patrick Buckland and John Belchem. Liverpool: Liverpool University Press, 1992, 1–10.

Fitzpatrick, David. *Oceans of Consolation: Personal Accounts of Irish Migration to Australia*. Cork, Ireland: Cork University Press, 1994.

Floyd, Janet. *Claims and Speculations: Mining and Writing in the Gilded Age*. Albuquerque: University of New Mexico Press, 2012.

Foner, Eric. *Free Soil, Free Labor, Free Men: The Ideology of the Republican Party before the Civil War*. London: Oxford University Press, 1971.

Foner, Eric. *Politics and Ideology in the Age of the Civil War*. New York: Oxford University Press, 1980.

Foner, Eric. *Reconstruction: America's Unfinished Revolution, 1863–1877*. New York: Perennial Classics, 2002.

Foner, Philip. *American Labor Songs of the Nineteenth Century*. Urbana: University of Illinois Press, 1975.

Foner, Philip, ed. *Mother Jones Speaks*. New York: Pathfinder, 2003.

Forbes, Bertie C. *Men Who Are Making the West*. New York: B. C. Forbes, 1923.

Foster, James C. "Western Miners and Silicosis: 'The Scourge of the Underground Toiler,' 1890–1943." *Industrial and Labor Relations Review* 37 (1984): 371–385.

Fox, William. "Patrick Edward Connor, 'Father' of Utah Mining." Master's thesis, Brigham Young University, Provo, UT, 1966.

Fredrickson, George M. *The Arrogance of Race: Historical Perspectives on Slavery, Racism, and Social Inequality*. Middletown, CT: Wesleyan University Press, 1988.

Geary, Laurence M., and Margaret Kelleher, eds. *Nineteenth-Century Ireland: A Guide to Recent Research*. Dublin: University College Dublin Press, 2005.

Gerard, David. "Transaction Costs and the Value of Mining Claims." *Land Economics* 77 (2001): 371–384.

Gilbert, Frank T. *Resources, Business, and Business Men of Montana, 1888*. Walla Walla, WA: Historic Publishing Company, 1888.

Gilfoyle, Timothy J. *City of Eros*. New York: W. W. Norton, 1994.

Gogan, Robert. *130 Great Irish Ballads*. Dublin: Music Ireland, 2006.

Golab, Caroline. *Immigrant Destinations*. Philadelphia: Temple University Press, 1977.

Goldman, Marion S. *Gold Diggers and Silver Miners: Prostitution and Social Life on the Comstock Lode*. Ann Arbor: University of Michigan Press, 1981.

Goodman, David. *Gold Seeking: Victoria and California in the 1850s*. St. Leonards, NSW: Allen and Unwin, 1994.

Gorn, Elliott J. *Mother Jones: The Most Dangerous Woman in America*. New York: Farrar, Straus and Giroux, 2015.

Gossett, Thomas F. *Race: The History of an Idea in America*. Dallas: Southern Methodist University Press, 1963.

Graham, Brian J. *An Historical Geography of Ireland*. London: Academic Press, 1993.

Grattan, Thomas Colley. *Civilized America*, vol. 2. London: Bradbury and Evans, 1859.

Graves, Jackson A. *Seventy Years in California, 1857–1927*. Los Angeles: Times Mirror, 1927.

Green, Archie. *Wobblies, Pile Butts, and Other Heroes: Laborlore Explorations*. Urbana: University of Illinois Press, 1993.

Greenway, John. *Folklore of the Great West: Selections from Eighty-Three Years of the Journal of American Folklore*. Palo Alto: American West, 1969.

Greever, William S. *The Bonanza West*. Norman: University of Oklahoma Press, 1963.

Griswold, Don L. *History of Leadville and Lake County, Colorado: From Mountain Solitude to Metropolis*. Denver: Colorado Historical Society, 1996.

Guglielmo, Thomas A. *White on Arrival: Italians, Race, Color, and Power in Chicago, 1890–1945*. New York: Oxford University Press, 2004.

Hagerty, Thomas J. *Economic Discontent and Its Remedy*. Cincinnati: Standard Publishing Company, 1902.

Halaas, David Fridtjof. *Boom Town Newspapers: Journalism on the Rocky Mountain Mining Frontier, 1859–1881*. Albuquerque: University of New Mexico Press, 1981.

Hall, William. *Progressive Men of the State of Montana*. Chicago: A. W. Bowen, 1902.

Hammett, Dashiell. *Red Harvest*. New York: Alfred A. Knopf, 1929.

Hand, Wayland D. "The Folklore, Customs, and Traditions of the Butte Miner." *California Folklore Quarterly* 5, no. 1 (January 1946): 1–25.

Handley, James E. *The Irish in Modern Scotland*. Cork, Ireland: Cork University Press, 1947.

Hanly, John, ed. *The Letters of Saint Oliver Plunkett, 1625–1681*. Bucks, UK: Colin Smythe, 1979.

Hannefin, Daniel. *Daughters of the Church: A Popular History of the Daughters of Charity in the United States, 1809–1987*. Brooklyn: New City Press, 1990.

Hardesty, Donald L. *The Archaeology of Mining and Miners: A View from the Silver State*. Pleasant Hill, CA: Society for Historical Archaeology, 1988.

Harris, Ruth-Ann Mellish. *The Nearest Place That Wasn't Ireland: Early Nineteenth Century Irish Labor Migration*. Ames: Iowa State University Press, 1994.

Haywood, William D. *Bill Haywood's Book: The Autobiography of William D. Haywood*. New York: International Publishers, 1929.

Helper, Hinton R. *The Land of Gold: Reality versus Fiction*. Baltimore: Baltimore, Pub., 1855.

Herbermann, Charles George. *The Catholic Encyclopedia: An International Work of Reference on the Constitution, Doctrine, Discipline, and History of the Catholic Church*. New York: Encyclopedia Press, 1914.

Hickey, Patrick. *Famine in West Cork: The Mizen Peninsula Land and People, 1800–1852, a Local Study of Pre-Famine and Famine Ireland*. Cork, Ireland: Mercier, 2002.

Higham, John. *Strangers in the Land: Patterns of American Nativism, 1860–1925*. New Brunswick, NJ: Rutgers University Press, 2002.

Hittell, John S. *Mining in the Pacific States of North America*. New York: John Wiley, 1862.

Hoerder, Dirk, ed. *American Labor and Immigration History, 1877–1920s: Recent European Research*. Urbana: University of Illinois Press, 1983.

Houston, Cecil J. *Irish Emigration and Canadian Settlement: Patterns, Links and Letters*. Toronto: University of Toronto Press, 1990.

Howard, Joseph Kinsey. *Montana: High, Wide, and Handsome*. Lincoln: University of Nebraska Press, 1983.

Huntley, Henry V. *California: Its Gold and Its Inhabitants*. 2 vols. London: Thomas Cautley Newby, 1856.

Hutchinson, Edward Prince. *Immigrants and Their Children, 1850–1950*. Census Monograph Series. New York: Wiley, 1956.

Hynding, Alan A. "The Coal Miners of Washington Territory: Labor Troubles in 1888–89." *Arizona and the West* 12 (1970): 221–236.

Ibson, John Duffy. *Will the World Break Your Heart? Dimensions and Consequences of Irish American Assimilation*. New York: Garland, 1990.

Idaho State Historical Reference Series. *Placer Mining in Southern Idaho*, no. 166. Boise: Idaho State Historical Society, 1980.

Idaho State Historical Reference Series. *Rocky Bar Mines*, no. 199. Boise: Idaho State Historical Society, 1996.

Illustrated History of North Idaho: Embracing Nez Perces, Idaho, Latah, Kootenai and Shoshone Counties, State of Idaho. Racine, WI: Western Historical Publishing Company, 1903.

Irish Emigration Database, Letters (Emigrants). http://ied.dippam.ac.uk/.

Jacobson, Matthew Frye. *Special Sorrows: The Diasporic Imagination of Irish, Polish, and Jewish Immigrants in the United States*. Berkeley: University of California Press, 2010.

Jacobson, Matthew Frye. *Whiteness of a Different Color: European Immigrants and the Alchemy of Race*. Cambridge, MA: Harvard University Press, 1998.

James, Ronald M. "Defining the Group: Nineteenth Century Cornish on the North American Mining Frontier." *Cornish Studies* 2 (1994): 35–45.

James, Ronald M. *The Roar and the Silence: A History of Virginia City and the Comstock Lode*. Reno: University of Nevada Press, 1998.

James, Ronald M., and C. Elizabeth Raymond, eds. *Comstock Women: The Making of a Mining Community*. Reno: University of Nevada Press, 1998.

Jensen, Vernon H. *Heritage of Conflict: Labor Relations in the Nonferrous Metals Industry up to 1930*. Ithaca, NY: Cornell University Press, 1950.

Johnson, Susan Lee. *Roaring Camp: The Social World of the California Gold Rush*. New York: W. W. Norton, 2000.

Jones, Mary Harris. *Autobiography of Mother Jones, American Labor: From Conspiracy to Collective Bargaining*. New York: Arno, 1969 [1925].

Julia, Sister. *Annunciation Parish: Leadville, Colorado, a History*. [No publishing information], 1953.

Karabel, Jerome. *The Chosen: The Hidden History of Admission and Exclusion at Harvard, Yale, and Princeton*. Boston: Houghton Mifflin, 2005.

Karson, Marc. "The Catholic Church and the Political Development of American Trade Unionism (1900–1918)." *Industrial and Labor Relations Review* 4 (1951): 527–542.

Kelly, James, ed. *The Cambridge History of Ireland*, vol. 3: *1730–1880*. Cambridge: Cambridge University Press, 2018.

Kelly, John S. *The Bodyke Evictions*. Scariff, Ireland: Fossabeg Press, 1987.

Kemmerer, Donald Lorenzo, and C. Clyde Jones. *American Economic History*. New York: McGraw-Hill, 1959.

Kenny, Kevin. *The American Irish: A History*. New York: Longman, 2000.

Kenny, Kevin. *Making Sense of the Molly Maguires*. New York: Oxford University Press, 1998.

Kinnersley, Thomas Harold. *Nevada, 1859–1881*. Ann Arbor, MI: University Microfilms, 1978.

Klein, Robert F. *Dubuque during the California Gold Rush: When the Midwest Went West*. Charleston, SC: History Press, 2011.

Knobel, Dale T. *Paddy and the Republic: Ethnicity and Nationality in Antebellum America*. Middletown, CT: Wesleyan University Press, 1986.

Knox, Celia I. "The Patriot Priest—Father Eugene Sheehy: His Life, Work, and Influence." PhD diss., University of Sussex, England, 1998.

Korson, George. *Coal Dust on the Fiddle: Songs and Stories of the Bituminous Industry*. Hatboro, PA: Folklore Associates, 1965.

Lang, Herbert O. *A History of Tuolumne County, California: Compiled from the Most Authentic Records*. San Francisco: B. F. Alley, 1882.

Lapp, Rudolph M. *Blacks in Gold Rush California*. New Haven, CT: Yale University Press, 1977.

Larsh, Ed, and Robert Nichols. *Leadville, U.S.A.* Boulder: Johnston Books, 1992.
Lazure, Joe. "Hobo Miner." *Miners' Magazine* 6, no. 94 (March 13, 1905): 13.
Lee, Joseph, and Marion R. Casey, eds. *Making the Irish American: History and Heritage of the Irish in the United States*. New York: New York University Press, 2006.
Levy, Jo Ann. *They Saw the Elephant: Women in the California Gold Rush*. Hamden, CT: Archon Books, 1990.
Lewis, Marvin, ed. *The Mining Frontier: Contemporary Accounts from the American West in the Nineteenth Century*. Norman: University of Oklahoma Press, 1967.
Limerick, Patricia Nelson. *The Legacy of Conquest: The Unbroken Past of the American West*. New York: W. W. Norton, 1987.
Ling, Huping. *Surviving on the Gold Mountain: A History of Chinese American Women and Their Lives*. Albany: State University of New York Press, 1998.
Lingenfelter, Richard E. *The Hardrock Miners: A History of the Mining Labor Movement in the American West, 1863–1893*. Berkeley: University of California Press, 1974.
Link, Arthur S., ed. *The Papers of Woodrow Wilson*, vol. 63. Princeton, NJ: Princeton University Press, 1994.
Linn, Meredith B. "From Typhus to Tuberculosis and Fractures in Between: A Visceral Historical Archaeology of Irish Immigrant Life in New York City 1845–1870." PhD diss., Columbia University, New York, NY, 2008.
Long, Priscilla. *Where the Sun Never Shines: A History of America's Bloody Coal Industry*. New York: Paragon House, 1989.
Luebke, Frederick C., ed. *European Immigrants in the American West: Community Histories*. Albuquerque: University of New Mexico Press, 1998.
Lukas, J. Anthony. *Big Trouble: A Murder in a Small Western Town Sets off a Struggle for the Soul of America*. New York: Simon and Schuster, 1997.
Macdonald, Alexander. *In Search of El Dorado: A Wanderer's Experiences*. G. W. Jacobs, 1907.
Macdonald, Fergus. *The Catholic Church and the Secret Societies in the United States*. New York: United States Catholic Historical Society, 1946.
Mac Gabhann, Micí. *Rotha mór an tSaoil*. Dublin: National Publications, 1953.
MacGowan, Michael. *The Hard Road to Klondike*. Translated by Valentin Iremonger. Cork, Ireland: Collins, 2003.
MacRaild, Donald M. *Culture, Conflict, and Migration: The Irish in Victorian Cumbria*. Liverpool: Liverpool University Press, 1988.
MacRaild, Donald M., and Enda Delaney, eds. *Irish Migration, Networks and Ethnic Identities since 1750*. New York: Routledge, 2007.
Makley, Michael J. *The Infamous King of the Comstock: William Sharon and the Gilded Age in the West*. Reno: University of Nevada Press, 2006.
Malone, Michael P. *The Battle for Butte: Mining and Politics on the Northern Frontier, 1864–1906*. Seattle: University of Washington Press, 2012.
Malone, Michael P., and Richard B. Roeder. "1876 in the Gulches: Mining." *Montana: The Magazine of Western History* 25 (1975): 20–27.

Mann, Ralph. *After the Gold Rush: Society in Grass Valley and Nevada City, California, 1849–1870*. Stanford: Stanford University Press, 1982.

Mann, Ralph. "Frontier Opportunity and the New Social History." *Pacific Historical Review* 53 (1984): 463–491.

Martinelli, Phylis Cancilla. *Undermining Race: Ethnic Identities in Arizona Copper Camps, 1880–1920*. Tucson: University of Arizona Press, 2009.

McCarthy, Thomas B. "From West Cork to Butte: The Irish Immigration to Montana, 1860–1900." Master's thesis, Washington State University, Pullman, 1987.

McConnell, William John. *Early History of Idaho*. Boise, ID: Caxton, 1912.

McCracken, Donal P. *The Irish Pro-Boers, 1877–1902*. Johannesburg: Perskor, 1989.

McGloin, John B., and Martin Francis Schwenninger. "A California Gold Rush Padre: New Light on the 'Padre of Paradise Flat.'" *California Historical Society Quarterly* 40 (March 1961): 49–69.

McGrath, Roger D. *Gunfighters, Highwaymen and Vigilantes: Violence on the Frontier*. Berkeley: University of California Press, 1984.

Mclean, Evalyn Walsh. *Father Struck It Rich*. New York: Little, Brown, 1936.

Mellinger, Philip J. *Race and Labor in Western Copper: The Fight for Equality, 1896–1918*. Tucson: University of Arizona Press, 1995.

Mercier, Laurie. "'We Are Women Irish': Gender, Class, Religious, and Ethnic Identity in Anaconda." In *Writing the Range: Race, Class, and Culture in the Woman's West*, edited by Elizabeth Jameson and Susan Armitage. Norman: University of Oklahoma Press, 1997, 311–333.

Merrifield, Robert B. "Nevada, 1859–1881: The Impact of an Advanced Technological Society upon a Frontier Area." PhD diss., University of Chicago, Chicago, IL, 1958.

Miller, David W. "Irish Catholicism and the Great Famine." *Journal of Social History* 9 (1975): 81–98.

Miller, Joaquin. *An Illustrated History of the State of Montana*. Chicago: Lewis, 1894.

Miller, Kerby A. *Emigrants and Exiles: Ireland and the Irish Exodus to North America*. New York: Oxford University Press, 1988.

Miskell, Louise. "Custom, Conflict and Community: A Study of the Irish in South Wales and Cornwall, 1861–1891." PhD diss., University of Wales, Cardiff, 1996.

Montgomery, David. *The Fall of the House of Labor: The Workplace, the State, and American Labor Activism, 1865–1925*. New York: Cambridge University Press, 1987.

Mooney, Bernice Maher. *Salt of the Earth: The History of the Catholic Church in Utah, 1776–2007*. 3rd ed. Salt Lake City: University of Utah Press, 2008.

Morris, Patrick F. *Anaconda, Montana: Copper Smelting Boom Town on the Western Frontier*. Bethesda, MD: Swann, 1997.

Murdoch, Angus. *Boom Copper: The Story of the First US Mining Boom*. New York: Macmillan, 1943.

Murphy, Mary. *Mining Cultures: Men, Women, and Leisure in Butte, 1914–41*. Chicago: University of Illinois Press, 1998.

Murphy, Mary. "A Place of Greater Opportunity: Irish Women's Search for Home, Family and Leisure in Butte, Montana." *Journal of the West* 31, no. 2 (1992): 73–78.

Murphy, Mary. "Women on the Line: Prostitution in Butte, Montana, 1878–1917." Master's thesis, University of North Carolina, Chapel Hill, 1983.

Myres, Sandra L. *Westering Women and the Frontier Experience, 1800–1915*. Albuquerque: University of New Mexico Press, 1999.

Nestor, Sandy. *Silver and Gold Mining Camps of the Old West: A State by State American Encyclopedia*. Jefferson, NC: McFarland, 2007.

Nevada Legislative Assembly. *The Journal of the Assembly of the Seventh Session of the Legislature of the State of Nevada, 1875*. Carson City, NV: John J. Hill, State Printer, 1875.

Nevada Legislative Senate. *The Journal of the Senate of the Sixth Session of the Legislature of the State of Nevada, 1873*. Carson City, NV: Charles A. V. Putnam, State Printer, 1873.

Nevada State Mineralogist. *Biennial Report of the State Mineralogist for the State of Nevada 1872–1873*. Carson City: State Printing Office, 1873.

Nic Congáil, Ríona, Máirín Nic Eoin, Meidhbhín Ní Úrdail, Pádraig Ó Liatháin, and Regina Uí Chollatáin, eds. *Litríocht na Gaeilge ar Fud an Domhain, Imleabhar I: Cruthú, Caomhnú agus Athbheochan*. Dublin: LeabhairComhar, 2016.

Noonan, Alan J. M. "From Ireland to Montana: A Study of the Frontier 1860–1900." M.Phil. thesis, University College Cork, Ireland, 2008.

Noonan, Alan J. M. "'Oh Those Long Months without a Word from Home': Migrant Letters from Mining Frontiers." *The Boolean* 2 (2011): 135–142.

Noonan, Alan J. M. "Progressive Men of the State of Montana: A Kaleidoscope of Irish Montanans." *Journal of the Trinity Postgraduate Seminar Series* 1 (2008): 12–19.

Noonan, Alan J. M. "Real Irish Patriots Would Scorn to Recognise the Likes of You." In *Locked Out: A Century of Working Class Life*, edited by David Convery. Dublin: Irish Academic Press, 2013, 57–73.

Noonan, Alan J. M. "Wandering Labourers: The Irish and Mining throughout the United States, 1845–1920." PhD diss., University College Cork, Ireland, 2013.

O'Connell, Michael. "'What a Pity at the Very Source of Wealth': Strikes and Emigration, Berehaven Mining District, 1861–c1900." *Saothar* 34 (2009): 7–18.

O'Connell, Michael. "Emigration from the Berehaven Copper Mining District to the United States of America, 1840–1900." Master's thesis, University College Cork, Ireland, 2007.

Ó Dubhda, Seán. *Duanaire Duibhneach: Bailú D'ampanaidh Agus De Phiosaibh Eile Filidheachta a Ceapadh Le Tuairim Céad Bliain i gCorca Dhuibhne*. Dublin: Government Publications Office, 1933.

O'Dwyer, Riobard. *Who Were My Ancestors? Genealogy (Family Trees) of the Allihies (Copper Mines) Parish, County Cork, Ireland*. Self-published, 1976.

O'Dwyer, Riobard. *Who Were My Ancestors? Genealogy (Family Trees) of the Bere Island Parish, County Cork, Ireland*. Self-published, 1976.

Ogden, Rollo, ed. *Life and Letters of Edwin Lawrence Godkin*. New York: Macmillan, 1907.

O'Hanlon, John. *The Irish Emigrant's Guide for the United States*. New York: Arno, 1976.

O'Leary, Paul B. "Anti-Irish Riots in Wales." *Llafur* 5 (1991): 27–36.

O'Neil, Timothy M. "Miners in Migration: The Case of Nineteenth Century Irish and Irish American Copper Miners." *Éire-Ireland: A Journal of Irish Studies* 36, nos. 1 and 2 (Spring–Summer 2001): 124–140.

Palais, Hyman. "Black Hills Miners' Folklore." *California Folklore Quarterly* 4, no. 3 (1945): 255–269.

Pascoe, Peggy. *Relations of Rescue: The Search for Female Moral Authority in the American West, 1874–1939*. New York: Oxford University Press, 1993.

Paul, Rodman W. *Mining Frontiers of the Far West, 1848–1880*. Austin: Holt, Rinehart and Winston, 1963.

Payton, Philip. *The Cornish Overseas*. Cornwall, UK: Alexander Associates, 1999.

Peck, Gunther. *Reinventing Free Labor: Padrones and Immigrant Workers in the North American West, 1880–1930*. Cambridge: Cambridge University Press, 2000.

Perrigo, Lynn I. "Law and Order in Early Colorado Mining Camps." *Mississippi Valley Historical Review* 28, no. 1 (June 1941): 41–62.

Perrigo, Lynn. "A Social History of Central City, Colorado, 1859–1900." PhD diss., University of Colorado, Boulder, 1936.

Peterson, Richard H. *The Bonanza Kings: The Social Origins and Business Behavior of Western Mining Entrepreneurs, 1870–1900*. Lincoln: University of Nebraska Press, 1977.

Pickett, Evelyne Stitt. "Hoboes across the Border: A Comparison of Itinerant Cross-Border Laborers between Montana and Western Canada." *Montana: The Magazine of Western History* 49, no. 1 (Spring 1999): 18–31.

Reel, Guy. *The National Police Gazette and the Making of the Modern American Man, 1879–1906*. New York: Palgrave Macmillan, 2006.

Report of the Board of Trade, Sea Fisheries Act, 1868. Main Papers 531. Journal Office, Parliament Office, Records of the House of Lords, Parliamentary Archives. London: UK Parliament, 1868.

Report of the Director of the Mint upon the Production of the Precious Metals in the United States. Washington, DC: US Government Printing Office, 1883.

Richards, Benjamin B. *California Gold Rush Merchant: The Journal of Stephen Chaplin Davis*. San Marino, CA: Huntington Library, 1959.

Richter, F. E. "The Copper-Mining Industry in the United States, 1845–1925." *Quarterly Journal of Economics* 41, no. 2 (February 1927): 236–291.

Richter, F. E. "The Copper-Mining Industry in the United States, 1845–1925." *Quarterly Journal of Economics* 41, no. 4 (August 1927): 684–717.

Robbins, William G. *Colony and Empire: The Capitalist Transformation of the American West*. Lawrence: University Press of Kansas, 1994.

Robertson, David. *Hard as the Rock Itself: Place and Identity in the American Mining Town*. Boulder: University Press of Colorado, 2006.
Rohe, Randall E. "After the Gold Rush: Chinese Mining in the Far West, 1850–1890." *Montana: The Magazine of Western History* 32 (1982): 2–19.
Rohrbough, Malcolm J., and American Council of Learned Societies. *Days of Gold: The California Gold Rush and the American Nation*. Berkeley: University of California Press, 1997.
Rosenbloom, Joshua L. "Looking for Work, Searching for Workers: U.S. Labor Markets after the Civil War." *Social Science History* 18 (Autumn 1994): 377–403.
Ross, Dudley T. *Devil on Horseback: A Biography of the "Notorious" Jack Powers*. Fresno: Valley Publishers, 1975.
Ruggles, Steven, J. Trent Alexander, Katie Genadek, Ronald Goeken, Matthew B. Schroeder, and Matthew Sobek. Integrated Public Use Microdata Series [IPUMS]: Version 5.0 [Machine-readable database]. University of Minnesota, Minneapolis, 2010.
Rutter, Michael. *Upstairs Girls: Prostitution in the American West*. Helena, MT: Farcountry, 2005.
Samito, Christian G. *Becoming American under Fire: Irish Americans, African Americans, and the Politics of Citizenship during the Civil War Era*. Ithaca, NY: Cornell University Press, 2009.
Sarbaugh, Timothy J. "The Irish in the West: An Ethnic Tradition of Enterprise and Innovation, 1848–1991." *Journal of the West* 31, no. 2 (1992): 5–8.
Sarbaugh, Timothy J., and James P. Walsh. *The Irish in the West*. Manhattan, KS: Sunflower University Press, 1993.
Scanlon, Gretchen. *A History of Leadville Theatre: Opera Houses, Variety Acts and Burlesque Shows*. Charleston, SC: History Press, 2012.
Schmitz, Christopher. "The Rise of Big Business in the World Copper Industry 1870–1930." *Economic History Review* New Series 39 (1986): 392–410.
Schrier, Arnold. *Ireland and the American Emigration, 1850–1900*. Chester Springs, PA: Dufour Editions, 1997.
Seagraves, Anne. *Soiled Doves: Prostitution in the Early West*. Hayden, ID: Wesanne, 1994.
[Searchable database for US Census details]. Ancestry.com.
[Searchable database for US Census details]. Familysearch.com.
Shannon, James P. *Catholic Colonization on the Western Frontier*. New Haven, CT: Yale University Press, 1957.
Shea, Debbie Bowman. *Irish Butte*. Mount Pleasant, SC: Arcadia, 2011.
Shepperson, Wilbur S. *Restless Strangers; Nevada's Immigrants and Their Interpreters*. Reno: University of Nevada Press, 1970.
Shinn, Charles Howard. *Mining Camps: A Study in American Frontier Government*. New York: Charles Scribner's Sons, 1885.

Shoebotham, H. Minar. *Anaconda: Life of Marcus Daly, the Copper King*. Harrisburg, PA: Stackpole, 1956.

Silber, Irwin, ed. *Songs of the Great American West*. New York: Macmillan, 1967.

Smith, Duane A. *Mining America: The Industry and the Environment, 1800–1980*. Lawrence: University Press of Kansas, 1987.

Smith, Grant H. *The History of the Comstock Lode 1850–1920*. Reno: University of Nevada and Nevada State Bureau of Mines, 1943.

Smith, Kathleen, and Lane Parker. *Smartsville and Timbuctoo*. Charleston, SC: Arcadia, 2008.

Smith, Robert Wayne. *The Coeur d'Alene Mining War of 1892: A Case Study of an Industrial Dispute*. 2nd ed. Gloucester, MA: P. Smith, 1961.

Smith, Rodney H. *Austin, Nevada, 1862–1888*. Reno: University of Nevada Press, 1963.

Smith-Rosenberg, Carroll. *Disorderly Conduct: Visions of Gender in Victorian America*. New York: Alfred A. Knopf, 1985.

Soloman, Barbara Miller. *Ancestors and Immigrants: A Changing New England Tradition*. Cambridge, MA: Harvard University Press, 1956.

Spence, Clark C., ed. *The American West*. New York: Crowell, 1966.

Spence, Clark C. *Mining Engineers and the American West: The Lace-Boot Brigade, 1849–1933*. New Haven, CT: Yale University Press, 1970.

Staley, W. William. "Gold in Idaho." Pamphlet 68. *Idaho Bureau of Mines and Geology*. Moscow: University of Idaho Press, 1946.

Stewart, John C. *Thomas F. Walsh: Progressive Businessman and Colorado Mining Tycoon*. Boulder: University Press of Colorado, 2007.

Stickney, Mary M. "Mining Women of Colorado." *The Era: An Illustrated Monthly Magazine of Literature and of General Interest* 9. Philadelphia: Henry T. Coates, 1902: 24–32.

Sullivan, Timothy D. *Speeches from the Dock; or, Protests of Irish Patriotism*. Dublin: M. H. Gill and Son, Ltd., 1945.

Swibold, Dennis L. *Copper Chorus: Mining, Politics, and the Montana Press, 1889–1959*. Helena: Montana Historical Society Press, 2006.

t'Hart, Marjolein. "'Heading for Paddy's Green Shamrock Shore': The Returned Emigrants in Nineteenth Century Ireland." Master's thesis, University of Groningen, the Netherlands, 1981.

Tannacito, Dan. "Poetry of the Colorado Miners: 1903–1906." *Radical Teacher* 15 (March 1980): 1–8.

Taylor, Lawrence. "Bás in Éirinn: Cultural Constructions of Death in Ireland." *Anthropological Quarterly* 62, no. 4 (1989): 175–187.

Terrar, Toby. "Catholic Socialism: The Reverend Thomas McGrady." *Dialectical Anthropology* 7 (1983): 209–235.

Thomas, William, and Florian Znaniecki. *The Polish Peasant in Europe and America*. Urbana: University of Illinois Press, 1996.

Thomes, William Henry. *A Gold Hunter's Adventures: or, Life in Australia*. Chicago: Laird and Lee, 1890.

Thurner, Arthur W. "Western Federation of Miners in Two Copper Camps: The Impact of the Michigan Copper Miners' Strike on Butte's Local No. 1." *Montana: The Magazine of Western History* 33 (1983): 30–45.

Todd, Arthur Cecil. *The Cornish Miner in America: The Contribution to the Mining History of the United States by Emigrant Cornish Miners, the Men Called Cousin Jacks*. Glendale, CA: Arthur H. Clark, 1967.

Todd, Arthur Cecil. "Cousin Jack in Idaho." *Idaho Yesterday* (Winter 1964): 6–11.

Toms, Don. *Tenderloin Tales: Prostitution, Gambling and Opium on the Gold Belt of the Northern Black Hills, 1867–1915*. Pierre, SD: State Publishing Company, 1997.

Travis, Merle. *Sixteen Tons*. Hollywood: American Music, Inc., 1947. Vinyl record.

Turner, Frederick Jackson. *The Frontier in American History*. New York: H. Holt, 1920.

Twain, Mark. *Roughing It*. Hartford, CT: American Publishing Company, 1891.

United States Congress, Immigration Commission. *Reports of the Immigration Commission: Abstracts of Reports of the Immigration Commission with Conclusions and Recommendations and Views of the Minority*, vol. 1. Washington, DC: US Government Printing Office, 1911.

United States Congress, Immigration Commission. *Report of the Immigration Commission, Occupations of the First and Second Generations of Immigrants in the United States*, vol. 65. Washington, DC: Government Printing Office, 1911.

US Bureau of Labor. *Report on Labor Disturbances in the State of Colorado, from 1880 to 1904*. Washington, DC: US Government Printing Office, 1905.

US Bureau of the Census. *Census of Population, 1910*, vol. 1: *General Report and Analysis*. Washington, DC: US Government Printing Office, 1913.

US Bureau of the Census. *Census of Population, 1910*, vol. 2: *Reports by States, Alabama–Montana*. Washington, DC: US Government Printing Office, 1913.

US Bureau of the Census. *Census of Population, 1910*, vol. 3: *Reports by States, Nebraska–Wyoming, Alaska, Hawaii, and Porto Rico*. Washington, DC: US Government Printing Office, 1913.

US Bureau of the Census. *Census of Population, 1920*, vol. 2: *General Report and Analytical Tables*. Washington, DC: US Government Printing Office, 1922.

US Bureau of the Census. *Census of Population, 1920*, vol. 3: *Composition and Characteristics of the Population by States*. Washington, DC: US Government Printing Office, 1922.

US Bureau of the Census. *Census of Agriculture, 1920*, vol. 5: *General Report and Analytical Tables*. Washington, DC: US Government Printing Office, 1922.

US Census Office. *Census of Population, 1850: The Seventh Census of the United States*, Washington, DC: US Government Printing Office, 1853.

US Census Office. *Census of Population, 1860: Population of the United States*. Washington, DC: US Government Printing Office, 1864.

US Census Office. *Census of Population, 1870*, vol. 1: *Statistics of the Population of the United States*. Washington, DC: US Government Printing Office, 1872.

US Census Office. *Census of Population, 1880*, vol. 1: *Statistics of the Population of the United States*. Washington, DC: US Government Printing Office, 1883.

US Census Office. *Census of Population, 1890*, vol. 1: *Report on Population of the United States, Part 1*. Washington, DC: US Government Printing Office, 1895.

US Census Office. *Census of Population, 1900*, vol. 1: *Population, Part 1*. Washington, DC: US Government Printing Office, 1901.

US Census Office. *Census of Population, 1900*, vol. 2: *Population, Part 2*. Washington, DC: US Government Printing Office, 1902.

US Industrial Commission. *Report of the Industrial Commission on the Relations of Capital and Labor in the Mining Industry*, vol. 12. Washington, DC: US Government Printing Office, 1901.

Van Tilburg Clark, Walter, ed. *The Journals of Alfred Doten, 1849–1903*. 3 vols. Reno: University of Nevada Press, 1973.

Vapnek, Lara. *Elizabeth Gurley Flynn: Modern American Revolutionary*. Boulder: Westview, 2015.

Walsh, Henry L. *Hallowed Were the Gold Dust Trails: The Story of the Pioneer Priests of Northern California*. Santa Clara: University of Santa Clara Press, 1946.

Walsh, James Patrick. "Michael Mooney and the Leadville Irish: Respectability and Resistance at 10,200 Feet, 1875–1900." PhD diss., University of Colorado, Boulder, 2010.

Waters, Mary C. *Ethnic Options: Choosing Identities in America*. Berkeley: University of California Press, 1990.

Watts, Gerard. "James Larkin and the British, American and Irish Free State Intelligence Services: 1914–1924." PhD diss., National University of Ireland, Galway, 2016.

Wedge, Frederick Rhinaldo. *Inside the I.W.W., by a Former Member and Official: A Study of the Behavior of the I.W.W., with Reference to Primary Causes*. Berkeley: F. R. Wedge, 1924.

West, Elliott. *The Contested Plains: Indians, Goldseekers, and the Rush to Colorado*. Lawrence: University Press of Kansas, 1998.

West, Elliott. "Five Idaho Mining Towns: A Computer Profile." *Pacific Northwest Quarterly* 73 (1982): 108–120.

White, Katherine A. *A Yankee Trader in the Gold Rush: The Letters of Franklin A. Buck*. Boston: Riverside Press Cambridge, 1930.

White, Richard, *"It's Your Misfortune and None of My Own": A History of the American West*. Norman: University of Oklahoma Press, 1991.

Whitley, Colleen K., ed. *From the Ground Up: A History of Mining in Utah*. Logan: Utah State University Press, 2006.

Whittaker, David J. *Mining the American West: A Bibliographical Guide to Printed Materials on American Mining Frontiers in the British Library*. London: Eccles Centre for American Studies, British Library, 1996.

Williams, George, III. *The Red-Light Ladies of Virginia City Nevada.* [No publication information], 1984.

Williams, George, III. *Rosa May: The Search for a Mining Camp Legend.* Riverside, CA: Tree by the River Publications, 1980.

Williamson, Ruby G. *Gold, God, the Devil, and Silver: Leadville, Colorado, 1878–1978.* Gunnison, CO: B&B Printers, 1977.

Willison, George F. *Here They Dug the Gold.* New York: Brentanos, 1931.

Wilson, Rodman W. *Mining Frontiers of the Far West, 1848–1880.* Albuquerque: University of New Mexico Press, 2001.

Wilson, Woodrow. *A History of the American People.* New York: Harper Brothers, 1902.

Wolf, Nicholas M. *An Irish-Speaking Island: State, Religion, Community, and the Linguistic Landscape in Ireland, 1770–1870.* Madison: University of Wisconsin Press, 2014.

Wolff, David A. *Industrializing the Rockies: Growth, Competition, and Turmoil in the Coalfields of Colorado and Wyoming, 1868–1914.* Boulder: University Press of Colorado, 2003.

Writers' Program of the Work Projects Administration in the State of Montana. *Copper Camp: Stories of the World's Greatest Mining Town, Butte, Montana.* New York: Hastings House, 1999.

Wyman, Mark. *Hard-Rock Epic: Western Miners and the Industrial Revolution, 1860–1910.* Berkeley: University of California Press, 1979.

Wyman, Mark. *Hoboes: Bindlestiffs, Fruit Tramps, and the Harvesting of the West.* New York: Hill and Wang, 2010.

Young, Otis E. *Black Powder and Hand Steel: Miners and Machines on the Old Western Frontier.* 1st ed. Norman: University of Oklahoma Press, 1976.

Yung, Judy. *Unbound Feet: A Social History of Chinese Women in San Francisco.* Berkeley: University of California Press, 1995.

Yung, Judy, Gordon Chang, and Him Mark Lai, eds. *Chinese American Voices: From the Gold Rush to the Present.* Berkeley: University of California Press, 2006.

Zanjani, Sally. *Goldfield: The Last Gold Rush on the Western Frontier.* Athens: Ohio University Press, 1992.

Zanjani, Sally. *A Mine of Her Own: Women Prospectors in the American West 1850–1950.* Lincoln: University of Nebraska Press, 1997.

Zhu, Liping. *A Chinaman's Chance: The Chinese on the Rocky Mountain Mining Frontier.* Boulder: University Press of Colorado, 2000.

Index

Page numbers followed by *f* indicate figures.
Page numbers followed by *t* indicate tables.